Policy, Finance & Management for Public-Private Partnerships

Edited by

Akintola Akintoye

School of the Built and Natural Environment
University of Central Lancashire

&

Matthias Beck

The York Management School
University of York

WILEY-BLACKWELL

A John Wiley & Sons, Ltd., Publication

This edition first published 2009
© 2009 Blackwell Publishing Ltd

Blackwell Publishing was acquired by John Wiley & Sons in February 2007. Blackwell's
publishing programme has been merged with Wiley's global Scientific, Technical, and Medical
business to form Wiley-Blackwell.

Registered office
John Wiley & Sons Ltd, The Atrium, Southern Gate, Chichester, West Sussex, PO19 8SQ,
United Kingdom

Editorial offices
9600 Garsington Road, Oxford, OX4 2DQ, United Kingdom
2121 State Avenue, Ames, Iowa 50014-8300, USA

For details of our global editorial offices, for customer services and for information about how
to apply for permission to reuse the copyright material in this book please see our website at
www.wiley.com/wiley-blackwell.

Library of Congress Cataloging-in-Publication Data

Policy, finance & management for public-private partnerships / edited by Akintola Akintoye
& Matthias Beck.
    p.   cm. – (Innovation in the built environment ; 1)
  Includes bibliographical references and index.
  ISBN 978-1-4051-7791-7 (hbk. : alk. paper)   1. Public-private sector cooperation.   2.
Infrastructure (Economics)–Management.   I. Akintoye, Akintola.   II. Beck, Matthias, 1964-
  HD3850.P59 2008
  332.1068–dc22

A catalogue record for this book is available from the British Library.

Set in 10/12 pt Sabon by Aptara® Inc., New Delhi, India
Printed in Singapore by Utopia Press Pte Ltd

1   2009

Innovation in the Built Environment *(vertical side title)*

Series advisors

Carolyn Hayles, *Queen's University, Belfast*
Richard Kirkham, *Liverpool John Moores University*
Andrew Knight, *Nottingham Trent University*
Stephen Pryke, *University College London*
Derek Thompson, *Heriot Watt University*
Sara Wilkinson, *University of Melbourne*

Innovation in the Built Environment is a new book series for the construction industry published jointly by the Royal Institute of Chartered Surveyors and Wiley-Blackwell. It addresses issues of current research and practitioner relevance and takes an international perspective, drawing from research applications and case studies worldwide.

- presents the latest thinking on the processes that influence the design, construction and management of the built environment

- based on strong theoretical concepts and draws on both established techniques for analysing the processes that shape the built environment – and on those from other disciplines

- embrace a comparative approach, allowing best practice to be put forward

- emonstrates the contribution that effective management of built environment processes can make

Forthcoming books in the IBE series

Pryke, *Construction Supply Chain Management*
Lu & Sexton, *Innovation in Small Professional Practices in the Built Environment*
Boussabaine, *Risk Pricing Strategies for Public-Private Partnerships*
Kirkham & Boussabaine, *Whole Life-Cycle Costing*
Proverbs et al., *Solutions to Climate Change Challenges in the Built Environment*

We welcome proposals for new, high quality, research-based books which are academically rigorous and informed by the latest thinking; please contact Stephen Brown or Madeleine Metcalfe.

Stephen Brown
Head of Research
RICS
12 Great George Street
London SW1P 3AD
sbrown@rics.org

Madeleine Metcalfe
Senior Commissioning Editor
Wiley-Blackwell
9600 Garsington Road
Oxford OX4 2DQ, UK
mmetcalfe@wiley.com

Contents

Foreword

When the editors of this book commenced their research on Private Finance Initiative (PFI) Projects in the late 1990s, Public-Private Partnership was in its infancy. In the United Kingdom the Private Finance Initiative had been launched in 1992. However, there had been considerable reluctance within the private sector to engage in partnerships, which had only been overcome by the late 1990s. This reluctance to engage in Public-Private Partnerships, in the UK as elsewhere, was rooted in a number of aspects. Firstly, PPP was new and as such risky. Secondly, PPP was complex in virtually every respect ranging from the bidding and negotiation process to financing and financial close. Thirdly, at least in some regions of the world, PPP was associated with anti-statist and neo-liberal ideologies, which saw PPP not as a complementary means for meeting infrastructure and service needs but rather as an instrument for diminishing the role of the public sector.

In the end, many of the teething problems of PPP were overcome by the sheer size of demand which has been associated, in particular, with the industrialisation of the non-Western world, and which has led to a massive global increase in the number of ongoing and completed PPP projects. This spread of PPP, needless to say, has not removed the conceptual, financial and managerial challenges associated with this approach to procurement. What it has done, however, is help us understand how different aspects of PPP can be approached in different contexts. If the purpose of this book can be summarised in one sentence it is, therefore, to disseminate some of the progress that has been made in our understanding of PPP procurement in different contexts and regions. This goal is reflected in the wide ranging disciplinary affiliations of its contributors, which include accountants; finance experts; engineers; construction, facilities and project managers; as well as those working in, and actively advising, public and private sector entities. It is also reflected in the broad geographic spread of contributors which includes, apart from the UK and US, authors from Australia, China, Greece, Ghana, Hong Kong, India, Nepal, South Africa, Taiwan and Turkey.

While it is, at this advanced stage in the development of PPP, impossible to write a truly comprehensive account, it is the hope of the editors that this book will prove useful to those who seek to expand their knowledge of PPP, whether it is for academic purposes or as practitioners. PPP as a topic of research is, and will remain, very much a moving target. In this sense this book should be seen, not only as a contribution in its own right, but also as an invitation to others to conduct research in this exciting and fast moving area.

Contributors

Dr Rifat Akbiyikli

Rifat Akbiyikli is an Assistant Professor at *Sakarya University* in Turkey. He holds an MSc in Civil Engineering from the Norwegian University of Science and Technology, an MSc in Construction Management from the University of Bath and a PhD from the University of Salford. His research interests are PFI/PPP in infrastructures, construction project management; technical and economical issues in transportation projects, productivity and performance in construction projects, innovation in construction, construction procurement, and health and safety in construction. His PhD thesis is on PPPs and he has published extensively in this field.

Professor Akintola Akintoye

Akintola Akintoye is a chartered surveyor and a chartered builder. He is Head of School and Professor of Construction Economics and Management. He was formerly Associate Dean of Research and Knowledge Transfer at the School of the Built and Natural Environment, Glasgow Caledonian University. Before his academic career, he worked as a quantity surveyor and a construction planner on major building and civil engineering projects. He was past chairman of the UK-based Association of Researchers in Construction Management (ARCOM) and the co-editor of the *Journal of Financial Management of Property and Construction*. His major research interests are in the economic formation, modelling and prediction of construction activities, construction risk management and procurement, construction estimating and modelling, construction economics, and construction inventory management. His is an active member of International Council for Research and Innovation in Building and Construction. He was Visiting Professor to the Asian Institute of Technology and Hong Kong Polytechnic University and a distinguished scholar of the University of Cape Town, South Africa. He is a co-editor of the most popular book on PPP titled: *Public Private Partnerships: Managing Risks and Opportunities*, published by Blackwell Science.

Dr Faisal Alsharif

Faisal Alsharif graduated with an architectural degree. He worked as a contractor for a number of years and was involved in several large housing projects. He then did an MSc in construction project management, followed by a PhD in financial modelling in PPP projects. The work focused on the UK construction industry and a computer-based financial model was developed and tested as a result. He has now moved back to industry and is currently working as main contractor on several large scale projects in Saudi Arabia, Sudan and other countries in the region.

Demos C. Angelides

Demos Angelides is Professor of Marine Structures in the Department of Civil Engineering at the Aristotle University of Thessaloniki, Greece. He has been Chairman of the same department since September 2005. Previously, he was with McDermott International, Inc., in the USA, having several engineering and management positions. He received his Dipl. Ing. Degree in Civil Engineering from the Aristotle University of Thessaloniki, Greece, and his SM and PhD both from the Civil Engineering Department of the Massachusetts Institute of Technology (MIT). He delivered several invited lectures in the USA and Europe. He has published extensively. He is a member of several scientific and professional societies. He is included in Who's Who publications of the Marquis Publications Board.

Aaron Maano Anvuur

Aaron Anvuur is a Ford Foundation International Fellow. He holds a BSc (First Class) in Building Technology from the Kwame Nkrumah University of Science and Technology (KNUST), Ghana, and an MSc(Eng.) (with Distinction) in International Construction Management and Engineering from the University of Leeds, UK. He is a lecturer in the Department of Building Technology, KNUST, on leave to the University of Hong Kong as a PhD candidate. He has consultancy, project management and contracting experience on a range of building and civil engineering projects. His research interests include construction procurement systems, human factors in project management, risk and value management.

Dr Darinka Asenova

Darinka Asenova is Senior Lecturer at the Division of Accounting, Finance and Risk, Glasgow Caledonian University. Having a scientific and economics background, her current research interests include a range of risk related issues such as risk management in PFI projects, behaviour of financial service providers in the PFU environment, communication of scientific risks as well as policy implications of the risk communication.

Guzide Atasoy

Guzide Atasoy has recently finished her MSc thesis about modelling project success using cognitive maps. She is currently conducting her PhD study, in the Civil and Environmental Engineering Department of Carnegie Mellon University, USA.

Professor Matthias Beck

Matthias Beck is Professor of Public Sector Management at the York Management School, University of York. He studied for a first degree in Architecture and Town Planning at the Universitaet Stuttgart in Germany.

In 1987 he participated in an exchange programme with the University of Kansas, where he graduated in 1989 with a Master in Architecture and a Master in Urban Planning degree. In the same year he began his doctoral studies in the Department of Urban Studies and Planning at the Massachusetts Institute of Technology, where he received a PhD in Urban and Regional Science in 1996. He had lecturing posts in economics and economic history at the University of Glasgow and in economics at the University of St Andrews. Before coming to York he was Professor of Risk Management and Director of the Cullen Centre for Risk and Governance at Glasgow Caledonian University. His main research interests are: risk management and risk regulation with a particular focus on the public sector; public-private partnerships; and state-business relationships in transitional and developed economies. He has been a co-principal investigator in a DETR/EPSRC LINK programme funded project titled *A Standardised Framework for Risk Assessment and Management of Private Finance Initiative Projects*. He is currently the principal investigator of a NHS, National Insitute for Health Research, Service Delivery and Organisation Programme into the The Role and Effectiveness of Public–Private Partnerships (NHS LIFT) in the Development of Enhanced Primary Care Premises and Services.

Professor M. Talat Birgonul

Talat Birgonul is a Professor and lectures in the Construction Management and Engineering Division of the Civil Engineering Department in the Middle East Technical University. His primary research interests include engineering economy, health and safety, construction planning, macroeconomic aspects of the construction industry and claim management. Apart from his academic activities, he acts as an expert witness in Turkish courts and Arbitral Tribunals and gives a claim management consultancy service to leading construction companies. Currently, he is acting as the director of Construction Management and Engineering Division of Civil Engineering Department in the Middle East Technical University.

Professor Philippe Burger

Philippe Burger is Professor of Economics and Head of Department of Economics at the University of the Free State, South Africa. During 2007 he was consultant to the OECD on PPPs, researching issues regarding the definition, budgeting and VFM of PPPs. In addition, in 1999 as member of a task team led by the national Department of Finance researching the viability of Public-Private Partnerships in South Africa, he co-authored an unpublished research report for the Department of Finance that entailed an economic and fiscal assessment of PPPs within the South African context.

Charles Y.J. Cheah

Charles Cheah is an Assistant Professor in the Division of Infrastructure Systems and Maritime Studies at NTU. He obtained his doctoral degree from the Massachusetts Institute of Technology (MIT) and also holds the Chartered Financial Analyst designation. Dr Cheah's research interests are broad and interdisciplinary in nature, generally focusing on corporate-level issues including strategy and finance. He has published more than 30 journal and conference papers in construction, finance, project management, business management and engineering education. His research projects include real option applications, catastrophic risk analysis of infrastructure systems and strategic analysis of large, global engineering and construction enterprises.

Dr Ezekiel Chinyio
Ezekiel Chinyio worked as a site quantity surveyor for 1 year before joining the academic sector 20 years ago. His research areas have included procurement, risk management and organisations. After spending 1 year as a visiting researcher at the University of Reading in the mid 1990s he went on to study for his PhD at the University of Wolverhampton. He has since then worked as a research fellow with Glasgow Caledonian University and senior lecturer with the University of Central England (now Birmingham City University). He is now in a senior lecturing post at the University of Wolverhampton.

Chris J. Clifton
Chris J. Clifton is currently a senior project manager with Multiplex Group where he has a particular focus on social infrastructure PPPs. He has recently completed a PhD investigating enhanced PPP frameworks utilising alliance techniques, and also holds degrees in Civil Engineering (Honours) and Commerce from the University of Melbourne.

Dr Irem Dikmen
Irem Dikmen is an Associate Professor and lectures in the Construction Management and Engineering Division of the Civil Engineering Department in the Middle East Technical University. Her primary research interests include risk management, knowledge management, strategic management of construction companies and use of IT to improve the construction value chain. She also conducts extensive research in the area of risk management of BOT projects in Turkey. In addition to her research activities, she gives continuing education seminars and consultancy services to construction professionals about international business development and construction risk management.

Dr Colin F. Duffield
Colin F. Duffield is an Associate Professor at The University of Melbourne and Academic Co-ordinator for postgraduate Engineering Project Management courses within the Department of Civil and Environmental Engineering. His research into efficient procurement of major projects has recently focused on the use of PPPs where the long-term sustainability of service outcomes is governed by the interaction between policy, technical matters, risk management, financing and contractual arrangements. He is a Fellow of Engineers Australia and a member of their National Committee for Construction Engineering, a member of the Australian Institute of Project Managers and a Registered Building Practitioner.

Professor David Eaton

David Eaton is a professor and chartered surveyor. His PhD is in Competitive Advantage in Construction from the University of Salford. He is currently Director of Management in Construction Research Centre. His research interests include risk and competition in construction, PFI procurement, innovation and improvement tools and techniques, risk management, procurement of major infrastructure projects, innovation and change in property and construction, quality improvements, management systems and competitive advantage in construction. He is currently Visiting Professor of Construction Management at University of Sakarya, Adepazari, Turkey.

Dr Rod Gameson

Rod Gameson has been involved in the construction industry as a practitioner, teacher and researcher for over 30 years. He spent a number of years working as a site engineer and site manager on UK construction projects and then completed a BSc (Hons) and a PhD at Reading University. Rod has held research and lecturing posts at Reading University, the University of Manchester Institute of Science and Technology (UMIST) (now part of the University of Manchester) and Newcastle University in Australia before joining the University of Wolverhampton.

Dr Michael J. Garvin

Michael J. Garvin is an Assistant Professor in the Myers-Lawson School of Construction at Virginia Tech. His research is geared toward fundamentally changing how institutional owners make real asset investment and financing decisions. His active research initiatives include a project to improve risk mitigation strategies for infrastructure projects where private finance is at risk, and a project to identify best practices for P3 arrangements. Dr Garvin is a recipient of the Presidential Early Career Award for Scientists & Engineers (PECASE), which is the highest honour bestowed by the US government on outstanding scientists and engineers beginning their independent research careers

Darrin Grimsey

Darrin Grimsey has worked on infrastructure projects variously as an engineer, project manager and financial advisor. He specialises in the delivery of infrastructure projects including commercial, strategic and financial advice, project structuring, risk identification and contract negotiations.

Dr S. Ping Ho

Dr Ho received his PhD degree from the University of Illinois at Urbana-Champaign in 2001. His research expertise includes game theoretic modelling, financial economics and strategic management. His research goals are to develop theories and policy/practical implications to serve as foundations for governance in construction and for the management of PPPs. He considers himself a scientist in economics and engineering management. Since Ho joined National Taiwan University, he has developed the Infrastructure Policy and Economic Research (IPER) group and maintained an active sponsored research programme, publishing papers in top-quality journals and advising governments and industry practitioners.

Professor Ammar Kaka

Ammar Kaka is Professor of Construction Economics and Management and has an international reputation for research on financial planning and control of construction projects. Author of more than 100 research papers, he has held several research grants that led to the development of a dynamic cash flow forecasting model, mapping of projects' financial management processes and occupancy cost prediction for buildings. He is currently leading EPSRC funded research projects aimed at studying innovative payment systems, the use of computer vision in the measurement of work in progress and the development of a best practice process map for FE/HE funded construction projects. He is a member of the CIB W55 and W65 commissions and editorial boards for three academic journals. He has been invited to give several keynote speeches and research workshops in several places in the UK and abroad.

Professor John Kelly

Professor Kelly, currently chairman of the consultancy Axoss Ltd and visiting professor at Nottingham Trent University, is a chartered surveyor with industrial and academic experience. His quantity surveying career began with a national contractor moving to a small architects' practice and later to an international surveying practice. His academic career began at University of Reading as a research fellow, moving to Heriot-Watt University as a lecturer and later senior lecturer and finally to Glasgow Caledonian University where he held the Chair of Construction Innovation until November 2007. His research into value management and whole-life costing began in 1983 and has been well supported by grants from both public and private sector. He has published four books and eight research monographs and technical manuals. John has facilitated a number of PPP/PFI value management workshops in which the option appraisal of capital spend and facilities management initiatives were of key concern. John has worked with the UK local authorities organisations COPROP and SCQS as a consultant to develop formalised approaches to option appraisal of construction projects.

Professor Mohan Maheswaran Kumaraswamy

Mohan M. Kumaraswamy, chartered civil engineer and chartered builder, is a Professor at the Department of Civil Engineering of The University of Hong Kong. After a BSc (Eng.) from Sri Lanka, he worked on designs, construction and construction management in Sri Lanka and Nigeria, before his MSc in Construction Management, and PhD, from Loughborough University. As a construction manager and then a director of the first construction project management company in Sri Lanka, he led many projects and internationally funded consultancies. He is active in professional bodies and is the Executive Director of the Centre for Infrastructure & Construction Industry Development based at the university.

Douglas Lamb

Douglas Lamb is a financial modeller in the Infrastructure Finance team at the Royal Bank of Scotland. He supports the team in developing and running models for infrastructure and energy projects. Douglas has published two books on topics covering value and risk analysis, private finance initiative and PPPs.

Professor Mervyn Keith Lewis

Mervyn Lewis is Professor of Banking and Finance at the University of South Australia. Previously he was Midland Bank Professor of Money and Banking at the University of Nottingham. He has been visiting professor at many international universities. In 1986 he was elected a Fellow of the Academy of the Social Sciences in Australia. Professor Lewis has authored or co-authored 20 books and over 120 research papers. His book, *Public Private Partnerships: the Worldwide Revolution in Infrastructure Provision and Project Finance* (2004), co-authored with Darrin Grimsey, won the 2005 Blake Dawson Waldron Prize for Business Literature.

Dr Florence Yean Yng Ling

Florence Ling is a tenured Associate Professor and Vice Dean (Admin and Finance). She was a quantity surveyor (1987–1995) and a Visiting Scholar at the University of California, Berkeley (2002). She teaches construction economics and has received five university level teaching excellence awards. Her research interest is in international construction. She has published some 60 journal papers, and 30 conference papers. She won the Emerald Literati Network 2006 Highly Commended Award. She is a member of the industry's Workplace Safety and Health Construction Advisory Sub-Committee.

Anthony Merna

Anthony Merna is a lecturer in the Project Management Division in the School of MACE at the University of Manchester. He teaches project finance, risk management and quality management to MSc and MBA students and supervises MSc, MBA, MPhil and PhD research students. The author of 15 books on topics covering project finance, Private Finance Initiative, risk management, BOOT strategies and dispute resolution and over 40 refereed publications. He is also Senior Partner of Oriel Group Practice, a multi-disciplinary research consultancy based in Manchester.

Professor Stephen Ogunlana

Stephen Ogunlana is currently Professor of Construction Project Management at School of the Built Environment, Heriot-Watt University, Edinburgh. Before this he was Professor of Construction Engineering and Infrastructure Management at the School of Engineering and Technology, Asian Institute of Technology, in Pathumthani, Thailand. His doctoral degree is from Loughborough University, UK. His research interests are in project management (risk management, stakeholder management, project simulation, etc.) public-private partnerships, human resources management (motivation, productivity, leadership studies, learning, training and empowerment, etc.), system dynamics simulation, and construction process improvement. He consults for several governments and conducts in project management and training. He is the editor of the book, *Profitable Partnering in Construction Procurement*, published by SPONS.

Dr Thillai A. Rajan

Thillai Rajan is currently an Assistant Professor in the Finance and Strategy area in the Department of Management Studies at the Indian Institute of Technology Madras. Before joining academia, he spent several years in the venture capital and IT services industry in leadership roles. His research interests include corporate finance, infrastructure finance, venture capital, and corporate strategy. He has a doctorate degree from Indian Institute of Management, Bangalore.

Dr Herbert Robinson

Herbert Robinson is a senior lecturer in construction economics and financial management at London South Bank University. He was recently involved in major research on knowledge transfer in PPP/ Private Finance Initiative (PFI) projects at Loughborough University. After graduating from Reading University with a degree in quantity surveying, he worked with international consultants Arup and in a World Bank funded project on public works infrastructure before returning to academia to pursue his research interests. He holds a PhD in infrastructure management and is the co-author of a recently published book, *Infrastructure for the Built Environment: Global Procurement Strategies*.

Jon Scott

Jon Scott is a chartered surveyor and has been working as a cost consultant for over 4 years with Cyril Sweett Limited, a leading consultancy that provides a whole range of PFI/PPP services including bid management, whole life cost advice and due diligence. Cyril Sweett has won Best Technical advisor at the Public Private Awards in 2001, 2004 and 2006. Prior to joining Cyril Sweett, he has worked for a variety of public sector organisations. He graduated in economics in 1993 and later gained a masters degree in quantity surveying from London South Bank University.

Sheetal Sharad

Sheetal Sharad has a Bachelor degree in Electronics & Telecommunications from Rajiv Gandhi Technical University, Madhya Pradesh, India. She subsequently completed her MBA with specialisation in Finance from the Department of Management Studies, Indian Institute of Technology Madras. She has interests in the field of corporate investment advisory and project financing. She currently works in the Global Clients Group, Corporate Banking Division of ICICI Bank in mid 2006.

Professor Jean Shaoul

Jean Shaoul is Professor of Public Accountability at Manchester Business School where she focuses on public accountability and social distributional issues in the context of business and public policy. She has written on: privatisation, particularly water and rail; the use of private finance in public infrastructure under the UK government's Private Finance Initiative and public-private partnerships, particularly in roads, hospitals and London Underground; the use of private finance in roads in Britain, Spain and internationally; international regulatory reform, e.g. WTOs GATS; and public expenditure

Dr Raju B. Shrestha

Raju Shrestha is a senior engineer in Nepal Electricity Authority. He obtained his MSc degree in Hydropower Engineering from People's Friendship University, Moscow and a PhD in Construction Engineering and Infrastructure Management from Asian Institute of Technology, Thailand. He has previously worked in construction of a major hydropower project in Nepal.

Sidharth Sinha

Sidharth Sinha has a Bachelor degree in Information Technology from the University of Delhi. He subsequently completed his MBA with specialisation in Finance from the Department of Management Studies, Indian Institute of Technology Madras. He has interests in the field of derivatives and likes to follow the Indian money markets. Currently he works for ICICI Bank in the Corporate Treasury division as a Forex Dealer.

Arthur L. Smith

Arthur L. Smith is Chairman of the US National Council for Public-Private Partnerships and President of Management Analysis, Incorporated, a consulting firm headquartered in Vienna, VA. Mr Smith has 30 years' experience in analysing and implementing PPPs. He has PPP experience on five continents, is the author of more than 30 articles on PPPs published in six languages, and is a frequent speaker and trainer on PPP-related topics. Mr Smith is currently leading an international team of consultants on a United Nations-funded 'Comparative Review of Public-Private Partnerships in Market and Transition Economies', and was a primary author of the United Nations Economic Commission for Europe's *Guidelines to Promoting Good Governance in Public-Private Partnerships*. Mr Smith holds an MS in Technology Management from the University of Maryland.

Professor Steven Toms

Steven Toms is Head of School and Professor of Accounting and Finance at the York Management School, University of York. As an undergraduate he read Modern History at Trinity College Oxford. He then qualified as a chartered accountant with Price Waterhouse in 1986. His PhD was on the finance and growth of the Lancashire cotton textile industry from the University of Nottingham in 1996. In the meantime he completed a PGCE and an MBA programme. Before joining York as Professor of Accounting and Finance in February 2004, he was Professor of Accounting and Business History at the University of Nottingham. In October 2007 he was appointed joint editor of Business History. His main current research interest is the role of accounting, accountability and corporate governance in the development of organisations, particularly from a historical perspective. He is interested in perspectives that integrate financial models with economic and organisational theory and corporate strategy. Specific applications range from business history, in particular cotton and other textiles trades to capital markets and social and environmental accounting.

Dr Yiannis Xenidis
Yiannis Xenidis received his PhD from the Aristotle University of Thessaloniki. His thesis was on 'Risk Analysis in Build-Operate-Transfer Projects with the Use of Fuzzy Theory'. He has published several papers and has made several presentations in conferences with regard to issues related to PPP. Currently, he is an adjunct lecturer at the Department of Regional Planning and Development and a Research Associate at the Department of Civil Engineering at the Aristotle University of Thessaloniki. He is a member of scientific and professional organisations and an associate editor at the journal of *Integrated Environmental Assessment and Management* (publisher: SETAC).

Dr Sudong Ye
Sudong Ye is an Associate Professor at Beijing Jiaotong University. His research interests include the areas of project finance, contracting strategies for infrastructure projects (such as PPP/BOT) and project management, and he has published articles in such journals as *Journal of Construction Engineering and Management, Construction Management and Economics*, and *Journal of Financial Management of Property and Construction*. In addition, he has served as an advisor to China Society for WTO Studies.

Dr Xueqing Zhang
Xueqing Zhang is an Assistant Professor in the Department of Civil Engineering, The Hong Kong University of Science and Technology. He holds BEng and MEng degrees from Hohai University, a PhD degree from The University of Hong Kong, and a PhD degree from The University of Alberta. He has presented papers at many international conferences and published widely in top international journals in the areas of construction engineering and management, project financing, and infrastructure development and management. He has worked with province and ministry level government departments in China for several years and also served as an editor of the *Journal of Soil and Water Conservation in China*.

Acknowledgements

The editors would like to thank Dr Sally Brown and Dr Deborah Fitzsimmons of The York Management School for their help in editing several chapters of this book. Thanks also go to a number of staff members at the University of York who conducted peer reviews of individual chapters.

Dedication

Many thanks to our families that just never stopped supporting us in so many ways, directly and indirectly.

In particular this book is dedicated to Dr Caroline Hunter-Beck who left this life three years ago. In her short time with us, Caroline lived life to its fullest, coping with each day, no matter how painful. Caroline showed us all how to live with pain and not miss out on living each day, with love and giving.

Introduction: Perspectives on PPP Policy, Finance and Management

Akintola Akintoye and Matthias Beck

Background and Purpose

Around the world, public-private partnerships (PPPs) have become an increasingly popular means for procuring public services and infrastructure. Much of this is due to the fact that PPPs allow governments to secure much-needed infrastructure without immediately raising taxes or borrowing (The World Bank, 2005). Today many governments view PPPs as a win–win option for meeting their investment needs (The World Bank Institute, 2006). These views are based on a number of rationales. Firstly, it is often thought that PPPs provide budgetary room without prejudice to the sustainability of the government's financial position (Heller, 2005). Secondly, there is a presumption that the fiscal space created via PPPs will boost medium-term growth and thereby generate fiscal revenue in the future (The World Bank, 2005). Thirdly, it is often assumed that PPPs will reduce government risk exposure by transferring those risks to the private sector, which is better able to bear or manage them (The World Bank, 2005). Lastly, there is an expectation that the involvement of the private sector in the financing of infrastructure and services will increase accountability and transparency, reduce corruption and create incentives for the prudent management of public expenditure (The International Monetary Fund, 2005).

Although it is assumed that PPPs, at least in theory, will bring benefits to their host governments, there is increasing evidence that the practical implementation of PPPs is not without managerial, technical and even fiscal problems (Erhardt and Irwin, 2004). The problems stem partially from the complexity inherent in many PPP projects and the related increased demand for skills amongst PPP participants. These issues affect not only public sector clients for whom intense collaboration with private sector parties will often be a novelty, but also private sector parties for whom PPPs have often presented unique and unfamiliar challenges (Ezulike et al., 1997).

This edited book seeks to contribute to the debate and the understanding of PPP in a number of ways. Firstly, it is our intention to examine the unique challenges which PPP, as a policy, presents to participants and stakeholders in different regions and continents. Secondly, this book examines state-of-the-art approaches to PPP finance with a view towards highlighting the broad range of options available to those involved in designing and managing the financial structures and parameters which underpin PPP. Lastly, the book maps out a number of approaches for the improved management of PPP with a focus on such issues as innovation, risk assessment and management and costing.

Structure and Summary of the Chapters

Part one: PPP policy

Part 1 of this books focuses on PPP as a policy and investigates issues such as PPP development, practices, trends and the inherent contradictions which characterise some of these approaches. It presents both theoretical and empirical evidence drawn from developed and developing countries. Throughout the chapters the need to develop teamwork and good organisational structures at firm and government levels is emphasised. This part has eight chapters.

Chapter 1 by Ezekiel Chinyio and Rod Gameson reviews the performance of the 'Private Finance Initiative' (PFI) in terms of service operation. Through PFI, public services are delivered using new or refurbished facilities which are maintained throughout a concession period. The benefits of PFI are identified. It is noted that many clients are happy with PFI because they do not have to maintain buildings, and soft services are provided by other parties. The private sector too is happy because they get a steady income that lasts decades and a high return on their investment. As several PFI schemes are now in their operational phases, an examination of the quality of service provided from the perspective of the users is worthwhile. It is noted that PFI schemes can be complex and several clients will use this procurement approach only once. To each new client, the process can present a steep learning curve. Indeed, the hospital sector has instances where some decisions made in the brief could not foresee the precise consequences; and in construction rectifying mistakes can be very costly. A means of exploiting the accumulated knowledge in the PFI domain can therefore be very helpful to new clients.

Chapter 2 by Jean Shaoul discusses some of the weaknesses in the appraisal methodology commonly used for justifying the use of private finance *ex ante*. It reviews *ex post* evidence as it relates to the claims for private finance in the UK. In the context of building to time and budget, robust project specification, the use of penalties to incentivise good performance, the financial cost of PFI, risk transfer and affordability for road and hospital schemes, and the additionality argument, evidence is found to indicate that outcomes rarely match initial expectations. Meanwhile, PFI schemes frequently result in hidden transfers of wealth to financiers.

Chapter 3 by Darinka Asenova and Matthias Beck focuses on accountability and transparency in PFI projects. It explores the meaning and the significance of these concepts in relation to the activities of the financial services providers who, apart from their decisive function as providers of capital, play multiple roles in PFI procurement. In theory, PFI should enhance democratic accountability, at least through the embedded mechanism for option evaluation and value for money (VFM) tests. However, the ability of PFI to deliver accountable solutions for public service provision has often been questioned, with critics pointing at the opaque nature of PFI contacts and the lack of democratic oversight. The authors argue that there is another, perhaps less obvious way in which the accountability and transparency of PFI is constrained. These constraints arise from the fact that the powerful global financial institutions which dominate the PFI scene impose risk–return criteria on PFI projects which severely restrict the options available to the public sector client and other stakeholders. The chapter utilises two case studies of UK PFI projects, one housing accommodation project and one waste management project, in order to demonstrate how this operates in practice.

Chapter 4 by Steven Toms, Darinka Asenova and Matthias Beck addresses the issues of profitability of PFI to the UK private sector. The profitability of PFI projects to the private sector remains one of the key areas of debate in the UK. In recent years this dispute has intensified as a consequence of the negative publicity associated with UK PFI refinancing deals which have opened some private sector protagonists to allegations of excessive profiteering. Moreover, the financial aspects of PFI contracts are often concealed, usually justified by 'commercial confidentiality', so that in the absence of verifiable data, the debate about profitability remains even further from resolution. This chapter investigates four issues. Firstly it carries out a sector by sector comparison of refinancing profits. Secondly it examines the profits from refinancing of one firm to another. Thirdly, it examines trends of profitability on refinancing contracts. Lastly, it analyses the relative public sector shares of refinancing profits by sector and compares them to private sector profits. One of the striking results of the cross-sector comparison is the excessive returns obtained in health, which highlights a need for greater adeptness on the public sector's part in negotiating PFI deals.

Chapter 5 by Philippe Burger considers the theoretical rationale for PPPs and discusses the creation of a dedicated PPP unit in South Africa. The chapter discusses the role and operation of the unit as well as its future challenges. With the dawn of democracy in South Africa in 1994, the new South African government decided to restructure the management of state assets. Following the introduction of PPPs in the UK in the early 1990s, the South African government in the late 1990s explored, and ultimately implemented, a framework that allows for the use of PPPs in South Africa. At the heart of the South African PPP structure is the National Treasury's PPP Unit constituted in 2000. The South African dedicated PPP unit plays a key role in the creation of PPPs. In particular the unit has the final authority in the approval of PPP agreements. It also provides technical assistance to government departments and provinces initiating PPPs.

Chapter 6 by Thillai Rajan, Sheetal Sharad and Sinha Sidharth explores the Indian experiences in PPP development. The government of India has recognised the importance of creating adequate infrastructures to achieve economic growth. To accelerate the process of creating infrastructure capacity, many sectors were thrown open for private sector investment. This study analyses the experiences of Cochin International Airport Limited (CIAL), the first commercial airport in India to come under the PPP format. Built at a cost of INR3.15bn, CIAL became operational in 1999. Supported by the growth in airline traffic, CIAL started generating profits within 2 years of operation.

Since CIAL was the first airport in India to have private sector participation, several innovative financing methods were tried out. The success of CIAL indicates that successful smaller projects can be helpful in attracting subsequent investment in larger projects. In an emerging economy like India, where there is no track record on private sector investment, political risk management is very important for a successful implementation of large infrastructure projects. A supportive bureaucracy can play a very important role in project implementation during political regime changes.

Chapter 7 by Akintola Akintoye is concerned with PPP in developing countries. The chapter presents an overview of PPP and discusses PPP in developed countries. General information on the use of PPP in developing countries, and various initiatives that have been developed to encourage these countries, as well as the extent to which PPP for infrastructure development has emerged in developing countries, are discussed. The chapter concludes with a discussion of the key elements which create enabling environments for the use of PPPs in developing countries. These include government commitment, increased private interest, move to competitive processes, greater availability of information, acceptable prices and high developer returns as incentive to the private investors and large size of projects.

Chapter 8 by Mohan Kumaraswamy, Florence Ling and Aaron Anvuur provides an introductory overview of the changing focus of recent PPPs. Previous needs for attracting private finance to public infrastructure are being superseded by pressures for better value services. This requires a shift in mind sets and skill sets of teams that can properly handle the increasingly wide-ranging and far-reaching PPP projects. Recent initiatives to develop a new wave of PPPs in Hong Kong and Singapore are explored, and compared with some other regions. Specific comparisons focus on lessons learned when selecting teams. A general conceptual framework is developed to indicate how appropriate teams may be chosen and developed in line with special PPP needs.

Part two: PPP Finance

Part 2 focuses on financial aspects of PPP and discusses the investment, modelling and accounting practices associated with PPP projects. This part has seven chapters.

Chapter 9 by Demos Angelides and Yiannis Xenidis discusses lessons learned, with emphasis on financing issues, from the systematic private participation in infrastructure development for almost 20 years. In the recent past, both developing and developed countries have been engaged in partnerships with the private sector in order to develop the required infrastructure in different sectors of the economy (e.g. power generation, transportation, etc.). These PPPs were implemented with several variations aiming, in all cases, at enhancing economic and social growth, while, at the same time, minimising requirements of public funds. Currently, this project delivery scheme seems to lack the strong support that was demonstrated both by the public and the private sector in the 1990s. In this chapter the potential for a new flourishing of PPPs as a means for infrastructure development is discussed. The aim is to provide decision makers with useful ideas on achieving sustainable financial structures for PPP projects.

Chapter 10 by Sudong Ye focuses on four components of project financing which are central to the development of PPP projects. These are the optimisation of capital structure, the design of organisational structure, the design of contractual structure and the enhancement of creditworthiness. Each component has various options, and their combination forms a financing pattern. According to the type of organisational structures, patterns of financing PPP projects can be classified into three general patterns, namely mono-entity structure, dual-entity structure, and multi-entity structure. The choice of financing patterns depends on various factors. Of them, the complexity of construction and the characteristics of fund providers are two key determinants. This analysis provides a thinking tool for designing an optimal project financing for PPP projects.

Chapter 11 by Arthur Smith focuses on PPP financing in the US. It notes that the PPP market in the US is complicated by its fragmented nature. Government agencies are increasingly turning to PPPs to accelerate infrastructure acquisition and maintenance. The transportation sector has been particularly active, with the federal government implementing a succession of new initiatives to facilitate private participation in transportation projects. State and local governments have also been active in this regard. However, unlike many other countries, the US has no single federal agency with oversight of PPP policy and issues. Authority to undertake PPPs is typically granted to agencies by Congress on an agency-specific basis, or even a function-specific or a project-specific basis, and there is no standard approach to federal PPPs for infrastructure. At the state level, the market is similarly fragmented, with each state enacting its state-specific laws. This chapter focuses on financing of the transportation PPPs and discusses four cases of recent transportation PPPs: Massachusetts Route 3; Chicago Skyway; Indiana Toll Road; and Pocahontas Parkway. Although there has been a number of highly successful PPPs, encouraging broader utilisation of this approach, success has not been not universal.

Chapter 12 by Ammar Kaka and Faisal Alsharif focuses on the financial management of PFI projects. The chapter commences with a review of the important literature in the area of financial management in the construction industry in general and PFI in particular. It argues that, although there has been limited work in financial modelling of PFI projects, extensive work in construction cost modelling and cash flow forecasting can provide the basis for the development of a financial model in PFI projects. The chapter reports on a survey of current industrial practices in appraising PFI projects and proposes a computer-based model that will assist both clients and project teams in assessing and/or tendering for PFI projects. Whilst the model has been developed for UK schools projects, the methodologies applied could be deployed across other sectors.

Chapter 13 by Charles Cheah and Michael Garvin investigates the use of real options theory as a means for PPP investment modelling. In PPPs where private finance is at risk, economic feasibility analysis is clearly significant. Many infrastructure investments possess option-like features such as deferment or staged investment. Frequently, decision makers intuitively account for such options during the decision process, but typical investment decision methods cannot quantify the value of opportunities or contingencies. This circumstance, combined with the reality that project stakeholders often independently appraise the economic viability of a project investment, each from their own perspective, can make it difficult to define the 'true' economic value of a privately financed project. Indeed, the independent and subjective derivation of the economic value of these arrangements is often an obstacle to negotiating the concession agreement, attracting equity investors, and securing long-term financing. The industry needs reasonable approaches for recognising and quantifying the value of such 'real' options to enhance the strategic consideration of value and risk in PPP investments. Real option theory is gradually gaining acceptance in many industries as an approach that can capture managerial and operating flexibility. Opportunities to transfer this theory to the PPP infrastructure project environment abound. However, the unique development context of construction projects and infrastructure systems typically poses multiple challenges when modelling option-like features.

Chapter 14 by Raju Shrestha and Stephen Ogunlana assesses the financial implications of power purchase agreements (PPAs). PPAs are the most important contract underlying the construction and operation of independent power productions (IPPs). In designing PPAs a variety of questions have to be answered concerning the concessions to be provided to IPPs at various stages of the partnership. Based on the design of PPAs, the clauses may have direct financial implications on the revenue stream of the project for the stakeholders. This chapter analyses the financial implications of PPAs using the example of Nepalese case studies. The analysis shows that 'take or pay' clauses, purchase guarantees of excess energy, supply guarantees of minimum energy, and 'allowance of third party sales' can significantly affect the revenue stream of a project.

Chapter 15 by Ping Ho introduces two theoretical models and assesses their policy implications on PPPs: bid compensation and financial renegotiation. These two issues are closely associated with the success of PPP projects, but often overlooked. The major problem of bid compensation is that it has never been proven effective, and if bid compensation is ineffective and governments are not aware of its ineffectiveness, governments will lose their chances of adopting other approaches to improving bid quality or concept development. Financial renegotiation refers to the rescuing financial subsidy negotiation due to project distress. The real problem of financial renegotiation is that the expectation that governments will bail out a distressed project through renegotiation can cause serious opportunism problems. A case study of Taiwan High Speed Rail is used to illustrate the renegotiation model and to illustrate potentially costly lessons.

Part three: PPP Management

Part 3 focuses on management issues associated with PPP and discusses risk, value and appraisal processes and practices. It covers issues such as whole-life cycle costing, design innovation, preparation of winning bids, risk assessment and allocation, VFM assessment, payment mechanism and concession period determination. The part has nine chapters.

Chapter 16 by David Eaton and Rifat Akbiyikli discusses issues associated with innovation in PPP. The chapter recognises that PPP in itself is an innovation in public procurement, but the public sector must decide what gives the best scope for the private sector to add value and in all cases adhere to key principles such as whole-life costing, VFM and optimum risk allocation. The chapter identifies incentives and impediments to creativity in PFI. The sources for competitive advantage in PFI/PPP road projects which the chapter identifies include investment innovation (financial model), VFM (value adding), partnering, performance-related output, superior service product and high-quality project management.

Chapter 17 by Colin Duffield and Chris Clifton focuses on how consortia seek innovative solutions to demonstrate that they offer a VFM solution in response to an invitation to bid for a PFI/PPP. Innovations range from technical advancement, creative design that leads to whole of life efficiency and functionality, optimised risk allocation (or for some governments – maximum risk transfer), corporate structures, operational improvements and efficiency and financial engineering to the most cost-effective outcome. Discussion on design innovation draws from a workshop convened in conjunction with The Royal Australian Institute of Architects in 2006. The chapter considers the financing options available within the maturing of the PFI/PPP market prior to discussing the relative merits of design innovation as it relates to the preparation of winning proposals. It concludes with a commentary on the importance of combining design and finance to produce winning proposals.

Chapter 18 by John Kelly discusses key attributes associated with the staging of whole-life value projects. The chapter examines four distinct attributes, including the identification of a project and its place within the strategies and programmes of a client organisation, the definition of the project in explicit functional terms, the value criteria by which the project will be judged a success and, finally, the method of calculation for determining which of the competing options best satisfy the functional values defined. These options have to be judged in terms of their value to the client and their whole-life cost, and these two evaluations jointly form a whole-life value evaluation.

Chapter 19 by Irem Dikmen, Talat Birgonul and Guzide Atasoy discusses the Turkish experience with build, operate, transfer (BOT) projects, especially in the transportation sector, and present two cases which highlight the complexity of procurement process in transportation investments. The first case study is the Izmit Bay Crossing project, which was cancelled as a result of the court cases, and the second is the Gocek Tunnel project which happens to be the first successfully implemented BOT project realised by the General Directorate of Highways (GDH) in Turkey. In the light of lessons learned from these two case studies, it is concluded that there is no single formula to be utilised during the evaluation of tenders. A best-value procurement approach where the evaluation criteria are determined according to the needs of a client organisation is proposed. The GDH's experience is used to demonstrate how best-value procurement can be employed in BOT projects to determine the ranking of companies which will be invited to negotiations. The proposed methodology consists of two parts: calculation of net present value of costs to the public, and the assessment of risks using a multi-attribute rating technique.

Chapter 20 by Tony Merna and Douglas Lamb outlines a quantitative approach to the analysis of risk and discusses how this approach can be applied through a case study. The case study outlines how risks can be assessed and applied in the VFM assessment. Many countries utilise private finance/PPPs to encourage investment in public services and have formed stringent economic assessments to appraise the validity of private investment in public services. Central to the assessment is the VFM and the associated transfer of risk. In order to form this assessment a public sector comparator (PSC) and private finance alternative (PFA) is created. The chapter concludes that as current practices of identifying the key inputs to VFM differ according to country and sector, the approach presented has been designed to operate worldwide allowing the PSC to aid in future negotiation up to the point of financial close.

Chapter 21 by Darrin Grimsey and Mervyn Lewis develops a five-stage framework for procurement options analysis which is then illustrated by a representative case study of a hospital redevelopment. In the evolving market place for public procurement there now exists a range of delivery models for infrastructure, comprising traditional construction-based procurement methods, PPPs of various forms, and hybrids of them, that can meet different infrastructure service requirements. The decision as to which procurement option to employ is necessarily determined on a case-by-case basis, but the choice can be aided greatly by adopting a systematic approach applicable to a wide variety of different projects.

Chapter 22 by Jon Scott and Herbert Robinson examines the role of the payment mechanism in providing 'value for money' in the delivery of public services. Public sector bodies put forward a VFM case for procuring a project through the PFI route which rests upon risk transfer and efficiency in service delivery. The payment mechanism puts into financial effect the allocation of risk and service performance and ensures that the public sector client's objectives for PFI projects are delivered as set out in the output specification and monitored through a performance measurement system. Using a case study methodology and interviews with key stakeholders of operational PFI projects from the public sector and private sector organisations, the authors found that subjectivity in the output specification and complexity in the performance measurement system affects the effectiveness of the payment mechanism as a risk allocation tool. The chapter concludes that there is a need for improving output specifications to reduce subjectivity, simplifying performance measurement systems so that they are more transparent and, more significantly, to strengthen the logic and link between the output specification, performance measurement system and the payment mechanism.

Chapter 23 by Xueqing Zhang presents a framework for determining the concession period for PPP projects using the example of Hong Kong PPP projects. The Hong Kong government has been seeking innovative and flexible financing strategies to enhance the efficiency and cost effectiveness in the provision of pubic works and services and to stimulate economic activities in general. A wide scope of infrastructure projects has been developed through PPPs, including road tunnels, highways/bridges, an international exhibition centre, prisons, sewage treatment services and massive cultural district projects. The determination of the suitable length of the concession period for different types of projects is a key issue to be addressed in successful infrastructure development through PPPs, because the concession period demarcates the rights and obligations between public and private sectors in a project's lifecycle and it is also critical to the project's sustainable development. A case study of a hypothetical infrastructure project is provided to demonstrate the application of the proposed methodology, mathematical model and simulation techniques.

References

Erhardt, D. and Irwin, T. (2004) *Avoiding Customer and Taxpayer Bailouts in Private Infrastructure Projects: Policy Toward Leverage, Risk Allocation, and Bankruptcy.* World Bank Policy Research Working Paper, 3274, April.

Ezulike, E.I., Perry, J.G. and Hawwash, K. (1997) Barriers to entry in the PFI market. *Engineering, Construction and Architectural Management*, 4(3), 179–193.

Heller, P.S. (2005) *Understanding Fiscal Space.* IMF Policy Discussion Paper, March, http://www.imf.org/external/pubs/ft/pdp/2005/pdp04.pdf

The International Monetary Fund, Fiscal Affairs Department in Consultation with other Departments, the World Bank and the Inter-American Development Bank (2005) *Public Investment and Fiscal Policy – Lessons from the Pilot Country Studies*, http://www.imf.org/external/np/pp/eng/2005/040105a.pdf

The World Bank (2005) *Part III. Special Topic: PPPs-Fiscal Risks and Institutions.* http://siteresources.woldbank.org/INTECA/Resources/eu8-jul05-part 3.pdf

The World Bank Institute (2006) *Initiating a Global Network of Public Private Partnerships for Infrastructure.* Report of from the World Bank Institute's Public Private Partnership for Infrastructure Days, http://go.worldbank.org/7AQTOH&US0

Part One
PPP Policy

Private Finance Initiative in Use

Ezekiel Chinyio and Rod Gameson

1.1   Introduction

This chapter reviews the performance of the Private Finance Initiative (PFI) in terms of service operation. Since its implementation in 1992, several PFI schemes have now gone into their operational phase, so an examination of the quality of service provided under PFI from the perspective of the users would now seem appropriate.

When a PFI scheme is proposed it is usually possible to indicate what the expected benefits will be. However, practical realities do not sometimes match projections. Although user groups are often involved in the planning and provision of facilities under PFI, the truest test of satisfaction is to evaluate the perceptions of actual users or beneficiaries of services. Interviews with stakeholders are used to evaluate the feelings of users concerning the efficacy of services and facilities under PFI schemes.

To facilitate this evaluation, the PFI concept and process is first introduced, drawing from theory. Readers who are familiar with PFI may thus wish to skip this section. The second half of the chapter describes a recent survey and looks at the performance of some PFI schemes in the health and leisure sectors.

1.1.1   Public-private partnerships

Public-private partnerships (PPPs) are long-term alliances formed between the private sector and public bodies often with the aim of exploiting the private sector's resources and expertise in the provision and delivery of public services. In PPP schemes, resources and risks are shared between the public and private sectors for the purpose of developing a public facility to enhance the delivery of public services (Norment, 2002). There are several PPP options that depend upon the remit of the private sector and these are discussed further below.

PPP schemes are often financed and operated by the private sector partner in return for revenues received for the delivery of the facility and services.

This arrangement benefits from the ability of the private sector to provide more favourable long-term financing options and to secure such financing in a much quicker timeframe (NCPPP, 2003). PPP contracts are often made to last long, and are typically for a 25–30-year duration. PPP schemes are often larger in magnitude and face complex risks beyond the scope normally experienced in typical construction. PPPs are generally seen to be capable of coping with projects of such magnitude.

PPPs address the common faults often associated with public sector procurement, such as high construction costs, time overruns, operational inefficiencies, poor design and community dissatisfaction (Mustafa, 1999). They also deliver advanced facilities and services more quickly and efficiently through innovative means (Field and Peck, 2003). Worldwide, PPPs have been used extensively in projects such as roads, prisons, stadiums and tunnels (Jefferies, 2006).

1.2   The Private Finance Initiative

The Private Finance Initiative (PFI) is a type of PPP launched in the UK in 1992 by the Conservative government. The concept went through moderate changes and adjustments in its early years and was later adopted by the Labour government after their election in 1997 (Heald, 2003). Reviews of the PFI process in 1997 and 1999 led to the formation of two establishments:

- Partnerships UK (PUK), as a PPP developer with the objective of providing public bodies with expertise and financial backing
- Office of Government Commerce (OGC), responsible for procurement policy development

PFI is commonly used as a form of procurement (Owen and Merna, 1997). In a PFI arrangement, the private sector partner takes on the responsibility of providing a public service, including maintaining, enhancing or constructing the necessary infrastructure or facility, while the public sector partner specifies the type and quality of service desired. PFI must secure value for money (VFM) to the public sector client.

The UK government has been very keen for its public sector establishments to use PFI as it offers them the opportunity to use private finance, albeit at a risk. This is attractive to the private sector as it offers them good returns in the form of annual payments, referred to as unitary charges. PFI is appealing because more, or improved, infrastructure is needed as public sector establishments cannot meet the supply of infrastructure required due to unceasing population growth (Walker and Smith, 1995).

Fundamentally the aim of PFI is to bring the private sector's finance, management skills and expertise into the provision of public sector facilities and services (Katz and Smith, 2003). PFI, therefore, takes advantage of the management skills of the private sector in the delivery of public services. It is believed that the private sector is better equipped than the public sector to handle some types of service delivery. However, in view of the high transaction costs involved in setting up a PFI scheme, this arrangement is better suited to projects with a capital price in excess of about £50m.

The main responsibility for the design, building, financing and operation of the assets and the risks associated with these is transferred to the private sector. However, such risk transfer warrants a profit incentive to the private sector consortium (Grimsey and Lewis, 2002). Conjoint with risk transfer is the requirement that there must be effective private sector control, i.e. the public sector must not have the dominant influence in a PFI joint venture.

1.3　UK Government's Influence on the Use of PFI

PFI is one strategy for delivering high-quality public services that has become particularly important to the UK government. In assessing where PFI is appropriate, the government's approach is based upon its commitment to efficiency, equity and accountability and on the Prime Minister's principles of public sector reform.

> From 1st June 2000 all UK Central Government clients should limit their procurement strategies for the delivery of new works to PFI, Design and Build and Prime Contracting and from 1st June 2002 these procurement strategies should be applied to all refurbishment and maintenance contracts. Traditional, non-integrated, strategies should only be used where it can be clearly shown that they offer the best value for money. This means in practice that they will seldom be used. (Government White Paper)

Several public establishments have used PFI. As at December 2006, 794 PFI deals had been signed, worth over £54bn; and more projects are in the pipeline. £26bn of further PFI investments across 200 new projects are currently proposed and this includes the planned delivery of over 60 health facilities and 104 schools. These schemes are expected to close by 2010 and their uptake will make PFI one of the largest programmes worldwide.

In April 2007 HM Treasury undertook a large validation exercise and updated its database of PFI projects to reflect this and take into account projects that have:

- Been concluded or terminated
- Changed their contractual structures and, for instance, are no longer classed as PFI
- Been contractually merged

This HM Treasury exercise also identified that some departments have stopped collecting data on some very small projects in order to reduce reporting burdens. The leading users of PFI according to the Treasury's update are shown in Table 1.1.

In October 2007, HM Treasury (2007) published a working document containing information on current signed PFI Projects. This document lists 622 projects with a total capital value of almost £57bn. In addition this document provides data on the unitary payments for these projects (excluding figures for Scottish projects), from 1992–2046, of just over £180bn.

Part One

Table 1.1   An overview of the uptake of PFI (HM Treasury, 2007).

Sector/client	Number of projects	Expenditure to date (£)	Share of the market (%)
Department for Transport	49	22 496.77	42.1
Health	86	8 290.61	15.5
Ministry of Defence	47	5 644.45	10.6
Department for Employment and Skills	106	4 388.94	8.2
Scottish Executive	96	4 175.73	7.8
Department for Environment, Food and Rural Affairs	17	1 505.98	2.8
Home Office	41	1 375.43	2.6
Local governments	46	1 154.45	2.2
Others	103	4 371.80	8.2

1.4   Private Sector Tasks in PFI

Private consortia are usually contracted to design, build, finance and in some cases manage or operate a public service. The combinations of tasks applicable to PFI schemes are:

- Design, build, finance, operate (DBFO)
- Build, own, operate (BOO)
- Build, own, operate, transfer (BOOT)
- Build, operate, transfer (BOT)
- Turnkey

Amongst these, the DBFO option is popular and highly used. Figure 1.1 depicts its contractual links where the client contracts with a consortium which initially is known as a concessionaire. When the contract is signed, the concessionaire is referred to as a 'special purpose vehicle' (SPV). The SPV is normally represented by three to five companies and these would generally include a construction company, a facilities management firm and a financial institution. Depending on the nature of the service to be delivered, specialised firms may form part of the SPV, e.g. a waste management firm in a waste disposal service.

A consortium is necessary since no one company has the in-house expertise required to fund, design, build and operate the service (Carrillo *et al.*, 2006). The SPV is an independent legal entity, typically with its own business name. However, SPVs tend to maintain a very lean structure and carry out most of their contractual obligations by outsourcing, frequently to the parent companies that formed the SPV, for obvious reasons.

1.5   Establishing PFI Contracts

The process leading to a PFI contract is longwinded. Typically, PFI projects consist of 13 stages (Carrillo *et al.*, 2006):

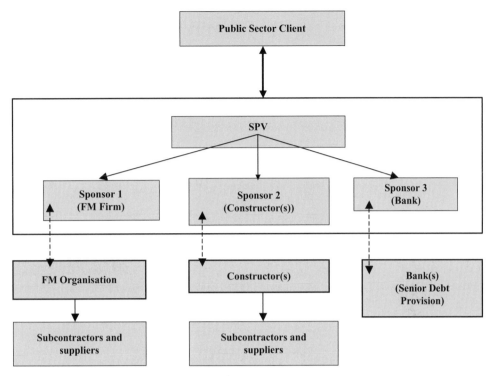

Figure 1.1   Participants in a PFI scheme.

1. Needs assessment
2. Strategic outline case
3. Outline business case
4. *Official Journal of the European Communities* (OJEC) advertisement
5. Pre-qualification questionnaire
6. Preliminary invitation to negotiate
7. Final invitation to negotiate
8. Final offer
9. Preferred bidder/final business case
10. Financial close
11. Construction
12. Operation and maintenance
13. Hand back

Table 1.2 and Figure 1.2 provide further details and an overview of this process. The period from the point of deciding to procure through to financial closure can last anywhere between 12 and 36 months. While early schemes used to take years to procure, financial closure is now reached in 12–18 months.

When a project is advertised and contractors express their interest, the client uses an iterative approach to screen the contractors. The intensity of scrutiny and amount of information involved increases with each subsequent iteration while the number of bidders is whittled down. Ultimately, a preferred

Table 1.2   The PFI process (HM Treasury, 2007).

Stage 1	Establish business need
	Consider key risks – outline risk matrix
Stage 2	Appraise the options
	Keep thinking about risks
Stage 3	Business case and reference project
	Work up reference project (embryonic PSC), risk matrix, costings, sensitivity and tentative transfers
Stage 4	Developing the team
Stage 5	Deciding tactics
Stage 6	Invite expressions of interest; publish *OJEC* Notice
Stage 7	Prequalification of bidders
Stage 8	Selection of bidders (i.e. shortlisting)
	During all of the above stages continuing to work up the PSC
Stage 9	Refine the proposal
	Review the PSC to ensure it is fully worked up before detailed bids are received from the private sector
Stage 10	The invitation to negotiate
	Publish the policy in relation to disclosure of PSCs
Stage 11	Receipt and evaluation of bids
	'Account' for all the risks
	Final check to see whether PSC needs to be revised because of availability of new data, but not for new ideas (picked up from PFI sides)
Stage 12	Selection of preferred bidder and the final evaluation
	Use this accounting to compare the PFI bids and the best PFI bid with the PSC which should be checked to ensure data and risk allocation are as accurate and comprehensive as possible
Stage 13	Contract award and financial close
Stage 14	Contract management
	Record details, share experiences and manage risk

bidder is selected and negotiations between the client and the preferred bidder result in a contract. A reserved bidder is typically appointed alongside the preferred bidder, so if the negotiations break down, the reserve bidder can be invited to step in.

The service operation period in a PFI project is long, often ranging from 15–35 years. After the service operation period, one of two things can happen:

- The provision of services and associated maintenance of assets reverts to the public sector client
- The public and private sector parties can renegotiate

When the first of these two options is selected, all aspects of assets or services below a set standard must be achieved before facilities are transferred to the public sector client (British Institute of Facilities Management, 2003).

1.6   Forms of Finance Used in PFI

PFI provides a way of funding major capital investments without immediate recourse to the public purse. Table 1.3 shows the funding options for PFI

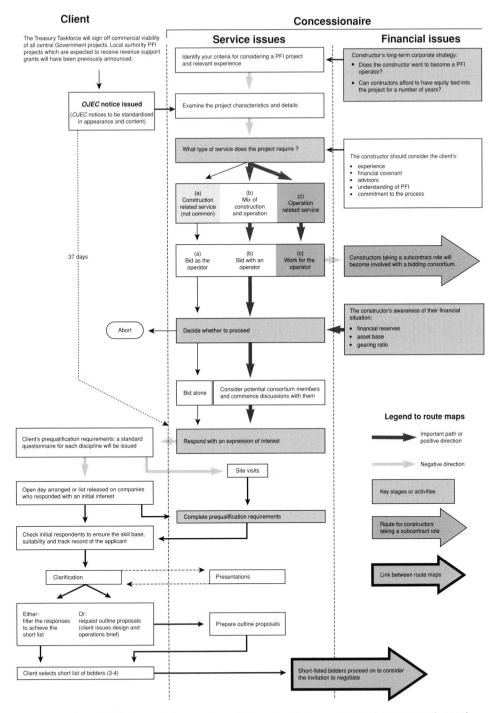

Figure 1.2   Towards formulating a PFI contract (Construction Industry Council, 1998).

Part One

Table 1.3 Funding Options.

Type	Usage
Bank debt	Frequently
Equity	Frequently
Bonds	Occasionally
Loan from shareholders	Occasionally
Mezzanine finance	Exceptionally

schemes. The private sector is required to invest equity in the project and a combination of equity and debt is often used. The debt to equity gearing is often 90:10 but can start from 95:5. In addition to equity, the SPV can source money from bonds, loans from outside the bank and mezzanine finance.

1.7   Performance of PFI Schemes – A Theoretical Perspective

1.7.1   Benefits of PFI

PFI yields certain benefits to the public or private sector or both.

Deregulation

Projects which had previously been delivered under the control of public bodies (e.g. prisons, hospitals, etc.) are now available to private sector organisations (PFP, 1995; Birnie, 1999).

Time savings

It has been reported that the construction period under PFI is shorter (Ward and Chapman, 1995; Price, 2000) with 80% of construction completions under PFI reported to be either within or on time, which is better than most other forms of procurement.

Cost savings

The whole-life price of a scheme procured by PFI is generally cheaper than for procurement by traditional means (PFP, 1995; Grubb, 1998) and this is a requirement for any UK PFI scheme. Before a PFI project is approved, the public sector client must prepare a public sector comparator (PSC) to show the advantage(s) of PFI. The client can also use PSC analysis to test whether another form of procurement will offer better VFM.

By taking advantage of private sector innovation, experience and flexibility, PPP and PFI schemes can deliver services more cost effectively than traditional approaches (Partnerships British Columbia, 2003). The lengthy negotiations preceding the formation of a PFI contract contribute immensely to driving down prices.

Reduction of public sector risk

The public sector bears very minimal risk in PFI projects as it is a requirement that most risk should be transferred to the private sector. Each risk should be allocated to the party best able to manage it and, in general, most of the project risks are better managed by the private partner.

Leeway on government spending

PFI projects have a reduced financial burden on the public purse (Beenhakker, 1997; Jones, 1998), at least initially, as government does not have to pay all costs up front. Through the unitary payments, clients and the government pay back the money invested in a scheme to the private sector. So the government can use its money for other projects while paying for PFI schemes over time. According to Partnerships British Columbia (2003) PPPs can reduce the government's capital costs, helping to bridge the gap between the need for infrastructure and financial capacity.

Further opportunity to make profit

For equity investors, PFI is perceived as a relatively low-risk investment as it is backed by government covenant, provides a stable long-term yield and many of the risks are sub-contracted. Unlike other areas of project finance, PFI has limited exposure to market risks (demand for the infrastructure, commodity prices, etc.). The trend in the secondary market is to develop reasonably large portfolios of yielding assets typically after the construction phase.

Opportunity to develop assets and/or infrastructure

Most PFI schemes involve the provision of new infrastructure. Where current stock is retained, it is often upgraded and maintained on a regular basis.

Enlargement of markets

Private sector participants utilise their skills and knowledge in a number of areas, e.g. finance, law, risk, insurance, facilities management. In this regard, PFI offers further trading opportunities to the private sector.

Innovative solutions

In PFI, the design solutions are not finalised completely until the end of the negotiations (Figure 1.2). During the competitive phase and subsequent

Part One

negotiations, the SPV refines the design and often uses this opportunity to introduce innovative solutions that will benefit the client.

Accounting for maintenance costs

To a client, PFI relieves them of the responsibility for maintaining facilities. If something goes wrong with a building the SPV must fix it within a specified time or be charged on the basis of a predefined penalty.

Curtailing cost escalations

Project services are provided at a predictable cost set out in the contract agreement (Partnerships British Columbia, 2003). Inflation should not affect what the client will pay.

Improved service delivery

This is achieved by allowing both sectors to do what they do best. For example, the private sector will provide high-quality food to hospital patients while the NHS is free to concentrate on treating those patients.

Optimal use of assets

Private sector partners are motivated to make optimal use of the facilities to maximise return on their investment. This can result in higher levels of service and reduced occupancy costs for the government (Partnerships British Columbia, 2003).

1.7.2   Downsides of PFI

Despite its numerous benefits, PFI has its downsides.

High transaction costs

The cost of bidding for PFI projects is quite high (Tiffin and Hall, 1998; Mustafa, 1999; Walker, 2000). Bidding costs for PFI schemes are estimated to be in £millions. The National Audit Office (NAO, 2007) has reported significant problems with tendering processes. In addition, governments can borrow money more cheaply than private firms (Jones, 1998), so to a public sector establishment the cost of financing PFI schemes is higher (Gaffney et al., 1999).

Demanding negotiations

When developing the contracts, the negotiations associated with PFI schemes are highly complex and very time consuming (Tiffin and Hall, 1998; Mustafa, 1999).

Bland products

There is the potential for innovative designs and construction methods to be inhibited as contractors may be wary of overruns (Mustafa, 1999; Birnie, 1999).

Unusual alliances

In the early days, the formation of project consortia was sometimes difficult as constituent members had differing objectives (Mustafa, 1999). An extension of this is the selling of stakes after the construction phase. By doing so, some companies have made profits and walked away from the risks.

Quantification of risks

High cost is ascribed to risk transfer (Gaffney *et al.*, 1999). As no PFI scheme has yet run out its life, it is argued that no one knows precisely the frequency of occurrence of risks and their associated impact.

Unusually high profits

Shareholders in PFI schemes can expect very high returns per year (Gaffney *et al.*, 1999); these returns can be perceived as unnecessarily high as this burden is passed on to the taxpayer.

Justification of PFI

According to Gaffney *et al.* (1999) the discounting method used to compare the 'present value' of different options is politically determined and set well above the government's interest rates. This favours PFI over other procurement options.

Inadequate prior knowledge of PFI

Most client organisations use PFI once so they have substantially fewer staff who fully understand the intricacies of PFI. In contrast, some private sector organisations have been involved with several PFI projects and therefore have significantly more experience (Robinson *et al.*, 2004).

Although there are issues with PFI, its advantages are many, visible and undeniable. It is these advantages that are sustaining PFI.

1.8   Improving the Performance of PFI

The National Audit Office reviews PFI projects, sometimes via case studies, and particularly to scrutinise whether projects have achieved VFM. In the UK, PFI was developed by the Private Finance Panel (PFP). Two reviews of the concept were carried out by Sir Malcolm Bates in 1997 and 1998. His second review was carried out in tandem with Peter Gershon's review of central government civil procurement (Her Majesty's Treasury, 1999). Gershon's review concentrated on the need to establish an integrated and strategic framework for the PFI procurement process with the aim of obtaining cost savings and establishing best practice in the procurement of all government projects. The government accepted the findings and recommendations of these reviews in 1999. Thereafter, the functions of the PFP were taken over by two bodies:

- PUK (Partnership UK) – responsible for coordinating and accelerating the development, procurement and implementation of PPPs. PUK works solely with and for the public sector.
- OGC (Office of Government Commerce) – responsible for ensuring best practices are achieved in PFI/PPP.

Following these developments, the Public Private Partnerships Programme (4Ps) Ltd has since emerged as an advisory body for local authorities. The NHS standard contract was formulated and later ProCure 21 (2007) was launched for hospital schemes.

The OGC developed and launched its Gateway Review Process in England in February 2001. They recommend that gateway reviews should be carried out at key decision points on all major capital projects including PFI/PPP projects. This should be done by a team of experienced people, independent of the project team. There are five review points during the lifecycle of a project; three occur before the contract award stage, while the other two concern service implementation and confirmation of the operational benefits. The OGC review gateways are shown in Figure 1.3.

1.9   Performance of PFI Schemes – An Empirical Review

Now that services are being delivered from PFI schemes it would seem appropriate to compare service delivery performance with delivery from schemes procured using traditional methods. In order to do so, users with experience in both PFI and traditional procurement methods were contacted to obtain their opinions. A questionnaire was designed to elicit their opinions concerning the operation of their facilities and services (see Appendix). The questionnaire was used as a basis for conducting interviews with the stakeholders.

Discussions were held with people operating or managing a PFI service in the hospital and leisure sectors. The discussions sought to find out how the services are faring.

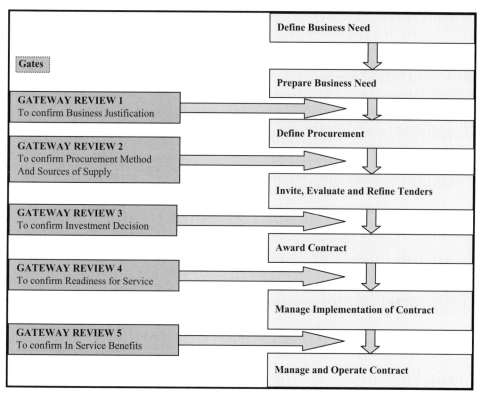

Figure 1.3   The gateway review template (OGC, 2001).

1.9.1   PFI In hospital projects

'The National Health Service (NHS) was set up in 1948 to provide healthcare for all citizens, based on need, not the ability to pay. It is made up of a wide range of health professionals, support workers and organisations' (National Health Service, 2004). The NHS is funded entirely by government, managed by the UK Department of Health, and serves over 800 000 patients per week. It is currently the largest organisation in Europe and employs over 1 million people in England. The advent of PFI was seen by the NHS as a means of improving its outdated system of procurement.

Where PFI is under consideration, an NHS trust is required to prepare an outline business case for approval by the NHS Management Executive and the Treasury. This outline business case indicates an estimate of the capital cost based on standard NHS costings. Final approval of a scheme is then dependent on the trust producing a full business case that will include an economic appraisal showing that the PFI option offers better VFM than a traditional funding stream.

In accordance with the guidance of the Department of Health, all major PFI schemes in the NHS should typically be arranged to cover the functions of: design, build, finance and operate (DBFO). However, in all these schemes, the core clinical services are provided by the NHS through the relevant primary care trust (PCT).

The first major UK PFI scheme was the Norfolk & Norwich University Hospital, which was commissioned in late 1997 (Boyle and Harrison, 2000). Since then, there has been a rapid increase in the number of major PFI hospital schemes that have been approved and are currently under construction or operation. PFI in the health sector is about ensuring new facilities that are 'as modern, efficient and cost effective as possible' (Boyle and Harrison, 2000). This is achieved by incorporating specialist teams from the private sector who are experts at delivering advanced facilities. The Department of Health sees PFI as a key instrument for improving the quality and cost effectiveness of public services. Accordingly, PFI is seen not only as a mechanism for financing capital investments, but an avenue to exploit the full range of private sector management, commercial and creative skills (Department of Health, 2003).

Recently, there has been a significant reduction in the number of PFI health-related projects, with volumes dropping by more than half compared with the same period a year ago. This drop has partly been offset by a rebound in the use of traditional funding approaches, which has helped to support a 6% rise in public non-residential new orders (Baldauf, 2007).

The health sector is one area where criticisms of PFI have been vociferous. According to Gaffney *et al.* (1999) clinical concerns are generally countered by assurances that the largely undisclosed price of PFI is worthwhile because schemes approved by the initiative offer better VFM than traditional public sector procurement. This claim is usually based on the fact that, for approval purposes, all PFI schemes are compared with a PSC. However, concerns remain about the precision with which the cash payments of each option are discounted, and the pricing of 'risk transfer'. These concerns have lingered because the appraisal methodology is prescribed by government guidance and is crucial in the justification of the choice of PFI for any scheme.

1.9.2   PFI In leisure projects

Projects in this sector usually involve the design, funding, construction, operation and maintenance of a leisure centre to replace an existing facility. Facilities include:

- Swimming pools
- Fitness gymnasiums
- Health suites and treatment rooms
- Sports/indoor games halls
- Flexible second sports halls/socialising spaces
- Meeting rooms/teaching spaces
- Dancing studios
- Café bars
- Crèches
- Therapy facilities (e.g. hydrotherapy)

Leisure projects fall under the remit of the Department for Culture, Media and Sport (DCMS), who review and approve business cases. A scheme would normally provide new facilities but may include the refurbishment of some existing facilities, usually for a local authority.

The client will consult with the public at all stages of the project, especially when the outline business cases and outline planning applications are being prepared. This is usually done using public workshops, displays and online interaction and establishes a number of things, principally, the range of facilities and activities to offer.

1.9.3   Performance of hospital and leisure schemes

As interviewees in this research were employees of their various organisations, they were understandably somewhat reserved in their discussions. While they were more willing to discuss positive achievements they were less willing to identify any negative issues.

Hospital schemes

In the survey of users about hospital buildings, most were appreciative of their new buildings which are continuously maintained. If paint is damaged, for example, the SPV will repaint the affected wall at short notice. They tend to maintain a presence in the hospital and, if anything goes wrong with the facilities, they can be contacted quickly.

Some doctors described the new facilities as having large windows that improved the lighting in the facility. Theatre facilities were described as purpose built and excellent. In some hospitals, public areas were described as impressive, spacious and comfortable. A doctor described the new service operation as planned and business-like. When any of the equipment becomes outdated or damaged, the SPV replaces it without additional charge to the PCT. So for the entire duration of the concession the PCT does not need to be concerned with the provision and maintenance of facilities. In some hospitals, equipment did malfunction after the services became operational, and the SPV replaced these at no cost to the PCT.

By contrast, one situation was described where the lobby was too large. Some doctors identified theatre changing rooms as being quite small and nurses mentioned some narrow corridors. Such features may be difficult to pick up from a design plan, especially for clinicians who are not experts in interpreting such documents. Many risks associated with projects and PFI schemes are large and it may not be feasible to address each of them. Therefore, in some schemes the inadequacy of the size of corridors or changing rooms may not be recognised prior to construction. Given that a PCT will almost certainly procure only one PFI scheme, it is plausible to experience such teething problems. Although a lot of planning and consultation goes into the design of PFI hospitals, it is still possible to fail to identify the adequacy, or inadequacy, of some areas. It is evident from such issues that the consultation

Part One

process can be improved and the use of visualisation techniques may assist in the briefing process.

In recent times the NHS in the UK has been under financial constraints. The government has been asking NHS trusts to make cuts in their budgets and, while this directive applies across the board, it affects PFI hospitals more. For example, for two hospitals, each with a £100m budget, one of these has buildings procured by means of PFI and the other does not. The one with PFI buildings is paying a £7m annual unitary charge to service their PFI scheme. If the Treasury asks the hospitals to cut £15m out of their respective budgets, the hospital with PFI buildings will be making a cut of £15m on £93m (16%) while the one without PFI buildings will be making a cut of just 15%.

A few users felt that the schemes were financially driven and too expensive for their PCT. This argument was used to explain why fewer car parks were now available in PFI hospitals and users were being charged more for using them. When two or three hospitals are merged to form one big hospital, some communities have to travel longer distances to get to the new hospital. For some accident and emergency (A&E) cases this increase in travel could be the difference between life and death. Also, when hospitals are merged, some senior personnel may become redundant. While all frontline staff may be needed in the new bigger hospital, a new ward may not need two managers or two deputy managers. Only a few staff are affected this way, however this had been noticed by some users. One doctor cited a situation where the design brief was scaled down midway into the construction phase, and some of the proposed new-build facilities were eliminated and replaced with the refurbishment of existing buildings.

Leisure schemes

In the leisure centres studied, the respondents were keen to discuss the benefits. The facilities are much more modern and attractive. The level of usage of one leisure centre was said to have increased from 600 users per month before PFI to 35 000 users per month after PFI. The respondents were generally very satisfied with their outcomes and the facilities were said to be commendable.

Attempts to identify negative issues in this sector were not successful. Either no problems were encountered in this sector or the projects studied did not encounter any significant difficulties.

The interviews with some respondents established that a proactive approach was often adopted and was seen to be worthwhile. If clients can be forthcoming with information or concerns, then solutions can be worked out amicably with the private sector. In one of the leisure centres, for instance, the client raised an issue concerning a club that was using their old swimming pool. Following discussions, it was agreed to allow the club use of the new pool for a specified period of time under terms that remained favourably unchanged. Without this special arrangement the club would have had to pay more. In another scheme, a hotline was set up to ensure constant access to the SPV where any issue of concern could be treated speedily.

In some of the projects the clients identified the major stakeholders and held consultations with them. The unions, communities and other stakeholders were consulted and their views taken into consideration. In one scheme where three swimming pools were replaced with one bigger swimming pool, the communities affected were consulted and in this respect the local authority client lowered the price of using the new pool and also made concessions in terms of the cost of commuting to the new pool.

1.10   A Generic Overview of PFI Schemes

Although a client gets new facilities via PFI, it was pointed out that one can acquire a new building by other means. It is not the use of PFI alone that yields new buildings or new construction facilities. However, the advantage that PFI offers over other forms of procurement is that the facilities are maintained at a high level of specification throughout the concession period and facilities are kept up to date. If a machine breaks down, and is beyond repair, it is replaced with one that has a similar or higher specification to the one that was damaged.

There is no uniformity in the specifics of PFI contracts. For example, in some projects, the SPV is responsible for some soft facilities management (FM) services like cleaning, laundry and portering; while in others they are not, so the client has the luxury of choice in PFI.

Clients have realised the importance of engaging the SPV in frank consultation. This helps them to achieve facilities that either match or surpass their requirements.

Some of the hospital schemes studied encountered teething problems whereas the leisure schemes did not and there are several reasons why this may be the case. Hospital schemes are bigger and much more complex, and so the chances of error therein are increased. PFI hospital schemes started earlier than leisure centres and lessons learnt in the former are taken on board in the implementation of the latter. In addition, the leisure centres had champions who knew what was expected in terms of facilities needed and how to generate their specifications. These champions were also able to identify their stakeholders and consult with them appropriately.

The foregoing findings are not an indication that one PFI sector is better than another, rather that the outcomes are transferable. Given the same conditions, some of the downsides identified in the health sector could manifest in the leisure sector. Likewise, with due care, the successes identified in the leisure centre can be duplicated in the health sector. It is getting the right conditions, personnel and resources that matters.

The downsides of PFI cannot be entirely eliminated, but need to be monitored to avoid or minimise their potential impacts. Other forms of procurement have their downsides too. PFI has lots of potential that can be exploited in a project. It seems that careful monitoring can enable a project team overcome most of the disadvantages of PFI. According to Liddle (2006): 'The industry continues to be distracted by the cascade of criticisms targeted at PFI. I am calling for a new attitude – one that embraces PFI and concentrates on the value it brings.'

1.11   Comparative Studies

PUK published a report into operational PFI/PPP projects in March 2006 (PUK, 2006). The report, which commented on the largest survey of PFI projects ever undertaken, contains a comprehensive review of the performance of PFI projects during their operational phase. The findings show that public sector managers and users are happy with the outcomes of their PFI/PPP projects. Specifically: 96% of projects in their survey are performing at least satisfactorily, with 66% of projects performing at the stated level of either good or very good standard; 89% of projects achieving contract service levels of either always or almost always; 80% of all users of PFI projects are always or almost always satisfied with the service being provided. In their survey, public sector managers believe that they have developed an effective partnership with the private sector to deliver services, wherein 97% believe that their relationship with their private sector partners is satisfactory or better. PUK's survey revealed that incentives within PFI contracts are working, with around 80% of public sector managers agreeing that the payment mechanism supports the effective contract management of the project.

According to Jefferies (2006), reporting on the Sydney superdome in Australia, which was procured by a DBFO arrangement, this project has certainly delivered an outstanding building and is an example of how both government and private industry can meet Australia's need for infrastructure in the new millennium. However, Jefferies's reflections indicate that more traditional economic infrastructure projects, such as roads, where there is a more defined revenue stream, appear to have been more successful than social infrastructure projects such as hospitals and schools.

1.12   Conclusion

PFI was introduced in the UK in 1992 to deliver improved services. Since then, its uptake has been significant with hundreds of signed deals having a capital value of almost £60bn. Whilst most public sector clients will undertake one or two PFI projects, the private sector side is often used to procuring that way. Thus, there tends to be an experience gap between the public and private sectors. The use of PFI ensures that services are delivered using new or refurbished facilities and these facilities are maintained throughout the service period. In projects procured by PFI the clients are happy because the benefits are many, e.g. they do not have to maintain buildings, soft services are provided by someone else, risks are transferred to the private sector, they obtain VFM, etc. The private sector too is happy because they get a steady income that lasts decades and a high return for their money.

Although the implementation of PFI is relatively complex and long winded, careful consultation and monitoring of the process can yield win–win outcomes. However, as the risks are many in PFI, an oversight can lead to an unpleasant outcome. As some of the hospitals have shown, an inadequate size of a room can affect the comfort of users. It can be difficult to get everything right in the PFI process and in this respect all the views of stakeholders should be accounted for.

Time is a teacher. With the passage of time, procurers of PFI are learning to avoid its pitfalls. Our study indicates a positive trend towards more of the advantages than a repetitive practice where mistakes are allowed to happen again. As most clients are usually new to PFI, it may be inevitable for mistakes to manifest in some PFI schemes. One way to avoid such is to use establishments like 4Ps who offer free advisory services to local authority clients. The National Audit Office, PUK and OGC can also extend their services to cater for new clients who need extra support in developing their PFI schemes. This, for instance, can be reflected in a revised OGC review gateway.

Despite its numerous advantages, any mistake in a PFI scheme is likely to stand out for a long time and be costly to correct, making it essential to get the outcome right first time.

References

Baldauf, M. (2007) New construction orders – February 2007: Expert analysis (Benchmark). *Contract Journal*, 26 April.

Beenhakker, H.L. (1997) *Risk Management in Private Finance and Implementation*. Quorum Books, London.

Birnie, J. (1999) Private Finance Initiative (PFI) – UK construction industry response. *Journal of Construction Procurement*, 5(1), 5–14.

Boyle, S. and Harrison, A. (2000) *PFI in Health: The Story so Far*. King's Fund, London.

British Institute of Facilities Management (2003) *Private Finance Initiative* [online]. Available from: www.bifm.org.uk/index.mhtml?lib/jargon.html+Private%20 Finance%20Initiative%20(PFI) (Accessed: 31 March 2004).

Carrillo, P.M., Robinson, H.S., Anumba, C.J. and Bouchlaghem, N.M. (2006) A knowledge transfer framework: the PFI context. *Construction Management and Economics*, 24(10), 1045–1056.

Construction Industry Council (1998) *Constructors' Guide to PFI*. Thomas Telford, London.

Department of Health (2003) *Public Private Partnerships in the National Health Service: The Private Finance Initiative*. Department of Health – NHS Executive, London.

Field, J.E. and Peck, E. (2003) Public–private partnerships in healthcare: the managers' perspective. *Health and Social Care Community*, 11(6), 494–501.

Gaffney, D., Pollock, A.M., Price, D. and Shaoul, J. (1999) PFI in the NHS – is there an economic case? *British Medical Journal*, 319, 116–119.

Grimsey, D. and Lewis, M.K. (2002) Evaluating the risks of public private partnerships for infrastructure projects. *International Journal of Project Management*, 20(2), 107–118.

Grubb, S.R.T. (1998) Private Finance Initiative – public private partnerships. *Proceedings, Institution of Civil Engineers*, 126, 133–140.

Heald, D. (2003) Value for money tests and accounting treatment in PFI schemes. *Accounting, Auditing and Accountability Journal*, 16(3), 342–371.

Her Majesty's Treasury (1999) *Modern Government Modern Procurement*. HM Treasury, London.

Her Majesty's Treasury (2007) *The Private Finance Initiative (PFI) Statistics: Signed projects list* (online). Available from: http://www.hm-treasury.gov.uk/documents/public_private_partnerships/ppp_pfi_stats.cfm (Accessed: 30 November 2007).

Jefferies, M. (2006) Critical success factors of public private sector partnerships – a case study of the Sydney SuperDome. *Engineering, Construction and Architectural Management*, 13(5), 451–462.

Jones, I. (1998) *Infrafin*. Final report of a project funded by the European Commission under the Transport RTD Programme of the 4th Framework Programme.

Katz, G. and Smith, S. (2003) Build-Operate-Transfer: The future of construction? *Journal of Construction Accounting and Taxation*, 12(1), 36–48.

Liddle, C. (2006) Setting the record straight. *The PFI Journal*, 54, 58.

Mustafa, A. (1999) Public–private partnership: an alternative institutional model for implementing the Private Finance Initiative in the provision of transport infrastructure. *Journal of Project Finance*, 5(1), 64–79.

National Audit Office (2007) *Improving the PFI Tendering Process*. The Stationery Office, London.

National Council for Public-Private Partnerships (NCPPP) (2003) *NCPPP White Paper*. The NCPPP, Washington DC.

National Health Service (2004) *How the NHS Works, its Structure and Organisation Types* (online). Available from: http://www.direct2communications.com/downloads/how_nhs_works.pdf (Accessed: 6 April 2007).

Norment, R. (2002) PPPs – American style. *The PFI Journal*, 39, 26.

Office of Government Commerce (OGC) (2001) OGC Best Practice and Operational Guidance: HM Stationary Office, London.

Owen, G. and Merna, A. (1997) The Private Finance Initiative. *Engineering, Construction and Architectural Management*, 4(2), 163–177.

Partnerships British Columbia (2003) *An Introduction to Public Private Partnerships*. Partnerships British Columbia, Vancouver.

Partnerships UK (PUK) (2006) *Report on Operational PFI Projects*. PUK, London.

Private Finance Panel (PFP) (1995) *Private Opportunity, Public Benefit: Progressing the Private Finance Initiative*. HMSO, London.

Price, J. (2000) Constructing success. *The PFI Journal*, 5(1), 6–7.

ProCure 21 (2007) *ProCure 21 – Achieving Excellence in NHS Construction* (online). Available from: http://www.nhs-procure21.gov.uk/content/home/home.asp (Accessed: 30 November 2007).

Robinson, H.S., Carrillo, P.M., Anumba, C.J. and Bouchlaghem, N.M. (2004) *Investigating Current Practices, Participation and Opportunities in the Private Finance Initiative*. Loughborough University, Loughborough.

Tiffin, M. and Hall, P. (1998) PFI – the last chance saloon? *Proceedings of the Institution of Civil Engineers*, 126, 12–18.

Walker, A. (2000) Equating FM. *The PFI Journal*, 5(1), 8–9.

Walker, C. and Smith, A.J. (eds) (1995) *Privatized Infrastructure: The Built Operate Transfer Approach*. Thomas Telford, London.

Ward, S.C. and Chapman, C.B. (1995) Risk management perspectives on the project lifecycle. *International Journal of Project Management*, 13(3), 145–149.

Part One

Part One

Appendix: Questionnaire used in the survey

Complete/provide answers in respect of current or recent place of work

Name of project?	
When buildings/facilities became operational?	
How long have you used these buildings or facilities?	
How often do you use the buildings? (e.g. daily, weekly, etc.)	
Your name	
Your designation	
Address	
Tel:	
Email:	

Rate the facilities on the basis of the satisfaction derived by you in the course of using them:

	Exc	Good	Slightly good	Neutral	Slightly poor	Very poor	Extremely poor	Comments
Buildings/facilities:								
Size (i.e. adequacy)								
Quality								
Functional efficacy								

Rate the facilities on the basis of the satisfaction derived by you in the course of using them

	Exc	Good	Slightly good	Neutral	Slightly poor	Very poor	Extremely poor	Comments
Durability								
Innovativeness								
Safety								
Security								
Compliance with DDA								
User-friendly								
Service delivery:								
Speed								
Quality								
Effectiveness								

Rate the performance of these facilities prior to PFI

	Exc	Good	Slightly good	Neutral	Slightly poor	Very poor	Extremely poor	Comments
Size (i.e. adequacy)								
Quality								
Functional efficacy								
Durability								
Innovativeness								

Part One

Rate the performance of these facilities prior to PFI

	Exc	Good	Slightly good	Neutral	Slightly poor	Very poor	Extremely poor	Comments
Safety								
Compliance with DDA								

Comment more on the facilities

Identify or comment on any positive feature of the PFI facilities, i.e. not captured above	Identify or comment on any negative feature of the PFI facilities, i.e. not captured above
Any other suggestion on how the facilities can be improved	

2

Using the Private Sector to Finance Capital Expenditure: The Financial Realities

Jean Shaoul

2.1 Introduction

Countries all over the world have turned to the private sector via PPPs to finance much needed investment in physical infrastructure, particularly in transport, water, energy and telecoms, and more recently in healthcare, education and prisons, the so-called human infrastructure.

There is no simple agreed definition of the term PPP, which covers several models of operation, including design, build, finance and operate (DBFO), build, own, operate and transfer (BOOT), build, operate and transfer (BOT), the Private Finance Initiative (PFI), concessions, sale and lease back arrangements, franchises and joint ventures between the public and private sectors, to name but a few variants. Furthermore, the terms are often used interchangeably. But essentially, there are two models: contractual relationship and joint ownership (HM Treasury, 2003). The policy encourages the involvement of the private sector in public infrastructure and service provision.

Under partnership arrangements, the private sector is responsible for constructing and operating the asset, providing the finance and assuming all or most of the risks associated with construction, operation and maintenance of that asset. Projects in the UK have typically been structured in one of several ways, although there are others:

- Under a contractual type arrangement, the public sector pays for the use of the asset and the services so provided under terms set out in a contract which may contain incentives for good and/or penalties for poor performance.
- In free-standing projects, the private sector charges the users directly via a system of road tolls or fees, as for example Britain's M6 toll road and National Air Traffic Services.
- Alternatively, there is some mix of both public and user funding for either the construction and/or the service element. One example is the Skye

Bridge (originally a free-standing project, where the government paid some of the construction costs and later subsidised the tolls before ultimately terminating the contract). Another is the London Underground PPP, a contractual arrangement, which receives a grant, in effect a subsidy to the private sector, and charges passengers.

■ Under joint venture (joint ownership) arrangements such as the Local Improvement Finance Trusts (LIFT), the partnership may charge either the public sector as in health and education, or the users (National Air Traffic Services).

The situation has become even more complex, however. For example, the UK government now calls the privatised railways a PPP (DfT, 2004). The railways are part funded by a system of operating subsidies to the private sector train operators who have a franchise to run designated services for a specified period of time. These subsidies are used by the operating companies to lease the trains from the rolling stock companies and access the track from Network Rail, the private not-for-profit network infrastructure company, as well as their own operating costs. There are also direct grants to Network Rail for capital expenditure.

PPPs in the UK now encompass most sectors and services across the public sector and all types of public bodies, national, local and non-departmental. They also involve working not just with the private for-profit sector but also the so-called third or not-for-profit sector. Under conditions where broader government policy is to include the private sector ever more directly in the provision of public services, one can expect an ever increasing diversity of hybrid forms of financing and funding.

While the UK has led the way in introducing partnership arrangements, within Europe there has long been a policy of concessions and management contracts for utilities and transport, particularly in Spain, France and Italy, and decentralised mixed-mode financing mechanisms, with the right to charge users directly. All these are now included under the umbrella of partnerships. With the increasing integration of the European economy via the EU, the EU has begun to formulate arrangements both in relation to the policy itself, which it broadly supports, and to its governance and reporting for national income accounting purposes (EC, 2004).

The contractual model between the public and private sectors, which is the focus of this chapter, involves a clearly defined project where the private sector finances and shares risks and rewards with the public sector over a 30-year period according to terms set out in the contract. Thus the policy carries with it long-term financial and legal commitments that bind future governments and gives private corporations a degree of control over the direction of future policy.

The private sector partner in such contractual relationships is usually, in the UK, a consortium, typically made up of a bank and construction, property and facilities management companies, constituted as the special purpose vehicle (SPV) that operates through a complex web of sub-contracting to sister companies. The SPV is a standalone company, financed predominantly by debt, and reliant on the revenue flows from this single project.

Should it experience financial problems, it has no recourse to its parent companies.

As with many policy innovations, the rationale has changed so much over time that even its proponents have described it as 'an ideological morass' (IPPR, 2001). In the UK, it was originally justified as a way of leveraging in the private finance the state could not provide – the so-called 'additionality' argument. In some countries, it is seen as a way of reducing public sector debt as the underlying asset and its corresponding debt may, if there is sufficient risk transfer, be treated as off balance sheet, thereby evading the strictures of the EU's Stability and Growth Pact. Now the policy is increasingly justified in terms of delivering value for money (VFM), in the form of lower discounted whole-life costs, including the cost of transferring some risks to the private sector, compared with conventional procurement as measured by a public sector comparator (PSC). This is known as the VFM or risk transfer argument that compensates for the higher cost of capital. More recently, the government has justified PFI on the basis that it delivers assets to time and budget (HM Treasury 2003). Other benefits are now believed to include:

- Introducing private sector expertise, innovation and efficiency
- Incentivising the private sector via the performance-related payments
- Ensuring that maintenance is carried out
- Lower whole-life costs because of the integration of construction, operation and maintenance
- Greater discipline at decision making about what the public sector is procuring, the outputs it expects, performance criteria, risk allocation and management
- A robust project's specification as a result of the independent due diligence carried out by the financiers of the project

But as others have noted, good research evidence to support the claims for superior private sector performance is lacking.

Within the UK, by December 2006, there were nearly 800 signed deals with a capital value of £55bn (HM Treasury, 2006a). The total amount of revenue expenditure committed for the next 30 years is unclear, since the Treasury has reported it after assumptions about the Corporation Tax yield (HM Treasury, 2003). The annual estimated payments are believed to be £6.9bn in 2006–07, rising to £8.9bn in 2016–17, before declining (HM Treasury, 2007). Between 1995 and 2034, total commitments are believed to be £204bn. However, since these projections necessarily omit the new deals yet to be signed, payments in later years of the largest scheme, the London Underground PPP, that are still to be negotiated, and increases in payments due to contract changes, these annual payments are set to increase. Thus, future payments will take an increasing amount of the key denominator, the annually managed public expenditure that is still spent 'in house', which is itself falling due to different forms of outsourcing (Pollock *et al.*, 2001).

The purpose of this chapter is to review the outcomes in terms of the claimed advantages, focusing in particular on the financial costs, including the cost of risk transfer, and hence VFM, and consider some of the wider implications of this policy for service delivery and control of public expenditure. There are, however, several important definitional points to be made. Firstly, while PPPs encompass both contractual (PFI/DBFO) and concessionary arrangements and joint ownership, this study excludes joint ownership schemes, due to the lack of financial evidence about joint ventures, which have even more diverse and less visible governance and reporting forms. Secondly, in focusing on the financial costs of using the private sector to finance investment, the assumption is made that the appropriate economic appraisal of the wider economic and social costs and benefits of such investment has been carried out. In other words, it is only the financing method, not the project *per se*, that is being evaluated. Thirdly, since private finance is inevitably more expensive than public finance, the additional financial costs must be borne by whoever funds the services and the underlying assets, either the state or users or some combination of the two. In other words, the vital distinction is made between the financing and funding.

The independent and empirical research into how long-term contractual arrangements (PFI, DBFO and concessions) are working in practice shows that they are costly and inflexible, create risks and liabilities for the taxpayers and must lead to some combination of higher taxes, cuts in service provision and user charges. Thus, the evidence undermines the claims made for the policy. As the European Investment Bank (2005) has argued, the sole evidence-based argument for private finance is that a project that would not otherwise proceed, gets built. Any rational government would therefore take note of independent and impartial evidence, abandon the policy, seek access to funding and return to the public financing of public infrastructure, which will reduce both the capital cost and the annual financial payments from both the capital and revenue budgets: a win–win situation.

The chapter is organised in several sections. First, it discusses the control of the policy and practice in the UK in order to understand how the assumed benefits are derived and the weaknesses in the appraisal methodology and process. This also determines in part at least the nature of any evaluative evidence. The second section reviews the evidence of how the policy is working in practice and the final section draws some conclusions.

2.2   The Control of PFI

For contractual arrangements that follow the PFI model to proceed, the project must demonstrate that it is likely to deliver VFM and be affordable (HM Treasury, 1997). This section considers each criterion in turn.

2.2.1   Value for money

VFM is dependent firstly upon appropriate arrangements to ensure competition for all aspects of the project, including financial advisors, so that competitive pressure will be exerted throughout the negotiation phase (NAO, 1997). But large-scale projects require and attract a limited number of highly experienced bidders so there is limited effective *ex ante* competition even in the best organised tendering processes (Estache and Serebrisky, 2004). It would indeed be highly unlikely to get more than three or four bidders for large projects as industry concentration means that there are few players. For example, just six infrastructure companies won 50% of the EU roads market and 16 had 90% of the market (Stambrook, 2005). Concentration in the construction industry has increased in recent years following takeovers and mergers and this has led to reduced competition in PPP procurement (Stambrook, 2005). This creates increased risk for the public sector because the companies are large and powerful enough to take on the regulators in the case of conflict and force contract renegotiation on more favourable terms (Molnar, 2003). Within the UK, the National Audit Office (NAO) (2007) and the Public Accounts Committee (2003) have also reported on the low and declining level of competition for PFI contracts. One in three PFI projects have attracted only two bidders, compared with one in six in earlier years. This means that the corporations are now in a position to exert the monopoly power that undermines the VFM argument and thus to control the direction of future policy in ways that privilege the few at the expense of the many.

Secondly, and this is the aspect that has attracted the most attention, VFM is demonstrated by identifying and discounting the whole-life costs of the project as financed under conventional procurement methods and known as the public sector comparator (PSC), which are compared against the discounted costs of the PFI option. The scheme with the lower cost is assumed to offer the greater VFM. The comparison also includes the costs of some of the risks associated with the construction and management of the asset and delivery of services. Since some of the risks are to be transferred to the private sector, for comparison purposes, the PSC needs to include the costs so transferred. It is argued that the PFI option will therefore provide greater VFM than a publicly financed alternative where the public sector bears all the risks. In effect, the proponents of PFI are arguing that the difference between the public and private sector cost of borrowing constitutes the risk premium, the price the public sector is paying for greater efficiency, expertise and innovation plus the cost of risk transfer.

But neither the appraisal methodology nor the control process is neutral. The highly technical VFM appraisal methodology, established by the Treasury, has been extensively critiqued in the research literature, although largely ignored in the corporate literature. It is not neutral but is itself biased in favour of the private sector option and has important wealth distributional implications (Shaoul, 2005). Conceptually and methodologically flawed, as the research evidence has demonstrated (Gaffney et al., 1999a,b,c; Pollock et al., 1999), such valuations encapsulated in VFM and set out in the projects' business cases are not generally, other than in health and education, in the public domain, for reasons of 'commercial confidentiality'. The hospital business cases that are in the public domain show that the VFM, resting upon uncertain projections of costs far into the future, relies overwhelmingly upon estimates of the cost of 'risk transfer' to the private sector, and is at best marginal (Pollock et al., 2002). In effect, the government created an in-built bias in favour of PFI, raising questions as to the degree to which the public agencies can and do reliably demonstrate that the higher cost of private finance is likely to deliver VFM as the NAO has acknowledged (NAO, 2000a). However, the government's response to critical research evidence has been to dismiss the scientific evidence, discredit and intimidate critics, and ultimately exclude and ignore it (Greenaway *et al.*, 2004).

Secondly, under conditions where private finance is the only game in town, then as the NAO has acknowledged, there are incentives to ensure that the case favours the private option. It is therefore almost unheard of for the business cases drawn up by the public sector's private sector financial advisors not to show that the private finance route is better VFM than a publicly financed option.

Thirdly, the key government department, the Treasury, both champions and controls the PFI process. The Treasury's Projects division was initially established in 1997 with a 2-year life, largely with staff on secondment from the private sector. This was later reconstituted as a PPP, Partnerships UK (PUK), whose mission is to help the public sector deliver: fast and efficient development and procurement of PPPs; strong PPPs that build stable relationships with the private sector; savings in development costs; and better VFM (PUK, 2003). Fifty-one per cent of the shares are held by private sector institutions, including financial services companies that have been involved in the financing of PFI projects, and others that have PFI contracts. Furthermore, the majority of the board members come from the private sector, with the public sector represented by only two non-executive directors and the public interest represented through an Advisory Council. The structure, ownership and control of PUK are important because they set the PFI agenda and reflect the conflict between policy promotion and policy control acknowledged by government (Timms, 2001).

Fourthly, the project and the case is managed and/or vetted by the Treasury, the Departmental Private Finance Units, PUK or 4Ps, all of whom are largely staffed by private sector secondees from firms with a commercial interest in the policy. This means that the control process is dominated by parties which have a vested interest in the policy's expansion (Craig, 2006). Under such circumstances, conflicts of interest abound.

One of the most egregious examples of the conflict of interests, the resultant poor financial advice and the cost to the public purse, is provided by the case of the National Air Traffic Services (NATS) PPP, which required a government bailout within 3 months of financial close in 2001. The Department of Transport had paid its advisors, one of whose tasks it was to evaluate and manage the risks to NATS' business, some £44m. This was £17m more than expected and at 5.5% of the proceeds of the sale, among the highest of all the trade sales examined by the NAO (2002a). But despite this, CSFB, the lead financial advisors, failed to evaluate the PPP correctly. It had ignored evidence and advice that did not fit with the government's and its own desired outcome: a signed deal. CSFB told the NAO that their prime motivation was to gain valuable experience of PPPs in order to win future contracts in this new and expanding market (NAO, 2002a).

Several further points should be noted. First, the VFM case is necessarily based on *estimates* of future costs and operates only at the point of procurement. Second, risk transfer is the crucial element in delivering whole-life economy since under PFI private sector borrowing, transactions costs and the requirements for profits necessarily generate higher costs than conventional public procurement. Third, the public sector retains the ultimate responsibility for essential and often statutory services for which there is usually no alternative. This, plus government commitment to the policy, means that the revenue streams are assured as the capital markets recognise (Standard and Poor's, 2003). Thus the ability to transfer risk may in practice be very limited.

The government claims that PFI represents VFM, but this is largely based upon the business case used to support the use of private finance. This is hardly an independent assessment as we have shown above. Apart from the London Underground PPP (NAO, 2000a), the NAO has not carried out any assessments of projects before financial close. While the NAO has carried out numerous VFM assessments after financial close, these were not independent in the sense that they collected new data. Instead the NAO scrutinised, and in many cases, criticised various aspects of the way the business cases were compiled and interpreted, questioning the degree to which the projects demonstrated VFM(NAO, 1997, 1998, 1999a, 2000a).

While the government has commissioned several surveys of PFI that purport to show that PFI represents VFM, these have been carried out by financial consultants with a vested interest in the policy. The first, the Andersen report, commissioned by the Treasury, is particularly important (Arthur Andersen/LSE, 2000) because it claims that PFI had 'saved' 17% on the cost of conventionally procured projects. However, this is based on a sample of 29 projects (out of a possible 400 projects), whose selection is not explained. Its evidence base is the business cases used to support a PFI deal over conventional procurement, rather than any independent analysis. But even more important, most of the savings come from just a few schemes as a result of the risk transfer to the private sector. Furthermore, about 80% of these savings came from just one project, the NIRS2 project for the Benefits Agency run by Andersen's sister company, Accenture, which has become a byword for failure. In other words, the study was based upon anticipated savings that were not achieved in practice. Despite this, the government has never repudiated the report.

The second report, commissioned from Pricewaterhouse Coopers (PWC) (2001), fails to provide even the most basic information that would enable the reader to assess the methodology and the value of the findings. It is based on the perceptions of senior managers responsible for commissioning 27 PFI schemes, not users, staff or project managers. While the report does not explain the sample choice or even provide any evidence about the nature or sector of the schemes, its author explained to this writer that PWC largely selected projects with which PWC had been involved as advisor to either the public or private sector, excluded IT projects and included the first eight DBFO road schemes (personal communication).[1] The report does not contain any supporting financial or other empirical data on service or volume levels.

A third widely cited report, authored by the Institute of Public Policy Research (IPPR, 2001), the think tank with the close relations with the Labour government, was sponsored by KPMG and other private sector companies with a vested interest in the use of private finance. It too used secondary, *ex ante* evidence. While the report had reservations about the use of PFI in health and education, it did endorse the turn to private finance via partnerships.

2.2.2   Affordability

The second criterion that a PFI must satisfy if a project is to proceed is that the annual payments are affordable, an issue that has largely been ignored in both the appraisal process and the wider public debate. The Treasury has not required a consistent reporting methodology that clearly describes and presents all the operating costs that enables an assessment to be made of the affordability of the scheme. Studies of PFI in hospitals have shown that affordability was indeed a problem (Gaffney *et al.*, 1999 a,b,c; Pollock *et al.*, 1999; Froud and Shaoul, 2001). The high cost of PFI in capital terms meant that the first wave of PFI hospitals were 30% smaller than the ones they replaced as trusts adjusted their plans downwards. The affordability gap was further reduced by subsidies from the Department of Health, land sales, a shift of resources within the local healthcare economy to the PFI hospital, and 'challenging performance targets' for the trusts' reduced workforce. Thus, PFI comes at the expense of both capacity and access to healthcare. The emphasis on VFM has served to disguise the high cost of PFI and downplay the importance of affordability, which in turn raises questions about VFM.

In summary then, VFM is based upon a flawed appraisal methodology and process for projects in an increasingly concentrated market of powerful international players. While the watchdogs have been critical of the business case for PFI projects, the government has commissioned reports supporting PFI from consultants with commercial interests in the development of the policy. As such, they do not constitute an independent unbiased source, one of the basic requirements for objectivity. But even accepting their findings, in the final analysis they all rest upon *expectations* or *estimates* of future VFM over the life of the project, and none of them address the second criterion, affordability, which the emphasis on VFM downplays.

2.3   Post-implementation Evaluation of PFI

There has as yet been little in the way of financial evidence as to how the turn to private finance is working out in practice. Indeed, Hodge's (2005) review of Australia's experience notes that there has been no comprehensive evaluation of PPPs; parliamentary enquiries have revealed 'a paucity of quantitative information relating to risk experience and weak financial evaluations' of the comparative performance of PPP and traditional mechanisms; and therefore that 'much of the political promise has not yet been delivered'.

In the absence of either a comprehensive evaluation of such claims or systematic evidence in the public domain that would enable such claims to be evaluated, the evidence presented here about how PFI is working in practice in relation to the claims used to justify private finance is drawn from a wide variety of both primary and secondary sources. These include NAO reports and academic, corporate and other commentaries.

2.3.1   Building to time and budget

The government claims that in contrast to conventional public procurement, PFI projects have been built to budget and on time. But first of all, this assumes that public procurement has been consistently late and over budget, and that this is greater than in the private sector. Good evidence on this is lacking, in part at least because so little was commissioned by the public sector after 1976. In the case of the NHS, cost overruns on the price agreed at financial close on conventional procurement in the early 1990s were of the order of 8%. Secondly, there are indeed well publicised examples of huge cost and/or time overruns on major projects, including the British Library, the Jubilee Line, and the Scottish Executive building. But similar examples can be given of such cost and time overruns in the private sector, such as the new Wembley Stadium. The most egregious example is the delay and escalation in cost of the upgrade of the West Coast Main Line which rose from an estimate of £2.5bn to £13bn under the privatised Railtrack, before being reined back by Railtrack's all but renationalised successor, Network Rail, to about £7.5bn (NAO, 2006). Thirdly, as Flyvbjerg et al. (2003) have pointed out, cost overruns are a common phenomenon in high-profile or megaprojects where political reputations and legacies are involved and occur whether publicly or privately financed. This is because everyone involved has an incentive to ensure that costs are underestimated and revenues inflated to ensure that the project gets the go ahead to proceed.

The government's case for building to time and budget under PFI rests upon on two reports by the NAO (2001, 2003a), which were surveys and consultations with project managers and were not backed up with any data on cost and time overruns, another study cited by the NAO (Agile Construction Initiative, 1999) and a Treasury report (2003), both of which contained neither data nor methodology. As Pollock et al. (2007) have shown, a fifth report (Mott Macdonald, 2002) contained so many flaws in the study design and methodology that the results are uninterpretable.

While the NAO reported that the aims of PFI had generally been met in the construction and design of the 11 hospitals built to date, this must be qualified by the widespread criticism of at least one hospital (it has corridors too narrow to permit more than one trolley) and problems in other hospitals. Other more strategic criticisms have been made of their design (Appleby and Coote, 2002; Worthington, 2002). In the context of schools, the Audit Commission's review of PFI schools (2003) found that PFI did not guarantee better buildings despite their higher cost. All this ignores the extent to which costs escalate during procurement, as others have shown in the context of new PFI hospital builds (Pollock et al., 2007). In the case of criminal justice contracts, court service projects have escalated in price, refuting the claim that PFI contracts deliver fixed prices (Centre for Public Services, 2002).

In the case of PFI, it should be noted that over the full planning period of a project the time taken for selection, bidding and contract negotiation processes under PFI may be months, or even years, longer than for Exchequer-financed schemes, introducing delay to the procurement process (NAO, 2007). The NAO (1998) also recognised that PFI is very costly in terms of legal and financial fees for both public and private sectors, compared to traditional procurement. Such costs incurred by private contractors on unsuccessful bids are likely to be recovered in future successful contracts, increasing the cost of subsequent PFI deals.

In other words, understanding the reality that underpins the rhetoric of 'on time and to budget' is not straightforward. It needs to be understood in the context of the costs of this achievement over the full planning period and not just the time period between financial close and project construction. The (high) costs associated with bidding have already resulted in fewer competing bids, and recouped or reimbursed costs for failed bids provide no VFM. In essence, it is difficult to quantify the benefit of finishing on time and to assess this against the increase in price that the contractor demands to carry the risk of timely completion, a cost that is shown below to be a high one. However, if this balance is a positive one, then such benefits are not exclusive to PFI, but could also be achieved with similar contractual arrangements for conventionally financed projects. Furthermore, these issues need to be considered in a holistic evaluation of PFI rather than in the context of individual projects.

2.3.2   Robust specification

While the Treasury (2003) and PWC (2004) argue that there will be greater discipline at decision making about what the public sector is procuring and that the independent due diligence carried out by the financiers of the project will ensure a robust project specification, this has not always turned out to be the case. Within the UK, the Channel Tunnel Rail Link PPP had to be renegotiated within months of signing. The National Air Traffic Services PPP collapsed within 3 months of financial close for reasons that were entirely foreseeable despite the official line that it was due to the collapse in transatlantic flights after the terrorist bombing of the World Trade Centre in 2001 (Shaoul, 2003). The Royal Armouries Museum deal had also to be bailed out, and the QEII Greenwich Hospital Trust is technically insolvent (PWC, 2005), in part at least due to the £9m extra costs resulting from PFI.

This is not a British phenomenon. Estache and Serebrisky (2004), in their overview of transport PPPs, note that such projects have not been uniformly successful. With a high cost of capital and lower than expected demand, 55% of all transport concessions implemented between 1985 and 2000 in Latin America and the Caribbean had to be renegotiated, a much higher proportion than all the other infrastructure sectors, and that such renegotiations took place within about 3 years. While governments gained in the short term from any proceeds and the low level of public investment, the renegotiations led to higher expenditure via up-front capital grants, subsidies and explicit debt guarantees to the private sector to make the schemes viable. New toll roads in Mexico were unsuccessful and had to be taken back into public ownership.

Boardman *et al.* (2005), in their review of private toll road cases in North America, report that even after refinancing and gaining tax-exempt status and extra ridership, the Dulles Greenway project was still making heavy losses. In the case of the Highway 407 Expressway, the Ontario provincial government had to assume the financing of a cost it had sought to transfer to the private sector, in order to make the road affordable to users. In the context of Spain, which has by far the longest experience of private finance in roads, three schemes had to be taken into public ownership in 1984, a large number of the foreign loans had to be renegotiated, state loans were made available, the remaining contracts had to be renegotiated and in some cases, public subsidies were given (Farrell, 1997). Hungary's M5 project had to be restructured within months of signing. In the case of the M6 toll road in Britain, where traffic flows are much lower than forecast and the concessionaire is unable to break even, this has led to the concessionaire lobbying for development in the region to promote traffic growth and paying for a new link road that will bring traffic to its toll road.

In short, the claims for robust project specification have not always been realised. At the very least, the robustness has served the private sector, particularly the banks, not the public sector, which to date have not lost out when projects have failed.

2.3.3   Penalties to incentivise operational performance

It is difficult to know the degree to which the penalty and incentive system operates to ensure satisfactory delivery of contracted services for several reasons. Firstly, the size of the penalties relative to the baseline payment below which the total payment cannot fall is not generally disclosed. One hospital for example reported that maximum deduction for poor service delivery was £100 000 on expected annual payments of £15m (Edwards *et al.*, 2004), which provides little effective sanction. Anecdotal evidence suggests that the scale of the penalties elsewhere while larger is, relative to the annual payments, small. Secondly, the public agencies neither report the standards of performance nor the amount deducted for poor performance.

There have been numerous adverse press reports in the UK of poor service delivery in hospitals under the contract, some of which are documented in evidence to the Health Select Committee (2002) and similar press reports of concerns about poor performance in schools projects. Metronet, which held the contracts for two of the three London Underground PPPs, was heavily criticised by London Transport and the Office of Rail Regulation for failing to meet the targets set for investment and maintenance and was reported to have overspent by nearly £1bn in its first 7.5-year contract due to not working economically, efficiently or in line with industry best practice. Nevertheless, there have, according to the credit ratings agency Standard and Poor's (2003), been few deductions on PFI contracts and these have been small, in part at least because of the complexity of the contracts that have proved difficult to enforce in practice. In many cases, the original contract negotiation team has moved on, making it difficult to know the assumptions and intentions underlying the contract.

A case study of an NHS trusts found that monitoring has turned out to be more costly than anticipated, performance indicators have been difficult to operationalise, due to the subjective nature of the outcome, and contracts changes have been time consuming and complex (Edwards *et al.*, 2004).

A report on prison performance noted that prisoners were confined to their rooms for longer periods and that their cells contained 'substantial ligature points' that 'rendered the cells unfit for use at all' (Chief Inspector of Prisons, 2000). HMP Altcourse at Fazakerley, the first PFI prison, was controversial from the start because of its poor planning, lack of scrutiny of costs, a flawed savings assessment, operational performance failures and, lastly, the refinancing scandal that saw the private sector refinance the deal in a way that generated an extra £11m for itself while at the same time increasing the risk to the public sector (NAO, 2000b). The NAO, in its investigation into PFI prison performance, reported that operational performance against contract had been mixed (NAO, 2003b). But PFI contracts, even when 'successful', have hidden costs to the rest of the public sector. Centre for Public Services (2002) found that the private sector paid lower wages to its prison staff than did the public sector and some of its workforce were paid such low wages that they qualified for working family tax credits, in effect a low wage subvention by the state to the private sector.

As is almost universally accepted, operational performance has been conspicuously poor in IT projects, and the payment mechanisms have failed to incentivise the contractor. Even where penalties could have been invoked, these were waived in the interest of good partnership working and/or not jeopardising the policy, as in the case of the Passport Agency (NAO, 1999b) and NIRS2 projects (Edwards and Shaoul, 2003). Indeed, the outcomes of IT projects in the benefits recording and payments systems, the criminal justice system and other administrative services have been so poor that even the government has had to admit that PFI may not be the best means of procuring IT services (HM Treasury, 2003) and PFI for IT has now been abandoned.

Thus once again, understanding the reality that underpins the rhetoric of 'incentivising the private sector' is not straightforward. Such evidence as exists suggests the scale of the penalties, the complexity of the contracts and the relative power of the partners do not provide the incentives that PFI's proponents claim, while simultaneously imposing additional costs on the public sector for monitoring and enforcing the contract.

2.3.4   Financial cost of PFI, risk transfer and affordability

There have been few studies that produce systematic financial evidence about the cost of PFI projects once they are operational. This section cites two, one in hospitals and the other in roads.

Hospitals

A study into the cost of the first 12 operational PFI hospitals in England as of 2001, which had capital costs of about £1.2bn, combined annual PFI payments of about £260m in 2005, and total payments of about £6bn over the 30-year life of the projects, found that in a number of cases, the actual payments to the private sector turned out to be considerably higher than originally estimated by the Department of Health (Health Select Committee, 2000). While the average increase was 20%, this was as much as 71% for North Durham, 60% for South Manchester and 53% for Bromley (Shaoul *et al.*, 2007). This may be due to some combination of: volume increases; inflation; contract changes; and failure to identify and/or specify the requirements in sufficient detail, e.g. the failure to specify marmalade for patients' breakfast led to an increased charge. But at the very least, such contract drift suggests that there will be further increases and that the total cost of PFI is therefore likely to be very much more than the £6bn predicted at financial close.

The hospital trusts' PFI charges, including both the availability and service elements, took 12% of income in 2005. The case of Dartford is particularly interesting because even after a refinancing deal that led to a reduction in their charges, PFI charges still took 17% of income. While the trusts received a 56% increase in funding (adjusted for any mergers) as well as in some cases a specific increase to cover some of the extra costs of PFI, PFI charges were still taking the same proportion of income, raising questions about the affordability of PFI. It is therefore difficult to avoid the conclusion that without the increase in funding, PFI was unaffordable.

Despite the increase in funding, the trusts' financial situation was neither stable nor robust, as indeed was the case for many non-PFI trusts. Without a detailed study of each trusts' caseload, it is difficult to determine the role of PFI as other factors have intervened. But two examples illustrate some of the problems. In the case of South Manchester, which had suffered a £7m deficit in 2003, this was because it was unable to shift a £20m caseload to other hospitals that had been part of a wider reconfiguration underpinning the original business case. The QEII Greenwich Trust, with one of the largest deficits – £9.2m in 2005 – declared that it was technically insolvent and was locked into a PFI deal that added £9m to its annual costs over and above that built under conventional public procurement (PWC, 2005). Without government support, its long-term financial situation was insoluble.

Irrespective of any causal role in the trusts' financial problems, PFI charges constitute a 'fixed cost' that cannot be reduced and are significant when margins are low due to other rising costs. This serves to reduce their flexibility in managing their budgets which must create affordability problems when the trusts have always struggled to break even.

The private sector companies, SPVs or consortia organised as brass plate companies, operate in a complex and opaque web of sub-contracting to their sister companies that increases the costs and complexity of monitoring and enforcing the contract, and makes it impossible to assess the parent companies' total returns. After paying interest on their debt, which was higher than the total construction cost and rising, of about 7–8%, the SPVs reported a post tax return on shareholders' funds in excess of 58% in 2005, after negative returns in the early years. The SPVs' high effective cost of capital (£123m in 2005) means that the annual risk premium, the difference between public and private sector interest as defined by the NAO (1998), was £51m, equivalent to 21% of income received from the trusts. It is unclear whether this represents VFM or indeed whether VFM can indeed be measured *ex post facto*. But irrespective of whether this represents VFM, this analysis raises questions about the affordability of PFI in practice, and future service provision, an issue which the emphasis on VFM downplays. It also underestimates the total leakages from the public purse since there are leakages in the supply chain that are not quantifiable in a systematic way: the contractors and sub-contractors' cost of capital, sub-contractors' income received directly from the public (parking, canteen and telephone/television charges which also represent lost income to the trusts), the proceeds of land sales and any refinancing of the SPVs' loans.

Consider next the impact of the annual *observable* leakages from all the trusts' budget, where leakages are about £51m a year on just 12 capital projects worth £1.2bn, on the cost of the PFI programme. The first wave of 18 projects, of which these 12 form a part, were expressly identified and progressed in order to create the model for PFI projects in the health sector (PWC, 2004). But if this experience is generalised across the entire PFI programme, although it could be argued that 'lessons have been learned' from these early deals, then the extra cost of private finance for the signed PFI capital programme in hospitals worth £8.67bn (HM Treasury, 2006b) is about £430m every year.

Roads

While the use of private finance in roads has been deemed a 'success', this was and is a consequence of very high payments to the private sector. Shaoul *et al.* (2006) examined the first eight DBFO contracts signed by the Highways Agency and paid for on the basis of shadow tolls. The study found that they are costing about £220m a year or £6bn over 30 years. The study found that the payments in just 3 years for which information is publicly available were £618m, more than the £590m cost of construction, refuting the claim that the government could not afford the capital cost.

After paying interest on their debt, which was higher than the total construction cost, of about 9%, the SPVs reported a post tax return on shareholders' funds of 29% in 2002. The additional cost of private over public finance (risk premium) was about £62m, more than half the cost of capital (£103m) and 40% of the income received from the Agency in 2002. With annual operation and maintenance costs of about £50–60m a year, or £1.8bn in total, this means that after paying interest on debt (about £1.8bn), itself more expensive than public debt, the Agency is paying nearly £1.8bn (out of a total of £6bn) for the major maintenance and private sector profits, a high price for risk transfer. Thus 'success' comes at the expense of affordability and VFM and must entail service cuts elsewhere. Indeed, a Highways Agency official said that annual payments for all its contracts are £300m a year, or 20% of its budget for 8% of its roads. The contract for the M25 will add a further £300m a year, meaning that 40% of the budget will be committed for a small proportion of the network (Taylor, 2005).

While the additional cost of private over public finance is attributable to the cost of risk transfer, it was difficult to see, given that the contracts involved roads that had already been designed and gone through all the planning stages, thereby reducing some of the main risks, how such a high 'risk premium' could be justified (Shaoul *et al.*, 2007)

Furthermore, this underestimated the total cost of private finance, since the private sector partners operate through a complex web of sub-contracting. Their parent companies therefore have additional, undisclosed sources of profit via sub-contracting the construction, operation, maintenance, financing and refinancing of the projects to related companies that make it difficult to establish the total cost of using private finance. These findings therefore rebut the arguments that the private sector would find the finance that the public sector could not (the macroeconomic or additionality argument) and that the additional cost of private finance would be counterbalanced by the risks transferred to the private sector (the microeconomic or VFM argument).

2.3.5   Risk transfer

Most of the additional cost of private over public finance is justified in terms of risk transfer, largely construction not operational risk. There is, however, no yardstick by which to measure whether this is a reasonable cost. For example, it is unclear why the cost of risk transfer is so high given that after completion of the construction phase, the companies have been able to refinance their deals. Furthermore, these refinancing deals carry with them the potential, as in the case of the refinancing of Fazakerley prison, for the companies to increase their profits at the expense of the public sector (NAO, 2000b, 2002b). This is because the private sector's debt repayment profile is restructured and the contract extended in order to accommodate this. The public sector could therefore find itself exposed to additional termination liabilities, should the contract be terminated for any reason. This increased exposure would occur when the private sector had received most of the benefits and be facing additional costs associated with long-term maintenance, thereby tempting the private sector in adverse circumstances to cut and run, as indeed has been the case with unprofitable rail franchises.

More fundamentally, the concept of risk transfer that lies at the heart of the rationale for partnerships is problematic, regardless of whether the project is 'successful' or not. If the project is successful, then the public agency may pay more than under conventional procurement: if it is unsuccessful then the risks and costs are dispersed in unexpected ways as a study of failed IT projects has shown (Edwards and Shaoul, 2003). Although a project may fail to transfer risk and deliver VFM in the way that the public agency anticipated, the possibility of enforcing the arrangements and/or dissolving the partnership is in practice severely circumscribed for both legal and operational reasons, with the result that a public agency may be locked into a partnership for better or for worse. This in turn undermines the power of the purchasing authority to incentivise its partner while strengthening the contractor's already powerful financial and monopolistic position, in circumstances where it is beyond the reach of public accountability and scrutiny. Under conditions where partnerships are the only means available to the public sector for procuring goods and services, then the VFM case is little more than a rationalisation for a decision already taken elsewhere. Thus, far from being a neutral policy-making decision tool, 'risk transfer' disguises its political and social consequences.

2.3.6   Additionality

Since the public sector repays the full cost of private finance via annual payments spread over 30 years, it does not access new forms or higher levels of funding than would otherwise be the case with public funding. Like buying a house, it simply spreads the cost over a longer period and ultimately pays at least three times the original cost. As others have noted, all capital spending over the period 1999–2002, and indeed since then, could have been replaced by conventional public procurement financed either through public debt without breaking either the so-called 'golden rule' or the Stability and Growth Pact. Furthermore, the current account surpluses in some years (£23bn for 2000–01 alone) could have more than covered the £14bn deals signed between 1997 and 2001. PFI has served to displace the burden of debt on to future generations.

In the context of hospitals, several further points emerge from the financial analysis. Firstly, while the government claims that PFI has led to the largest building programme in the history of the NHS, the first wave of PFI hospitals were so costly that they created an affordability gap, leading to asset sales, extra subsidies, charity appeals and cuts of up to 30% in bed provision (Gaffney et al., 1999a). In other words, they are smaller than the ones they replace. Secondly, the annual observable extra costs of private finance in hospitals, extrapolated across the whole hospital sector, shows that the programme is costing an extra £430m a year, equal to at least two major hospitals every year or 60 over the life time of the programme. Thirdly, irrespective of whether private finance represents VFM, PFI creates affordability pressures for the trusts, which have been cushioned to some extent by increased funding. This is not set to continue after 2008, and in the context of a new funding regime where money follows patients on the basis of average costs will create even further cost pressures for trusts that are locked into PFI contracts since they have essentially higher fixed costs than non-PFI Trusts, as the QEII Trust noted (PWC, 2005). At the very least, PFI creates budget inflexibilities that increase the pressure on the NHS to cut their largest cost, staff, and thus access to quality healthcare.

In the context of DBFO in UK roads, as the evidence above has shown, the £590m construction costs were paid for in 3 years, which shows that far from providing additionality, the new construction (and maintenance) comes at the expense of other Highways Agency projects.

2.4   Conclusion

These perverse results are not a purely British phenomenon, as the evidence on the hospital sector in Australia (New South Wales Auditor General, 1996;

Auditor General Western Australia, 1997; Senate Community Affairs References Committee, 2000), and privately financed roads in Spain (Acerete *et al.*, 2007) shows. There too the outcomes were inconsistent with the claims. At best, PFI has turned out to be very expensive with the inevitable consequences for service provision, taxes and user charges, not just today but for a long time to come. These projects may burden government with hidden subsidies, diversion of income streams and revenue guarantees whose impact on public finance may not become apparent for many years. When things go wrong, and this is not infrequent, the costs are diffused throughout the public sector and on to the public at large, a travesty of risk transfer.

This analysis has not only demonstrated that the outcomes do not match the claims but, even more importantly, has indicated the reason for this. The government's claims ignored the competing demands of the numerous stakeholders and the particular characteristics of public services: cash strapped with no excess capacity to enable 'surplus fat' to be trimmed without affecting service delivery. In these circumstances it was and is impossible to reconcile all the conflicting claims on the funds and protect both the taxpayers and users. PFI ensures a resolution of the distributional conflict in favour of the corporations and more particularly the financial sector, who are its chief promoters, under the guise of additionality, risk transfer, efficiency, incentives, etc. Thus while the government's case rested upon risk transfer, additional investment and private sector efficiency, and therefore benefits for all, the real effect was the redistribution of wealth to the financial and corporate sectors. The government, by focusing on a concept as ambiguous as VFM under conditions where no public finance would be made available, made the distribution issue invisible in order to justify a deeply unpopular policy.

Note

1. Personal communication in response to a request for further information from the authors of this paper

References

Acerete, J.B., Shaoul, J. and Stafford, A. (2007) Taking its toll: the cost of privately financed roads in Spain. Presented at the World Conference on Transport Research, Berkeley, California; to be published in 2009 in *Public Money and Management*.

Agile Construction Initiative (1999) *Benchmarking Stage Two Study*. Agile, London.

Appleby, J., and Coote, A. (2002) *Five Year Health Check: A Review Of Health Policy 1997–2002*. The Kings Fund, London.

Arthur Andersen and Enterprise LSE (2000) *Value for Money Drivers In The Private Finance Initiative*, Report commissioned by the Treasury Taskforce, January, available from the Treasury's website: treasury-projects.gov.uk/series_1/andersen.

Audit Commission (2003) *PFI in Schools: The Quality and Cost Of Buildings And Services Provided By Early Private Finance Initiative Schemes*. Audit Commission, London.

Auditor General Western Australia (1997) *Private Care for Public Patients: The Joondalup Health Campus*, Report No 9. Perth, Australia.

Boardman, A.E., Poschmann, F. and Vining, A. (2005) North American infrastructure P3s: examples and lessons learned. In: Hodge, G., and Greve, C. (eds) *The Challenge Of Public-Private Partnerships: Learning From International Experience.* Edward Elgar Publishing, Cheltenham, UK.

Centre for Public Services (2002) *Privatising Justice: The Impact of the Private Finance Initiative in the Criminal Justice System.* Sheffield.

Chief Inspector of Prisons for England and Wales (2000) *HM Prison Altcourse.* Report of a Full Inspection 1–10 November 1999. Home Office, London.

Craig, D. (2006) *Plundering the Public Sector.* Constable, London.

Department for Transport (2004) *The Future of Rail,* White Paper, CM 6233. The Stationery Office, London.

Edwards, P., and Shaoul, J. (2003) Partnerships: For Better, For Worse? *Accounting, Auditing and Accountability Journal,* 16(3), 397–421.

Edwards, P., Shaoul, J., Stafford, A., and Arblaster, L. (2004) *Evaluating the Operation of PFI in Roads and Hospitals,* Research Report no 84. Association of Chartered Certified Accountants, London.

Estache, A. and Serebrisky, T. (2004) *Where Do We Stand on Transport Infrastructure Deregulation and Public Private Partnership?* World Bank Policy Research Working Paper 3274. The World Bank, Washington, DC.

European Commission (2004) *Public-Private Partnerships and Community Law on Public Contracts and Concessions.* Green paper, COM (2004) 327 final. European Commission, BrusselsEurostat.

European Investment Bank (2005) *Evaluation of PPP Projects Financed by the EIB,* synthesis report. European Investment Bank, Luxembourg.

Farrell, S. (1997) *Financing European Transport Infrastructure: Policies and Practice in Western Europe.* Macmillan, Basingstoke.

Flyvbjerg, B., Bruzelius, N. and Rothengatter, W. (2003) *Megaprojects and Risks: An Anatomy of Ambition.* Cambridge University Press, Cambridge.

Froud, J. and Shaoul, J. (2001) Appraising and evaluating PFI for NHS hospitals. *Financial Accountability and Management,* 17(3), 247–270.

Gaffney, D., Pollock, A.M., Price, D. and Shaoul, J. (1999a) NHS capital expenditure and the Private Finance Initiative – expansion or contraction? *British Medical Journal,* 319, 48–51.

Gaffney, D., Pollock, A.M., Price, D. and Shaoul, J. (1999b) PFI in the NHS – is there an economic case? *British Medical Journal,* 319, 116–119.

Gaffney, D., Pollock, A.M., Price, D. and Shaoul, J. (1999c) The politics of the Private Finance Initiative and the new NHS, *British Medical Journal,* 319, 249–253.

Greenaway, J., Salter, B. and Hart, S. (2004) The evolution of a meta-policy: the case of a Private Finance Initiative and the health sector. *British Journal of Politics and International Relations,* 6(4), 507–526.

Health Select Committee (2000) *Public Expenditure on Health and Personal Social Services 2000: memorandum received from the Department of Health containing replies to a written questionnaire from the Committee,* HC 882, Session 1999–2000. The Stationery Office, London.

Health Select Committee (2002) *The Role of the Private Sector in the NHS,* HC 308, Session 2001-02. The Stationery Office, London.

HM Treasury (1997) *Step by Step Guide to the PFI Procurement Process.* HM Treasury, London.

HM Treasury (2003) *PFI Meeting the Investment Challenge.* HM Treasury, London.

HM Treasury (2006a) *PFI Statistics: Signed Projects List.* http://www.hm-treasury. gov.uk/documents/public-private-partnerships/ppp_pfi_stats.cfm. (Accessed 12 December 2006).

HM Treasury (2006b) *PFI Statistics: Signed Projects List*. http://www.hm-treasury. gov.uk/media/5CE/7B/pfi_signed_projects_list_december_2006.xls. (Accessed 23 March 07).

HM Treasury (2007) *Budget 2007: Building Britain's Long Term Future, Prosperity And Fairness For Families*. Treasury, London.

Hodge, G. (2005) Public private partnerships: the Australasian experience with physical infrastructure. In: Hodge,G. and Greve,C. (eds) *The Challenge of Public-Private Partnerships: Learning from International Experience*. Edward Elgar Publishing, Cheltenham.

Institute of Public Policy Research (2001) *Building Better Partnerships*, The Final Report of the Commission on Public Private Partnerships. IPPR, London.

Molnar, E. (2003) *Trends in Transport Investment Funding: Past Present and Future*. UNESCO and CEMT/CS/12.

Mott MacDonald (2002) *Review of Large Scale Procurement*. Mott MacDonald, London.

National Audit Office (1997) *The Skye Bridge*, Report of Comptroller and Auditor General, HC 5, Session 1997-98. The Stationery Office, London.

National Audit Office (1998) *The Private Finance Initiative: The First Four Design, Build, Finance and Operate Roads Contracts*, Report of Comptroller and Auditor General, HC 476, Session 1997–98. The Stationery Office, London.

National Audit Office (1999a) *The PFI Contract for the New Dartford and Gravesham Hospital*, Report of Comptroller and Auditor General, HC 423, Session 1998–99. The Stationery Office, London.

National Audit Office (1999b) *The Passport Delays of Summer 1999*, Report of Comptroller and Auditor General, HC 812, Session 1998–99. The Stationery Office, London.

National Audit Office (2000a) *The Financial Analysis for the London Underground Public Private Partnership*, Report of Comptroller and Auditor General, HC 54, Session 2000–2001. The Stationery Office, London.

National Audit Office (2000b) *The Refinancing of the Fazakerley PFI Prison Contract*, Report of Comptroller and Auditor General, HC 584, Session 1999–2000. The Stationery Office, London.

National Audit Office (2001) *Modernising Construction*, Report of Comptroller and Auditor General, HC 87, Session 2000–01. The Stationery Office, London.

National Audit Office (2002a) *The Public Private Partnership for National Air Traffic Services Ltd*, Report by the Comptroller and Auditor General, HC 1096 Session 2001–02 The Stationery Office, London.

National Audit Office (2002b) *PFI Refinancing Update*, Report of Comptroller and Auditor General, HC 1288, Session 2001–02. The Stationery Office, London.

National Audit Office (2003a) *PFI: Construction Performance*, Report of Comptroller and Auditor General, HC 371, Session 2002–03. The Stationery Office, London.

National Audit Office (2003b) *The Operational Performance of Prisons*, Report of Comptroller and Auditor General, HC 700, Session 2002–03. The Stationery Office, London.

National Audit Office (2006) *The Modernisation of the West Coast Main Line*, Report by the Comptroller and Auditor General, HC 22, Session 2006–07. The Stationery Office, London.

National Audit Office (2007) *Improving the PFI Tendering Process*, Report by the Comptroller and Auditor General, HC 149, Session 2006–07. The Stationery Office, London.

New South Wales Auditor General (1996) *Report for 1996; Volume I*. NSW Parliament, Sydney.

Partnerships UK (2003) *What is Partnerships UK?* http://www.partnershipsuk.org. uk/puk/index.htm. (Accessed 12 December 2006).

Pollock, A., Dunnigan, M., Gaffney, D. *et al.* (1999) Planning the 'new' NHS: downsizing for the 21st century. *British Medical Journal*, 319, 179–184.

Pollock, A., Price, D. and Player, S. (2007) An examination of the UK Treasury's evidence base for cost and time overrun data in UK value for money policy and appraisal. *Public Money and Management*, 27(2), 127–134.

Pollock, A., Shaoul, J., Rowland, D. and Player, S. (2001) *Public Services and The Private Sector – A Response To The IPPR*, Catalyst Working Paper. Catalyst, London.

Pollock, A., Shaoul, J. and Vickers, N. (2002) Private finance and 'value for money' in NHS hospitals: a policy in search of a rationale? *British Medical Journal*, 324, 1205–1208.

Public Accounts Committee (2003) *Delivering Better Value for Money from the Private Finance Initiative*, HC 764, Session 2002–03. The Stationery Office, London.

PWC (2001) *Public Private Partnerships: A Clearer View*. PWC, London.

PWC (2004) *Developing Public Private Partnerships in New Europe*. PWC, London.

PWC (2005) *Queen Elizabeth Hospital NHS Trust*, Public Interest Report. PWC, London.

Senate Community Affairs References Committee (2000) *Healing Our Hospitals*. Commonwealth of Australia, Canberra.

Shaoul, J. (2003) Financial analysis of the National Air Traffic Services public private partnership. *Public Money and Management*, 23(3), 185–194.

Shaoul, J. (2005) A critical financial appraisal of the Private Finance Initiative: selecting a financing method or reallocating wealth? *Critical Perspectives on Accounting*, 16, 441–471.

Shaoul, J., Stafford, A. and Stapleton P. (2006) Highway robbery? A financial evaluation of design build finance and operate in UK roads. *Transport Reviews*, 26(3), 257–274.

Shaoul, J., Stafford, A. and Stapleton P. (2007) Evidence based policies and the meaning of success: the case of a road built under design build finance and operate. *Evidence and Policy*, 3(2), 159–180.

Stambrook, D. (2005) *Successful Examples of Public Private Partnerships and Private Sector Involvement in Transport Infrastructure Development*, final report under contract with OECD/ECMT Transport Research Centre, Contract # CEM JA00028491. Virtusity Consulting, Ottawa, Canada.

Standard and Poor's (2003) *Public Finance/Infrastructure Finance: Credit Survey of the UK Private Finance Initiative and Public-Private Partnerships*. Standard and Poor's, London.

Taylor, G. (2005) Major roads works ahead: 10 years of the UK private finance initiative roads program. In: *Public Private Partnerships: Global Credit Survey 2005*. Standard and Poor's, London.

Timms, S. (2001) *Public Private Partnership, Private Finance Initiative*. Keynote Address by the Financial Secretary to the Treasury to Global Summit, Cape Town, December 6th.

Worthington, J. (2002) *2020 Vision: Our Future Healthcare Environments*, Report of the Building Futures Group. The Stationery Office, London.

Part One

3

Obstacles to Accountability in PFI Projects

Darinka Asenova and Matthias Beck

3.1 Introduction

Accountability is a central aspect of governance and a cornerstone of the ethos of the British public sector. Significantly, accountability and transparency are increasingly in vogue in the private sector, in the wake of gross failures in corporate governance. However, under public partnership, and more specifically in implementing Private Finance Initiative (PFI) projects, there exist myriad problems of accountability in public procurement. These have arguably led to some of the most fundamental objections to this new form of governance as they relate to wider concerns about declining democratic oversight and the public interest. Below, we seek to explore what accountability can and does mean under PFI by examining how financial services providers, in allocating risk under partnership, often stymie accountability and transparency, and arguably impact upon the actual provision of public goods and services.

Today, the preferred model of PPP in the UK is the PFI (DETR, 1998; HM Treasury, 1999). Under the PFI, the private sector undertakes to design, build, finance and operate physical assets in order to provide a required service demanded by a public sector body that itself is responsible for the ultimate delivery of the service (Kirk and Wall, 2001). A distinguishing characteristic of PFI schemes, as compared to other forms of contracting out, lies in the emphasis on the procurement of services, rather than the creation of physical assets (Merna and Smith, 1999). One objective for the PFI is to allow the private sector to decide how to provide certain services, and how to finance the required assets. PFI procurement, furthermore, differs from other forms of PPP, in that PFI schemes require full financial support from the participating private companies over the life of the project[1] (Private Finance Panel, 1995; Treasury Taskforce, 1997; Allen, 2001). Crucially, PFI procurement contractually commits public and private sector parties to a pre-negotiated allocation of risks (Glaister, 1999). Underlying the service and long-term partnership focus of PFI procurement is the assumption that allowing the private sector to introduce its own management and procurement strategies will lead to cost saving and efficiency-enhancing innovation (Birnie, 1999), and that both public and private sectors share the benefits of these efficiencies.

In recent years there has been widespread criticism of PFI procurement. Given the central role of private sector actors, much of this criticism has focused on various aspects of the procurement scheme which are, *prima facie*, not conducive to maintaining adequate levels of public accountability. A 2001 UNISON study noted:

> Most local authorities only go down the PFI route because they believe public sector capital will not be available. If they stated this, they would not qualify for government funding. This is because the approval process requires them to confirm that they have evaluated the PFI option against the public sector comparator and found it to be better value. In practice, most of the financial issues associated with PFI schemes are either kept from elected councillors or where they are shared they are so complex that they are not understood. Most local authorities have gone out of their way to avoid public scrutiny either by publishing no information or by publishing edited versions of the full business case. Public accountability is almost non-existent. (UNISON, 2001)

Whether most local authorities have sought to avoid public scrutiny is difficult to test. Judging by the nature of the legal formal requirements associated with PFI procurement, at least, other observers have taken an opposite view. Option evaluations and VFM exercises, whether they are carried out objectively or not, do at least force the public sector to justify its choices. In terms of accountability, this is arguably a step forward compared to what public decision makers' practices were pre-PFI. Yet it is fair to say that there remains imperfect democratic oversight of PFI and PPP by the public and their elected representatives (especially at the local level). Moreover, many groups that attempt to represent specific stakeholders, including trade unions, non-governmental organisations (NGOs) and independent parties, are rarely involved as partners in PFI.

In addition to issues of accountability, some critics have questioned the reliability of economic estimates underpinning individual PFI projects. Thus, a UNISON (1999) sponsored report on the Cumberland Infirmary, Carlisle questioned the level of financial diligence and accountability of the respective NHS trust. Specifically, this report stated:

> We conclude that the deal does not give the taxpayer value for money. We have shown that the interest rate assumption at the heart of the economic appraisal has been deliberately set to favour the private sector, and that after only a minor adjustment the alleged advantages of the PFI option disappear. However, in Carlisle's case, political manipulation alone was insufficient to make the economic case. Only major errors in the Trust's economic calculations could do that. If these were rectified, the PFI option would be seen to be a bad economic option, more costly than the public alternative by £11 million. On a proper economic appraisal, Carlisle's PFI should have never left the drawing board. (UNISON, 1999)

While criticisms of the accuracy of the VFM model of this project may be justified, the assertion of a breakdown of accountability leading to financial harm is debatable. Even if one accepts that business models are often inaccurate, it is necessary to recognise that a true evaluation of the financial performance of most PFI projects will probably only be possible some years after the project has entered the operational phase, perhaps even only after the full length of the contract.

Others have queried the ability of PFI schemes to meet public expectations of accountable and transparent decision making. Amongst the most cogent criticisms of PFI procurement are those by Kerr (1998) and Froud (2003). Based on an essentially Marxist analysis of the post-modern capitalist state, Kerr has argued that the rhetoric of PFI, as improving efficiency and public services, merely serves to mask the fact that PPP is actively depoliticising state-sponsored service provision and subjecting it to the rule of money. This depoliticisation, according to Kerr, is part of an effort by the state to disengage from investment while simulating capital accumulation. While we need not agree with Kerr's assumption that this process marks a deliberate policy trajectory, his comments on the impact of PFI on the governance of public service provision are intuitively appealing:

> PFI marks a fundamental transformation of traditional public sector procurement methods, one in which the traditional and clear distinction between public and private activities and spaces is becoming obscured... This means that the public sector is now forced to think more objectively about the services it requires and also has to develop techniques to evaluate the complex private sector bids which have to be shown to provide value for money and transfer of risk. The private sector also had to come to grips with new organisational forms and methods of appraisal... In this way then, the PFI is attempting to transform the 'public' service provision labour process in, at least two ways. Firstly, through the requirement to define and monetarise risk and to quantify future life-cycle costs and future service needs, the PFI is attempting to force greater objectification and 'marketisation' into the provision of 'public' services. Secondly, through displacing the service provision labour process from the public to the private sector, the PFI is attempting to subordinate that labour process more effectively to the rule of money. (Kerr, 1998)

This analysis ties in with a wider literature on the remorseless creep of the private sector into domains that were once the sole preserve of the public sector (Habermas, 1976), a process very much accelerated under globalisation (Teeple, 1995; Leys, 2001). Kerr's suggestion that PFI acts as an instrument of objectification of services and marketisation of labour raises important questions about accountability in PFIs. If PFI forces both the public and private sector to apply a criterion-based approach to decisions on service provision, then PFI might well bring gains in terms of accountability as long as the criteria applied are transparent and defensible. Much more difficult to assess in terms of accountability, however, is the marketisation of labour in service provision. Here it could be argued that introducing private sector criteria of commercial profitability potentially undermines whatever levels of accountability may have been gained elsewhere.

If we assume that the trajectory of PFI in terms of current macro-accountability is at best ambiguous, it is worth querying the future or long-term accountability potentials of PFIs. Investigating the impact of long-term contractual commitments typical of PFI contracts on the ability of the public sector to deliver services, Froud (2003) has questioned the compatibility of

PFI with traditional assumptions about the appropriate role and responsibilities of the public sector. Rather than focusing on the issues of accountability in PFI, Froud's analysis centres on the question as to how the contractual management of risks in PFI projects impacts on the ability of the state to respond flexibly to the current and future needs of the public. In this context, Froud argues that the contractual treatment of risk in PFI undermines the traditional role of the government as risk bearer of last resort:

> Under PFI, risk is seen as the chance of incurring increased costs and is managed by the application of an approach based on inter-firm contract relations such that, in principle, risks are distributed to those best able to bear them...There is little explicit recognition in this that government as a contracting party has particular characteristics that make it different from firms or individuals in terms of responsibilities, interests and modes of operation...it is clearly simpler to employ a technicist approach to consider the risks from and to a particular public sector business unit or project, than to evaluate the issue of risk and uncertainty at the level of a public service. But it denies the traditional nature of government in taking responsibility for planning, organising and monitoring public service provision and responding to internal and external change. (Froud, 2003: 585)

Froud's analysis, when applied to issues of accountability, further complicates issues of PFI procurement. Taking Froud's analysis literally, PFIs are not lacking in accountability because criteria such as VFM are too ambiguous to protect taxpayers, but rather because the rigid contractual framework which inevitably underpins PFI schemes will make it difficult for the state to flexibly fulfil its traditional role as service provider of last resort. Like Kerr's earlier criticism of PFI, Froud's analysis is intuitively credible, because it points to the possibility that the extensive usage of PFI will make the public sector more vulnerable to unforeseen demands and events.

While both marketisation and flexibility can be seen as accountability-reducing elements of the PFI regime, there is a third, perhaps more hidden, aspect of PFI which adversely affects public accountability. This aspect arises principally from the fact that the private sector provides most, or all, of the capital for the PFI projects.[2] Access to capital is never unconditional, but rather is premised on a project meeting the risk–return criteria of private sector financial services providers. The application of these overarching risk–return criteria not only modifies service provision in a way which is not subject to traditional accountability criteria, but also reduces the possibility of PFIs, in their full complexity, being subject to traditional public sector criteria of political accountability and transparency.

This argument is presented in four sections. The first section maps out some earlier theoretical contributions on the effects of finance capital on the decision making of institutions which are dependent on its support. The second and third sections present two case studies of PFI projects which highlight the 'hidden' deal-shaping role of financial institutions in the PFI context. The fourth section concludes with a tentative analysis of the contradiction between the reliance on private finance which characterises PFI projects, and the desire to deliver public services in an accountable and innovative manner.

This chapter is based on a series of interviews which were conducted in 2001–2002 in connection with a DETR/EPSRC LINK project. Interviewees included senior public sector managers who headed the respective PFI projects as well as senior bank and SPV representatives. Interviews were supplemented by contract documentation which were generously made available to the researchers by respondents.

3.2   Finance Capital and Institutional Decision Making

Theoretical arguments asserting the power of finance capital in shaping the policy of institutions have an extensive academic pedigree, which ranges from the writings of Hilferding (1910) to the more recent works by Glasberg (1989). As a general tenet, the finance capital literature proposes that there is a close relationship between the power of financiers to influence institutional decision making on the one hand, and the level of dependency of the client institutions under conditions of uncertainty on the other. As such, financial institutions are more likely to shape institutional decision making where their client is heavily dependent on their input and where the financial sector itself is taking risks in providing the requisite capital to that client. In one of the earliest formulations of this relationship, Hilferding noted that:

> The development of capitalist industry produces concentration of banking, and this concentrated banking system is itself an important force in attaining the highest stage of capitalist concentration in cartels and trusts. How do the latter then react upon the banking system? The cartel or trust is an enterprise of very great financial capacity. In the relations of mutual dependence between capitalist enterprises it is the amount of capital that principally decides which enterprise shall become dependent upon the other. (Hilferding, 1910: 223)

Similarly Aaronovitch (1961) argued that:

> In financial institutions, more than in industrial combines, great sums of money are under the control of a limited number of people who themselves own directly only a small fraction of the actual capital. When these individuals are closely linked with the dominant shareholders and controllers of industrial enterprises, or are or become those very people, the concentration of control is greatly increased. (Aaronovitch, 1961: 43)

He further argued that even in cases when industrial conglomerates explore different forms of self-financing, this does not decrease the level of control exercised by the financiers:

> While self-financing has grown in scale since 1945, a very considerable fusion of industrial and finance capital had already taken place before that date. Under these circumstances, self-financing has nothing to do with 'independence' but only with the policies pursued by the largest groups. (Aaronovitch, 1961: 47)

The author rightly observed that the dominance of financial institutions can take more diverse forms, including consultancy and advisory services:

Part One

In fact, there is hardly a large combine in Britain today which is not professionally advised by one of the merchant banks and which has not got insurance companies among its substantial shareholders. (Aaronovitch, 1961: 47)

If this analysis is applied to the PFI context, it points towards additional means for influence to be exercised by financial institutions. In recent years, many financiers that traditionally provided the main project capital (80–95% of the project requirements) have simultaneously become involved in the equity provision, thus having a degree of influence over the 'self-financing' component of the PFI projects. In most cases, these are not accidental developments as now some senior debt providers explicitly require participation in the equity stakes. Moreover, the majority of well established PFI financiers in the UK tend to be multifunctional. Thus, in some transactions they act as senior debt and/or equity providers for the private sector project company, while in others they are involved as advisors to public or private sector partners, thus having even greater influence over their decision making. It can be argued that the degree of influence is partially determined by the experience and availability of relevant in-house commercial skills. Therefore, it can be assumed that when involved as financial advisors to private consortia, financial institutions provide professional advice. On the other hand, when involved as financial advisors to relatively inexperienced public sector clients, financial institutions can be expected to have significantly more influence over the institutional decision-making process.

The mechanism of exercising control by financiers is described by Rochester (1936) as follows:

... when a bank advances business credits it may demand full information about the company's other obligations, its profits, payroll, position on the competitive market, etc. So long as a loan is outstanding, the bank has a whip-hand over the corporation. It may order wage cuts or technical reorganisation. The bankers may even agree to boycott a company and drive it to the wall. (Rochester, 1936: 106)

Rochester and Hilferding's analyses placed heavy emphasis on concentration and power as factors underpinning high levels of dependency. Over the last decades the growth of financial institutions, on an international scale, has been facilitated by the process of ongoing globalisation and the creation of powerful national and international interlocking networks between financial services companies (Aaronovitch, 1961). One driver of merger activities in the financial sector is the possibility of enhanced market power and market hegemony (Kane and Pennacchi, 2000). In the domestic context, Carroll and Alexander (1999) have suggested that these activities can have a negative impact on 'the coherence of national economies and thus, of nationally focussed finance capital', which can lead to fragmentation of national financial networks. Furthermore, it has been suggested that financial deregulation and the activities of supra-national institutions such as the World Bank and the European Central Bank, have strengthened the position of international finance capital. According to Tabb (1999:12) the broad acceptance of neo-liberal logic has led to a situation where 'all sorts of regulation ... become impediments to the efficient functioning market. ... Indeed, the international financial institutions have forced the rollback of a host of government pro-grams around the world'.

In the context of the UK PFI market the role of global finance cap-ital is evident through the activities of foreign banks such as Dresdner Kleinwort Benson, Deutsche Bank, Bayerishe Landesbank Girozentrale, Bank of America, Bank Gesellschaft, etc., all of which have established consider-able experience in PFI projects. Additionally, those 'UK' banks such as the Royal Bank of Scotland, HBoS, Barclays and Lloyds TSB which participate in PFI are trans-national, globalised financial institutions.

While Hilferding depicted conditions of oligopolistic banking which to a large degree apply to the UK PFI market, the contemporary application of Hilferding's analysis is limited by the fact that he did not foresee the possi-bility of client relationships being formed between state agencies and finance capital. Both control over private sector firms and over public sector institu-tions were studied by authors such as Rochester (1936) and Glasberg (1987, 1989). In this context Glasberg concluded that:

> The conventional view of the business structure is based on the assumption of free and open competition, presumed to ensure that only the most efficient forms would survive. Day-to-day consultation between banks, however, contradicts this view-point. The banking community resides in a structural arrangement that necessitates banks' cooperation instead of competition. (Glasberg, 1987: 325)

Glassberg further argued that:

> Finance capital is ultimately the most critical resource: it is the only resource for which there are no alternatives. . . . Moreover the structural hegemony of the bank-ing community, produced by the legal and financial necessity of lending consortia, erodes the competitive nature that may be present in material resource industries. Finally, finance capital is more than a resource: it is a relationship, that unlike the temporary alliances characteristic of material capital supplier arrangements, has long term consequences. . . Finance capital relationships cannot be broken without deleterious consequences (since banks typically recognise and honour each others' customer supplier relationships and since they are collectively the 'only game in town'). Hence reliance on finance capital as a resource is unique to all other re-source dependencies and should be considered specifically and separately from a general resource dependence model. (Glasberg, 1987: 327)

Glasberg's analysis stands, and can be applied to PFI, even without the assumption of bank collusion. PFI projects rely on private capital and all providers of private capital will, within a certain range, apply similar risk–return criteria to evaluate projects. Where projects only marginally meet the banks' expectations, adjustments to the content, guarantees, or payment mechanism selected will have to be made for the project to proceed. Structural dependence on private capital, in this sense, is an ever-present feature of PFIs, but it is only likely to become an explicit part of PFI negotiations where the expectations of banks are not fully met. Where expectations are met, some level of structural dependence remains implicit to the PFI deal, in the sense that the public sector client will already have anticipated what constitutes an acceptable deal and adjusted its service provision requirements accordingly.

The following two case studies investigate the involvement of financial services providers in PFI projects where it was difficult to ensure external financing. Both cases examine risk assessment by the brokering banks. The focus of this analysis is on the decisive role played by financial services providers in structuring the conditions of the deal and, in some instances, shaping the service mix provided to the public.

Case Study 3.1   On-Balance Sheet PFI Housing

Background

The client of this project was a local branch of a central government agency, with technical expertise, long-standing experience of contracting with the private sector, and familiarity with PFI procurement. The project was located in a remote and sparsely populated area, which affected the unique risk portfolio associated with the facility. It involved the design and construction of nearly 300 houses for government agency employees as well as facilities maintenance over a 25-year period. The requirement for the accommodation arose from an urgent relocation of employees, which imposed very tight project timescales. The project company selected to undertake the works included two construction companies and a bank which together established a joint venture. The capital requirement for the construction phase of the project was in the range of £26m, while the overall cost of the project over the concessional period was estimated at £72m. The SPV members supplied 10% of the capital requirement, while the remaining 90% was financed through senior bank debt. The bank provided both the senior debt and the equity for the development. In order to reduce the transaction risk associated with contracting with a separate company, the SPV members sub-contracted the facilities management services to a subsidiary of one of the construction companies.

As part of the project agreement, the unitary charge which had to be paid by the client on project completion was to be split into three parts. These included a charge for capital repayment, one for maintenance over the asset lifecycle, and one covering the equity which had been provided to the project. A relatively high proportion of the unitary charge, 90%, was fixed and only 10% was variable with a capped value related to the Retail Price Index (RPI).

One senior manager representing the client noted that the bidding process was largely conducted in accordance with detailed government guidance and therefore did not present any unexpected challenges to the experienced project team. However, the team was conscious of time restrictions and aimed to speed up the procurement process. Consequently, the negotiation period between the selection of the preferred bidder and the financial close took only about 10 months as compared with the average of 12 months on comparable PFIs. This was followed by a construction phase of 18 months resulting in the timely completion of the project.

From the outset the client had clear objectives which centred on the provision of standard accommodation units. It was decided that the private sector partner should bear the responsibility for securing suitable land. The client's specifications involved standard family-type housing which had been relatively loosely defined with reference to UK building practices. However, there were some specific requirements concerning the minimum size of the houses, internal spaces and gardens, as well as some features relating to site location and characteristics. For example, the client required the maximum commutable distance between the location and the workplace to not exceed 10 miles, a reasonably attractive site with access to social amenities such as schools, supermarkets and nurseries. In interviews the chief project manager of the construction firm noted that combining these requirements within the time limits was a relatively difficult task, wherein the selection of a suitable site constituted a major project risk. Being unable to allocate all required housing into a single site, the construction company, in consultation with the bank, decided to develop three sites. This plan carried cost implications involving a less efficient use of construction equipment, and the cost of moving equipment between sites, which eventually proved unavoidable.

Table 3.1   Distribution of some major risks between the public and the private sector partners.

Main risks allocated to the private sector	Main risks retained by the public sector
Construction risks including time and cost overruns	Interest rates before financial close
	Occupant generated damages
Availability and quality of facilities	Financial risk – the difference between the contractual swap rate and the real interest rate is paid by the client
Financial risk – the SPV gets fixed price bank debt through an interest rate swap	
Lifecycle costs	
Facilities management – fixed price sub-contract	
Old property available	
Provision of land	
Performance of the technical consultants	
Interest rates after financial close	
A degree of residual value	

The overall risk profile of the project and the contractual distribution of some of the major project risks are given in Table 3.1. During the construction phase the public sector client gave priority to risks associated with time and cost overruns, and other construction problems such as the suitability of the site land and adverse weather conditions. During the operational phase, attention shifted to risks associated with the availability and performance of the facilities as well as the possibility for escalating lifecycle costs. The client meanwhile, retained very few risks, primarily those over which they had some degree of control, such as the risk of internal damage during usage.

Risk identification and evaluation by the public sector

In its risk analysis the project team identified three key areas that were likely to determine off-balance sheet status for the project. These included demand risk, availability risk and residual value risk. The availability risk (i.e. whether the building is available for its stated use) was mitigated by developing a robust payment mechanism including a pre-agreed scoring system accounting for faults and defects. The demand risk was essentially political given the possibility that the client would relocate or close the facility in the long run. To reduce this risk, and ensure compliance with the bank's wishes, the client included a contractual clause stipulating that, after the 6th operational year, it could step down a maximum of ten houses per year. These houses would then become non-core stock owned either by the client or the SPV, which would reduce costs to the client. Coping with residual value risk proved problematic given the bank's unwillingness to accept this risk given the remote location and the lack of a buoyant housing market in that area. One senior public sector manager admitted that, even at current conditions, after the 5th year the costs of the buildings were likely to exceed their market value. Unsurprisingly the project's risk profile attracted considerable debate in government circles, particularly in relation to demand and residual value risks. After examination by the National Audit Office[3] (NAO), the relevant government authorities decided that the project could not be treated off-balance sheet.

From the public sector's perspective the main difficulties encountered in this project related to the achievement of off-balance sheet treatment in parallel with VFM. These problems were aggravated by changes in government accounting regulations introduced shortly before financial close. Specifically the Accounting Standard Board modified balance sheet treatment of risks, requiring the project team to re-evaluate their estimates. This had a significant impact on the client. As it became clear that the project could not be treated as an off-balance sheet transaction, there was a danger that the whole project would fail due to central government regulations. Once the project team was able to demonstrate considerable overall savings between the public sector comparator (PSC)[4] and the risk-adjusted private sector bid, however, the Treasury approved the project.

Risk assessment and management by the bank

The bank's team scoped the risk portfolio of the project upon publication of the *Official Journal of the European Communities* (*OJEC*) notice. Since each SPV member was a nationwide player, the bank believed that the location of the project was not a substantial concern. The bank was further reassured as the SPV's building and the operating companies had the same parent company. The bank's approach to risk assessment and mitigation centred on the financial model. This model assessed the risk probabilities using external experts. The main risk categories evaluated included lifecycle costs, land suitability, design planning and availability, and risks related to the transfer of old property. Lifecycle risks were to be distributed over time, as an accumulated cash reserve would smooth the lifecycle cost curve.

When investigating the land suitability and land condition risk, the bank recognised early on that the best site was owned by another construction company. This led to the decision to create a joint venture which involved this company. One risk, related to the land condition, which was omitted during the tender stage, related to the discharge of surface water. Since water discharged in a local loch, which was a 'site of scientific interest', water quality had to satisfy environmental standards, requiring a special filtration system to be added.

Design and planning risk was reduced as the client's brief was sufficiently prescriptive to reduce these risks to a negligible level. Meanwhile, availability issues (which arise because of latent defects) were a major concern. A points system was created in collaboration with the client where each particular defect was assigned certain weight and the building was deemed unavailable once the total score exceeded an agreed number.

One bank representative noted that a principal goal of his organisation was to ensure that the risks which were transferred from the client to the SPV were in turn transferred to the sub-contractors (construction and operational companies). Those risks that could not be transferred, such as the risk associated with ground conditions, had to be priced by the SPV. Only after the appropriate premiums were added, was the bank willing to consider these risks as being reasonably mitigated.

In this project, the consortium became the preferred bidder primarily on the basis of its ability to quote the lowest overall project cost in terms of net present value. To achieve this result, the pricing of risks had to be considered carefully within tight limitations of capital and operational prices. As a part of due diligence before financial close, the pricing provided by the SPV was vetted by the bank's external consultants. The model was re-examined on several occasions, especially at invitation to negotiate (ITN) and best and final offer (BAFO) stages. This provided comfort that both the structure and content of the model were robust. The most important input categories of the financial model comprised costs, revenues and economic inputs. The input costs were classified into two categories: up-front costs and ongoing costs (Table 3.2).

Other inputs included items from the bank's term sheet which set the length of the lending period, the up-front and ongoing fees. In addition, the model incorporated information on the revenues and how they were expected to behave over time. Special attention was paid to the correction factor, which was used to moderate the predetermined value of the RPI. This factor is often used by the client to mitigate the inflationary risk and/or to achieve more favourable price estimates. For example, if the long-term operational costs were inflated with a coefficient equal to the RPI (Figure 3.1), the shareholders' profit could grow very fast over time. Therefore, variable costs are usually inflated by a factor, proportional to the RPI[5]. The bank was closely involved in the calculation of this correction factor which ultimately had to satisfy both the shareholders and the public sector client.

Table 3.2   Main costs considered in the financial model.

Up-front costs	Ongoing costs
Construction costs	Lifecycle costs
Up-front facilities management costs	Ongoing facilities management costs
Up-front shareholders' costs	Ongoing shareholders' costs (e.g. audit, insurance)
Bank's due diligence costs	Ongoing bank costs
Up-front advisers' costs	

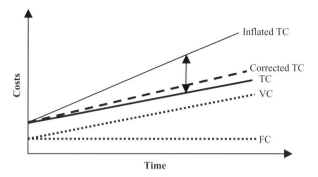

Figure 3.1   Development of project costs over time. (TC = total costs; VC = variable costs; FC = fixed costs).

Other inputs were based on assumptions about the future behaviour of variables such as corporation tax, the RPI and VAT, which accounted for the existence of systematic risk factors. The interaction between variables over time was assessed in the model to produce outputs such as a profit and loss account, a balance sheet and cash flow projections. The outputs estimated the level of retained cash, the corrected total cost curve, the ratio maps against base cover ratios, and graphs of shareholders' return. Some of these outputs were used to justify the expenses to the client.

Overall there was clear evidence that the bank's concerns with managing the risk–return profile of this project had a crucial impact on the way the public sector had to structure its approach. This was most visible when the public sector client had to consent to the bank's refusal to accept the transfer of the volume risk to the SPV. As a consequence, the client had to retain occupancy risk and was unable to document a sufficient level of risk transfer. This meant that the client was unable to achieve the off-balance sheet treatment conventionally available to PFI projects. In terms of accountability, the process by which this project was procured along PFI lines, but not accounted as such, must be considered unsatisfactory from the client's perspective, and thereby the wider public interest. However, much of the underlying negotiations never became public, due to the commercial confidentiality requirements.

Case Study 3.2:   Waste(ful) Management

Background

The responsibility for household waste management and disposal in the UK lies with local authorities. Despite technological developments in waste processing, the majority of local authorities dispose waste at landfill sites, with only 3% recycled. Following the EU Landfill Directive, which aims to move waste from landfill, two local authorities opted to develop more sophisticated and sustainable waste management solutions. Both authorities sought to use technology for retrieval of materials for recycling, composting, energy production, and using fuel produced to generate electricity. This required substantial up-front capital investment to finance the required processing equipment and the clients decided to utilise PFI procurement.

In early 1997, the client issued an *OJEC* notice and received a dozen applications from potential bidders. Six of these companies were invited to prepare more detailed formal submissions, and three were approved for the next selection stage. After careful consideration of these bids, in 1998 the clients selected the preferred bidder, which offered distinctly innovative solutions. The bid was submitted by a foreign company specialising in recycling and waste recovery. The company was seeking to enter the UK waste market and the management was prepared to take a higher level of risk, compared to UK companies, to establish a presence in Britain.

By partnering with a foreign company little known in the UK, the clients took a substantial risk. Moreover, accepting innovative but unproven technologies created additional uncertainties. The incentive for this active risk taking was that the proposal offered compliance with the EU Landfill Directive 8 years ahead of schedule. The proposed plant was designed to achieve 25% recycling, while the current recycling levels of this client were about 6%. According to the contract, it also had to reach 30–40% composting and about 35% for energy recovery. Therefore, it was processing on average three-quarters of the waste stream away from the landfill with 49% material recovery. In order to encourage the SPV to surpass these targets, the contract imposed further incentives which were based on an overall expected recovery level of around 70%.

The project would deliver the first fully integrated resource recovery centre located under one roof in the UK, including: a facility for sorting and separating over 135 000 tones of annual household and trade waste; organic waste compost control; recycling of textiles, plastics, glass, metals and paper; and converting part of the remaining waste into pellets to provide fuel for the production of electricity.

Negotiations between the public sector representatives and the bidding consortia began in 1998 and planning permission was granted 2 years later. The agreement was signed during the second half of 2000, specifying that the plant had to be fully operational in 2 years' time. The construction value of the contract is in the range of £33m and the concession period will run for 25 years, with the plant generating income mainly through waste gate fee, and a small proportion from recycled materials. The main parties comprising the SPV included the construction and engineering company (51%), the local authority's waste management company (20%) and an external equity provider (29%).

The capital requirement for this transaction was provided through a bank loan (87.5%) and some equity supplied by the project company and an external investor (12.5%). The local authorities also decided to participate with an equity stake through a specifically created company. This arrangement was designed to allow for a closer collaboration between public and private sector partners. The risk distribution achieved in this project is summarised in Table 3.3.

Risk management from the perspective of the senior debt provider

The bank's involvement with this project started with the publication of the *OJEC* notice, when a general letter of interest was provided to support the bid. At that time, the bank did not conduct any risk analysis, merely indicating potential interest. An internal team of four people was later established to work on the transaction. More serious involvement in the project developed after the short-listing stage.

According to the bank representative, the key risks in this particular project were anticipated from the outset because of the bank's existing experience in both PFI procurement and construction-type projects. The broad view adopted by the bank was that, under the PFI regime, the risks facing financiers were inevitably allied with the risks of the borrower. Regarding specific risk allocation, the bank was determined not to take any risks which their organisation could not control. Considering

Table 3.3 Risk distribution between the public and the private sector.

Risks transferred to the private sector	Risks partially retained by the public sector
Planning, design, construction time and cost overruns	The price of the electricity off-take
Latent defects and system failure	A degree of financial risk, particularly some limited recourse for the finance provision in case of force majeure conditions
Other performance-related risks	
The SPV will have increased tax obligation if landfill conversion targets are not fulfilled	A limited degree of waste stream composition risk that could affect the recycling targets (shared)
	Landfill tax
Risk from protestors' actions	Changes in specific legislation
Commercial and technical risks	
Changes in general legislation	
Financial risks	

Part One

the non-recourse features of the PFI borrowing and the limited resources commanded by the SPV, the bank therefore insisted that all major risks passing from the public sector were to be allocated to sub-contractors.

The bank's own risk identification process was predominantly based on project documentation and involved a combination of internal expertise and external advice. One bank representative noted that at the pre-qualification stage there were some commercial risks which had to be properly allocated in order to secure senior debt provision. These risks were well understood as 'deal breakers' and were identified without external advice. Commercial risks related largely to construction and operational activities. With construction risk the bank applied the general rule that the risks of time and cost overruns should be transferred to those involved with the construction. As there were four sub-contractors involved (some being divisions of the main engineering company), the bank anticipated problems with joint responsibility. To avoid such issues the bank insisted on establishing a contractual structure that would ensure a single interface to the borrower. Consequently, one company took the responsibility for all construction risks. Regarding operational risk, the bank insisted that the covenant of the SPV members be investigated in terms of their ability to manage risks, including their delivery track record. This information was used to determine levels of performance related liability.

According to the bank representative, during the BAFO stage, the bank involved legal advisors to investigate the contractual documentation, but did not require a detailed investigation of the project risks. A comprehensive risk analysis was conducted at the preferred bidder stage when, in addition to the legal advisors, the bank involved technical and financial advisors. The latter acted as auditors to the financial model and paid particular attention to compliance with legislation, reliability of calculations and correspondence of the model to the features of the commercial transaction. In order to judge the project's financial performance, the bank used the financial model to calculate some key financial ratios, such as the annual debt service cover ratio (ADSCR) and the loan life cover ratio (LLCR). The ADSCR was calculated as the cash revenue available for debt service divided by the amount of debt in the corresponding year. It was used to indicate the ability of the SPV to pay its debt. The LLCR is defined as the NPV of the sum of all future income for the life of the loan divided by the outstanding debt at a particular point of time. Both ratios were investigated in terms of critical values which had to be observed through the life of the project.

The bank investigated the construction and operations/maintenance sub-contracts. Besides commercial risks, the bank focused on financial risks. During negotiations the project team relied heavily on sensitivity analysis, which was used to assess the impact of different risk factors on the revenue stream and debt repayment. There were some additional factors peculiar to this waste management project. Thus, special attention was given to the risks affecting the recycling and electricity generation facilities, as they had to provide 20% of the payments stream.

The respondent noted that, under the PFI agreements the bank inevitably retained a degree of all risks that are contractually transferred to the SPV. Thus, for the financier, the risk mitigation process translates into reaching a reasonably comfortable pattern of risk sharing between the three parties. This involved the SPV passing some risks to the client and others to the sub-contractors. As the main construction and operational risks were off-laid, other important risks considered by the bank were addressed though due diligence and sensitivity analysis. These were used only after the main ('deal breaking') risks were sufficiently mitigated. Some specific risks investigated in this project related to possible changes in the electricity and recycling markets and landfill capacity risk.

The possibility for changes in the recycling market derived through sensitivity analysis indicated their likely impact in terms of waste quantity and composition, and corresponding effects on prices. The bank took independent advice about trends in this sector. Additional uncertainty was associated with future electricity prices, as the plant had to generate a proportion of its revenue from burning recycled waste. In recent years, the UK electricity market has undergone substantial changes as new regulations regarding the buying and selling of electricity have been introduced. The bank sought to ensure that, at least during the first years of the project, the price of electricity would be sufficient to support the revenue.

According to a bank representative, at the preferred bidder stage the bank received a version of the concessional agreement and commented on the proposed risk allocation. Afterwards, when the lawyers to the SPV drafted the sub-contracts, these drafts were also sent to the bank. After careful investigation the bank had concerns regarding the risk distribution. Most of the time the bank was not directly involved in the negotiations between the public and private sector partners. As a consequence, the iterative process of negotiating and agreeing contractual details took several months. Towards the end of the negotiations, the bank conducted due diligence procedures, which scrutinised the risk identification, evaluation and allocation by involving outside experts. Moreover, due diligence effectively served as a tool for off-laying part of the risks to the consulting companies involved. For example, operational problems with the financial model were the responsibility of the auditors. As a rule, external experts have an insurance cover for such risks and their liabilities are limited to a certain percentage of the damages.

According to one bank representative, construction and operation risks were mitigated primarily through the use of construction bonds. Construction bonds are payable from another bank acting as an insurer. They guarantee the availability of a certain amount of capital if the project company becomes insolvent, the construction contract is terminated and the bank has to incur additional costs to complete the project with another company. The senior debt provider received construction bonds not only from the main project company, but also from the major sub-contractors, which covered about 20–30% of the construction value of the transaction.

One of the major risks that the bank was trying to assess during the negotiations related to available landfill capacity. Awareness of this problem came from experience with similar types of projects and discussions were held to clarify the impact of any unforeseen difficulties that could jeopardise recycling waste targets. Subsequently, a special agreement was signed with a third party with a landfill license, which mitigated this risk.

Again, this project highlights the influence of financial service providers in determining the procurement. By opting for an innovative solution, this client created particular risks, which the senior debt provider was not willing to accept unless some specific and, by PFI standards, unusual, arrangements were made. These affected both the construction company and some of the sub-contractors, who were required to supply construction bonds in order to placate the financiers' concerns. The cost of these special arrangements were passed to the client and, ultimately, led to increases in the cost of the project to the public. In addition, the bank required the signing of a special third-party agreement regarding the use of an additional landfill site, which also increased project costs. Innovation combined with PFI hence came at a significant cost to the public, primarily because it was viewed as additional risk by the bank.

3.3   Conclusion

Despite differences between both schemes, the general approaches of the financial institutions to these projects were broadly similar. The key consideration of the financiers was that, under PFI procurement, their risks were allied with the risks of the borrower, the SPV. Therefore, a series of actions were taken to ensure an 'acceptable' approach to risk management. In this context, the banks scrutinised the contractual risk allocation while attempting to ensure that all important risks were passed through the SPV to the parties that had control over them. Meanwhile, capital providers largely avoided substantial risk taking. Very few residual risks were allowed to remain with the SPV and even then the financiers required strong evidence – in terms of past experience, skills and resources – of the ability of the partners to manage them. For the financiers, the 'proper' allocation of crucial commercial risks was a key criterion for determining the 'financability' of the particular transaction. Most risks were investigated on the basis of the full contractual documentation, including the main project agreement and the supplementing sub-contracts. Once the main project risks were mitigated, the residual risks were then assessed in detail in the course of the due diligence.

Past research has attributed comparatively little importance to the role played by financial institutions in PFI procurement. This omission is problematic for several reasons. Firstly, any study which underplays the role of financial institutions in PFI procurement is likely to ignore the genuine material considerations which make and break PFI deals. PFIs, viewed from an economic perspective, do not stand in a financial vacuum. Rather, their scope and feasibility are intrinsically linked to the expectations of financial markets. As such, the feasibility of PFIs in general depends on certain market conditions which currently favour the financing of PFIs, but may not necessarily do so in future. Secondly, decision making on PFI projects is not exclusively a political process. Rather, by scrutinising the role of financiers in PFI, it is clear that it is a process which is conditioned by the expectations and requirements of individual suppliers of finance capital. These expectations are likely to take precedence over other considerations, including the public sector's quest for innovative or high-quality services; all of which can only be addressed once the requirements of the financiers have been satisfied. In this sense, the relevance of research into PFI risk management by the financial sector arises not from its descriptive insight, but rather from the fact that it can demonstrate how the policies and preferences of financial sector companies can concretely constrain the range of possibilities available to public sector clients. As principal players in PFI projects, senior debt providers can determine what is commercially acceptable, dictating which measures other PFI participants must undertake in order to secure finance. This approval of a PFI project is often implicit in the process. However, where financial service providers' requirements are not met, finance will ultimately not be available.

Considering the issue of accountability in PFI projects, it is clear that the prerequisite of obtaining capital from the private sector will inevitably restrict openness and discussion of crucial financial and project parameters. What is relevant to them are not minute project details but rather questions about the allocation, mitigation and retention of key risks; and very little of that is ever discussed in public.

For the public sector to maintain adequate levels of accountability within these constraints is not an easy task. One way of ensuring accountability and transparency is for the public sector to have realistic expectations *vis-à-vis* financial services providers. Essentially, public sector clients need to be clear about their goals and not allow major decisions to change during final stages of negotiation. Lastly, accountability in PFIs requires public sector clients to possess the skills and capacity to undertake at least some of the analysis conducted by financial services providers if only to verify that the public's interests are served.

Notes

1. A recent publication by HM Treasury (2003) hints the distinctive role of private finance in general and financial institutions in particular in PFI procurement. Thus, in relation to senior debt provision this document notes that:

   > Typically, third party credit providers are more risk-averse than equity providers and provide the majority of the funding. The PFI approach and process thus leads banks and other financial institutions who lend to PFI projects to play an important role in ensuring that proper due diligence is performed, all important risks are identified and properly addressed and allocated to appropriate parties. They will seek to have robust and rigorous contractual undertakings from private sector participants in PFI scheme and this is one of the reasons the PFI process delivers projects on time and to budget. (HM Treasury, 2003: 40).

2. The PFI procurement option is predominantly used for capital intense projects, typically in the range of tens or hundreds million of pounds. Such capital requirements can be met only by large, well established and centralised financial institutions.

3. The National Audit Office (NAO) is an organisation independent of British government which acts on behalf of Parliament with an aim to audit and review public spending. The areas of investigation cover the performance of all kind of government departments, agencies and other public bodies. The NAO has produced a number of reports on the economy, efficiency and effectiveness with which government bodies have used public money.

4. In this case the PSC was audited by the NAO.

5. While the Treasury Taskforce (TTF) recommends the long-term value of the RPI, the public sector client can use a discretion regarding the value of the correction factor. Moreover, the value of this factor can be used to judge the level of competition in any particular project.

Part One

References

Aaronovitch, S. (1961) *The Ruling Class*. The Camelot Press Ltd, London.

Allen, G. (2001) *The Private Finance Initiative* (PFI), Research Paper 01/117. Economics Policy and Statistics Section, House of Commons Library, London.

Birnie, J. (1999) Private Finance Initiative (PFI) – UK construction industry response. *Journal of Construction Procurement*, 5(1), 5–14.

Carroll, W. and Alexander, M. (1999) Finance capital and capitalist class integration in 1990s: networks of interlocking directorships in Canada and Australia. *The Canadian Review of Sociology and Anthropology*, 36(3), 331–355.

DETR (1998) *Modern Local Government – In Touch with the People*. HM Stationery Office, London.

Froud, J. (2003) The Private Finance Initiative: risk, uncertainty, and the state. *Accounting, Organisations and Society*, 28(6), 567–589.

Glaister, S. (1999) Past abuses and future uses of private finance and public private partnerships in transport. *Public Money and Management*, 19(3), 29–36.

Glasberg, D. (1987) Finance capital markets and corporate decision-making process: the case of W. T. Grant Company bankruptcy. *Sociological Forum*, 2(2), 305–330.

Glasberg, D. (1989) *The Power of Collective Purse Strings: and the State, The Effect of Bank Hegemony on Corporations*. University of California Press, Berkley.

Habermas, J. (1976) *Legitimation Crisis*. Heinemann, London.

Hilferding R. (1910) *Das Finanzkapital; eine Studie ueber die juengste Entwicklung des Kapitals*. I. Brand, Vienna.

HM Treasury (1999) *Modern Government Modern Procurement*. HM Treasury, London.

HM Treasury (2003) *PFI Meeting the Investment Challenge*. HM Treasury, London.

Kane, E. and Pennacchi, G. (2000) Incentives for banking megamergers: what motives might regulators infer from event-study evidence? *Journal of Money, Credit and Banking*, 32(3), 671–706.

Kerr, D. (1998) The PFI Miracle. *Capital and Class*, 64, 17–28.

Kirk, R.J. and Wall, A.P. (2001) Substance, form and PFI contracts. *Public Money and Management*, 21(3), 41–46.

Leys, C. (2001) *Market Driven Politics: Neoliberal Democracy and the Public Interest*. Verso, London.

Merna, A. and Smith, N. (1999) Privately financed infrastructure in the 21st century. *Proceedings, Institution of Civil Engineers*, 132, 166–173.

Private Finance Panel (1995) *Private Opportunity, Public Benefit: Progressing the Private Finance Iinitiative*. HM Stationery Office, London.

Rochester, A. (1936) *Rulers of America: A Study of Finance Capital*. Lawrence & Wishart, London.

Tabb, W. (1999) Labor and the imperialism of finance. *Monthly Review*, 51(5), 1–13.

Teeple, G. (1995) *Globalisation and the Decline of Social Reform*. Garamond Press, Ontario.

Treasury Task Force (1997) *Treasury Taskforce Guidance, Partnerships for Prosperity, The Private Finance Initiative*. HM Stationery Office, London.

UNISON (1999) *The Only Game in Town. A Report on the Cumberland Infirmary Carlise PFI by UNISON Northern Region*. UNISON, London.

UNISON (2001) *Public Services, Private Finance, Accountability, Affordability and the Two-tier Workforce*. UNISON, London.

4

Refinancing and the Profitability of UK PFI Projects

Steven Toms, Darinka Asenova and Matthias Beck

4.1 Introduction

The profitability of PFI projects to the private sector remains one of the key areas of debate in the UK. In recent years this dispute has intensified as a consequence of the negative publicity associated with UK PFI refinancing deals which have opened some private sector protagonists to allegations of excessive profiteering. Moreover, the financial aspects of PFI contracts are often concealed, usually justified by 'commercial confidentiality', so that in the absence of verifiable data, the debate about profitability remains even further from resolution.

Nonetheless, ascertaining profitability is an important task. If the UK government and governments elsewhere are serious about involving the private sector in the delivery of public services on a large scale, profits must be at a level to offer sufficient incentive. From the point of view of the private sector provider, profits must be sufficient to compensate for perceived risk, particularly in the set-up and construction phases of the project. At the same time if public officials are to ensure that Value for Money (VFM) criteria are met, they must ensure that the private sector is not overcompensated, particularly in the case of risk pricing. Indeed, risk transfer has been identified as the most important determinant of VFM (Treasury Taskforce, 2000). In view of the inevitable asymmetric experience between public and private sector managers in drafting and negotiating commercial contracts, some bias in favour of the private sector is perhaps to be expected.

Wrapped up in the question of profitability therefore are a number of important questions relating to the public interest and the efficiency of resource allocation. Such issues are matters of serious concern for the National Audit Office (NAO) and the House of Commons Public Accounts Committee, both of which have commissioned statistical surveys of refinancing profitability. Whilst these surveys have been used to call officials responsible for the PFI programme to account (HM Treasury, 2007), there has been no further analysis of the detailed figures which have been generated as a result of this process.

It is the objective of this chapter to present such an analysis. In doing so, several interesting questions can be addressed. Because the PFI has been at work for longer in some sectors than in others, it might be expected that the degree of expertise also differs from one sector to another, impacting on the likely accuracy of risk pricing. It is also likely that the degree of risk varies from sector to sector. The first purpose is, therefore, to carry out a sector-by-sector comparison of refinancing profits. Any differentials in this respect are also suggestive of differential profits available to private sector operators, depending for example on their chosen portfolio of investment, the under-lying risks in the sectors chosen and the specialist knowledge of their staff. The second purpose is to compare the profits from refinancing of one firm to another. Because the public sector has traditionally lacked expertise in com-mercial contract negotiation relative to the private sector, it might be expected that there is a learning curve effect and that if there has been any mispricing, this is more likely to occur in early contracts rather than later ones. The third purpose is to examine the trends of profitability on refinancing contracts. One aspect of the 'learning curve' has been that on later contracts the public sector has negotiated a share of the refinancing gains. The detailed origins and relative scales of these gains have not yet been investigated. The fourth purpose is to analyse the relative public sector shares of refinancing profits by sector and compare them to private sector profits. An important reason why little analysis of this sort has been conducted hitherto is that, as noted above, the disclosure of information about the profitability of PFI contracts has been very limited. The information analysed here was collected by the NAO, but even so produced very different levels of response to the request. So much so that the PAC was forced to demand further information in specific cases of prior omission (HM Treasury, 2007). Differential disclosure levels provide the opportunity to examine which sectors and which companies are concerned with greater relative secrecy. Reluctance to respond to requests for informa-tion even from crown officials is perhaps suggestive of 'something to hide' and it might be supposed that if excessive profits are being made, then the level of voluntary first round disclosure and refinancing profits might be correlated. Investigation of these relationships is the fifth and final purpose of this chapter.

This chapter is structured as follows. The first section outlines the history of PFI with a focus on VFM criteria and the rationales for refinancing. The sec-ond section discusses the mechanics of PFI refinancing together with key UK government policies regarding these transactions. The third section presents an original financial analysis of refinancing returns which is based on financial details of 46 refinancing transactions reported in the 2006 NAO publication 'Update on PFI Debt Refinancing and the PFI Equity Market' (NAO, 2006). The chapter concludes with a discussion of the policy implications of this analysis.

4.2   PFI Finance and Value for Money

In PFI schemes, the private sector undertakes the design, building, financing and operation of public sector assets, in return for long-term payments from the government. PFI procurement differs from other forms of PPP, in that PFI schemes require full financial back-up from the participating private companies over the lifespan of the project (PFP, 1995; Treasury Taskforce, 1997; Allen, 2001). Another unique feature of PFI procurement is that it contractually commits public and private sector parties to a pre-negotiated allocation of risks (Glaister, 1999).

PFI-type arrangements in Britain can be traced to the early 1980s when a few projects such as toll roads were 'purchased' under 'PFI-like' terms under the so-called Ryrie Rules (Hall, 1998; Allen, 2001). The Ryrie Rules established that in public sector projects, private capital investment could not supplement public expenditure. This requirement, in addition to the prerequisite for unambiguous evidence for VFM, limited the use of private funding in public sector infrastructure development. Later on, some of the terms of the Ryrie rules were relaxed to allow certain projects to be financed through user charges. These 'pre-PFI' schemes sought to introduce private investments into public services, allegedly without affecting an overall reduction in the level of government direct investment (Allen, 2001). PFIs proper were introduced in England, Wales and Scotland in 1992, initially with a view towards increasing public service provision within existing borrowing constraints (Grout, 1997; Glaister 1999; HM Treasury, 1999). The support for PFI schemes was widened during the mid 1990s to include projects which were not subject to borrowing constraints, ostensibly on account of potential efficiency gains resulting from private sector involvement (Robinson, 2000).

Following the official launch of PFIs in Britain, the initial uptake for this type of procurement was relatively low, primarily due to a combination of hesitancy by the public authorities and of scepticism by private sector companies (Glaister, 1999; Beck and Hunter, 2002). In 1995, the government reaffirmed its commitment to PFI by introducing a series of 'priority projects.' At that time, a new government guidance document entitled 'Private Opportunity, Public Benefit: Progressing the Private Finance Initiative' advocated the expansion of PFIs (HM Treasury, 1995). Thereafter, soon after coming to power, the New Labour government emphasised its strong commitment to PFI (Timmins, 2001), while stressing that PFI procurement at all costs was not appropriate (Treasury Taskforce, 1997). Over the next years, the New Labour government commissioned a series of reports which were aimed at assisting the public sector in PFI based procurement (Bates, 1997, 1999; HM Treasury, 1999, 2000). In addition, in 1997, the Treasury Taskforce was established, which was tasked with facilitating the development of PFI expertise among the public sector managers. As a consequence of these measures, PFI projects have gradually become more widespread in Britain, with 450 projects having been signed with a total capital value of over £20 billion by September 2001 (Allen, 2001; National Audit Office, 2002) and more than 780 projects with a total capital value of £53 billion by the end of 2005 (International Financial Services, 2006).

Throughout this period of massive expansion of PFI-based procurement in the UK, there has been one striking continuity, namely the pronounced attempt by virtually all key UK governments to present PFI as a 'purely technical procurement process' (Dawson, 2001: 479), or in other words to disassociate

PFI from other policy agendas such as the Conservative's support for New Public Management or New Labour's Modernisation Agenda. This approach has been somewhat puzzling, since, at least on a theoretical level the economic case for PFI is by no means clear cut. Economic analyses of PFI have noted relatively early on that, since the public sector could borrow more cheaply than the private sector, the advocacy of PFI had to rest on other than pure public finance considerations (Spackman, 2002; Jenkinson, 2003). One potentially feasible argument was that PFI was politically, rather than economically, attractive not because it was cheaper but because it allowed government authorities to, firstly, ease macroeconomic constraints on account of the off-balance sheet treatment of PFI projects and, secondly, in so doing allowed them to bypass control on public investment (Spackman, 2002). While historically these considerations may have provided a rationale for expanding PFI procurement, more recent analyses indicate that the justification of PFI as a means of averting constraints on public sector borrowing has become, for a number of reasons, largely spurious (Dawson, 2001; Jenkinson, 2003). An alternative set of economic arguments in support of PFI has focused on the notion that incentives for efficient performance are inherently stronger in the private sector and that therefore any public service which utilises privately owned and managed assets, and in particular private finance, is likely to provide higher-quality and/or more cost-efficient services (Grout, 1997; Dawson, 2001).

In the UK, the notion that the private sector must demonstrate that efficiency gains in excess of its higher cost of capital is operationalised through the VFM requirement for PFI projects. This requirement evolved historically when, in 1997, the New Labour government abandoned the rule that all public sector projects were to be considered for PFI (HM Stationery Office, 2000). According to a new set of rules, the PFI option now had to be applied only if 'a robust assessment of the options in each set of circumstances confirms that the private sector proposal demonstrates considerable advantages over the public estimates' (HM Stationery Office, 1998). This link between PFI and VFM was explicated in the Treasury Taskforce's paper 'Partnerships for Prosperity' (1997) which specified that 'PFI solutions should be pursued where they are likely to deliver better VFM'. Moreover, this and consecutive Treasury papers, emphasised that the achievement of VFM centred crucially on a combination of competitive tendering processes and optimum risk transfer to the public sector, with the latter ensuring appropriate incentives for private sector PFI partners. Specifically the Treasury recommended that two VFM exercises be undertaken in order to evidence compliance with Best Value criteria. Accordingly, an initial VFM assessment is to be conducted once the business need has been identified. This assessment serves two purposes. Firstly, it provides an estimate for potential savings arising from the PFI procurement option, and secondly, it gives an assessment of the likelihood that these savings will materialise (Treasury Taskforce, 1997). In addition to this, the public sector client is expected to demonstrate that the project lends itself to the PFI option on the basis of pre-set conditions. These conditions include a clear operational need, scope for sufficient risk transfer, and the availability of adequate market interest from potential private sector bidders to ensure genuine competition. A complete estimate of whether a particular project fulfils this requirement has to be made later, on the basis of the full business case, the bids received and the outcome of final negotiations (NAO, 1999; Akintoye *et al.*, 2001).

In order to establish a benchmark for assessing private sector bids, the public sector client is expected to develop a financial model known as a public sector comparator, which reflects all revenues and costs associated with a particular project (Treasury Task Force, 1998). While the public sector comparator is a key measure for ensuring VFM, it is not the only criterion. In addition, government guidance recommends that other project aspects, such as service quality, risk transfer, and wider policy objectives be investigated (Treasury Taskforce, 1998).

In line with this emphasis on 'voluntary partnering', consecutive government guidelines have given the public sector client an increased degree of flexibility (HM Stationery Office, 2000). Before financial close, the client can, on the basis of the information collected during the PFI procurement process, reverse the decision to implement a project via the PFI option. Moreover, any project modifications during the construction and operational phases which affect the risk allocation, can repeal the off-balance sheet treatment of the project.

Earlier guidance by the Treasury Taskforce as well as more recent reports by the NAO (Treasury Taskforce, 1997, 1999; NAO 1999, 2000; HM Stationery Office, 2000) have tended to emphasise the need for accuracy in VFM exercises together with the need for clients to manage projects effectively through the implementation and operational phase. What has virtually escaped the attention of these guidance documents has been any requirement for the public sector clients to monitor the private sector income shares and especially to ensure a public sector share in the potentially very significant gains from refinancing during the later project stages. This omission is now being at best half-heartedly addressed through guidance which encourages the public sector either to restrict certain types of refinancing practices or to contractually ensure an 'appropriate' public sector share in refinancing gains (Office of Government Commerce, 2001; HM Treasury, 2004). It is in part due to this omission that there is widespread unease about private sector profits associated with many UK PFI projects.

Underlying these changes in approach to refinancing and PFI profitability in general is the perception that excessive private sector profits inevitably occur at the expense of the public purse, and that high private sector profits deprive the public from infrastructure and services that genuinely represent value for money (Ball *et al.*, 2001; Pollock *et al.*, 2002). Additionally, it has been argued that the excessive private sector return associated with some PFI projects undermines the efficiency rationales under which these schemes have been introduced, in a manner that puts into question the very reasoning and motivation of UK government policy (Froud and Shaoul, 2001; Edwards and Shaoul, 2003). Put differently, there is, among many observers concerned with private sector PFI profits, a suspicion that the current UK government continues to favour PFI schemes not because they have proven themselves as a cost-effective means for procuring high-quality infrastructure and services, but rather because it is politically opportune to allow the mostly UK-based private sector participants in PFI to reap potentially excessive profits (Gaffney and Pollock, 1999).

As an alternative to this hidden subsidy argument, it has been suggested that the continuing commitment of the UK to PFI in light of significant private sector profits, is not based on a desire to subsidise certain industries indirectly, but rather on the inability of government watchdogs and agencies to restrain what they do in fact perceive as excessive profits (Allen, 2001). This view of a 'lock-in' of UK PFI into undesirable practices is supported above all by publications of the House of Commons Committee on Public Accounts which, together with the UK NAO, have consistently criticised specific PFI projects for resulting in excessive private sector profits as well as condemning the failure of the private sector to adequately share gains from PFI refinancing with the public sector (NAO, 2006; House of Commons Committee on Public Accounts, 2007).

4.3   Mechanisms and Policies of Refinancing

PFI projects are financed through a large proportion of bank debt or bonds (typically around 90% of the capital requirement) and equity finance, primarily in the form of subscription to shares in the project company. The precise financial structure of these projects is usually shaped by private sector companies and the party which acts as the project's loan arranger or financial advisor to the project company. The objective of these arrangements is to ensure such financial arrangements which will guarantee that the project's financial requirements will be met and the shareholders will receive a return on their investments. The financing costs of the PFI projects are determined by various factors, such as the project's scope and scale, economic and market conditions, the credit reputation and rating of the borrower etc., and are closely associated with the project's risk profile.

The key risks involved in a PFI project include the risk of the project collapsing before the debt repayment and the risk of inaccurate revenue forecasts (Allen, 2001). Due to a less than obvious interplay of factors and difficulties in measuring these risks, cost evaluations associated with different methods of borrowing can be very imprecise, which is one of the factors which gives rise to the possibility of refinancing.

As a financial transaction, refinancing has become popular in the UK primarily on account of the willingness of government organisations to make it work. However, the very fact that refinancing exists paradoxically raises concerns about the nature of PFI contracts. Among other things, the large investor returns which were generated by the refinancing of some early PFI projects indicate that such basic fundamentals of PFI transactions as the pricing structure are frequently inaccurate (HM Treasury, 2007).

Today, refinancing is considered particularly suitable for projects where the construction phase has been completed and the operational phase is demonstrably successful. The risk profile of such projects is significantly less critical and revenues can be forecast more accurately. Refinancing, however, often increases the risk borne by the public sector, for example, in situations where equity is replaced by debt.

Technically, the refinancing of PFI projects involves a reconsidering of the project features according to which the loan was initially provided. The HM Treasury's Standardisation of PFI Contracts (Version 3) defines refinancing as follows:

> During the life of the Project, the Contractor may wish to replace, augment or change the structure, nature or terms of the financing solution that it put in place at Financial close for the purposes of financing the Project. Where such restructuring changes will have the effect of increasing or accelerating distribution to investors or of reducing their commitment to the Project, these effects are individually or collectively referred to as Refinancing Gains. (HM Treasury, 2004:254)

Since 2004, the UK Treasury has tended to encourage public authorities to approve and endorse refinancing arrangements. Thus, the above-cited Treasury document states that:

> Refinancing of PFI projects is one way in which both the Authority and investors in the Contractor can share in the benefits of a successful project. Accordingly, Authorities should be receptive to proposals from the Contractor to refinance, and are encouraged to consent to such proposals. (HM Treasury, 2004:255)

Despite this generally positive attitude towards refinancing and the private sector gains associated with these practices, there are some aspects which appear to be of concern to the UK Treasury. These include, in particular, increases in senior debt which can be crucial for ensuring the involvement of the private sector companies in refinancing activities. While Treasury guidance documents have been discouraging of such practices, they have not been prohibitive:

> Increases in Senior Debt for a PFI project, whether through the Contractor or otherwise having security (or other rights) over and/or recourse to the assets, contracts or cash-flows of the Contractor, beyond the original capital value of the Project should not be approved by the Authority without it first seeking appropriate professional advice. (HM Treasury, 2004:256)

The key components of refinancing can include alterations in financial parameters, such as interest rates, repayment dates, margins and/or the level of senior debt, at which the loan was provided in the original contract as well as the release of contingent junior capital (Treasury Task Force, 1999; HM Treasury, 2004).

Following bad publicity in relation to some earlier refinancing arrangements (Anon, 2006; HM Treasury, 2006; Rozenberg, 2006; Settle, 2006; Timmins, 2006; Hencke, 2007; Russell, 2007) the government has attempted to provide guidance on how these transactions should be structured. References to refinancing were first made in the publication 'Guidance on the Standardisation of PFI Contracts' (OGC, 1999) and subsequently revised (OGC, 2001). As a minimum, this early guidance recommended that the client's consent be gained prior to refinancing. At the same time, the guidance maintained that, even in the absence of formal provisions, the public sector should share benefits. According to the Office of Government Commerce

(OGC, 2001), refinancing profits should compensate parties for risks taken during the construction phase. Therefore the benefits should be shared on equal basis between the public and the private sectors.

The requirement for equal shares (50:50) has been disputed by many private sector companies, which believed that they should receive higher proportion from the profits (e.g. 75:25) as a reflection of the actual risk distribution. In 2001, when an attempt was made to establish new regulations regarding the conditions for refinancing, approval by private sector parties became subject to intensive negotiations. In July 2002 the body commissioned with this task, the OGC, published for the first time its revised guidance to the public sector authorities stipulating that in all new PFI contracts the refinancing gains should be equally distributed between the public and the private sector partners (HM Treasury, 2003).

The second important outcome of these consultations was that the private sector agreed to adopt a voluntary code for sharing the gains from earlier PFI contracts (i.e. contracts signed up to 30 September 2002) which otherwise did not contain explicit arrangements regarding the refinancing. The introduction of this voluntary code was considered an important achievement in light of the fact that early PFI projects were likely to have the greatest potential for refinancing (NAO, 2006). According to this voluntary code, the public sector was entitled to no more than 30% share of the refinancing gains, which was considered 'the best it (OGC) could have achieved' (HM Treasury, 2003: 7). The argument put forward for this decision reflects the government's own perception that the balance of power within the PFI scene is distributed overwhelmingly in favour of the private sector:

> To have sought more would have increased the risks that the private sector would not agree to the code or would seek to avoid complying with it. (HM Treasury, 2003:7)

Overall there is little evidence of a willingness of the private sector to voluntarily share refinancing gains with the public. Thus the Treasury's 'Twenty-Second Report on Refinancing' (HM Treasury, 2003) noted that prior to June 2000 only 4% of all PFI contract contained provisions to share half (50%) of refinancing gains with the public sector. This percentage had not increased by June 2001. For contracts providing for a share of 30% going to the public sector, meanwhile, there had been an increase from 4% for the period prior to June 2000 to 23% for the year to June 2001. For contracts providing for less than 30% share going to the public sector, lastly, there had been an increase from 18% for the period prior to June 2000 to 27% for the year to June 2001. Overall prior to June 2000 only 26% of PFI contracts included requirements for a sharing of refinancing gains, while for the year to June 2001, this figure increased to 54%.

Despite industry agreements to a voluntary regulation of refinancing deals, these deals continue to be, by the government's own admission, riddled with problems. One of the most important recent allegations relates to the level of disclosure of the precise scale of the refinancing benefits. Information in the press quotes Edward Leigh, Chairman of the Commons Public Accounts Committee referring to 'obscene' rates of return (Timmins, 2006) as well as his concern that due to the lack of transparency and full disclosure the real profits can be 'even more grotesque'. Leigh warned that some contractors may have to be called to provide an explanation of controversial refinancing deals to the Members of Parliament. Perhaps unsurprisingly, some public sector authorities have been equally reluctant to provide details on refinancing deals. For example, in November 2006, NHS Lothian refused to disclose the financial details related to the refinancing of one of its early flagship PFI projects – the Edinburgh Royal Infirmary. In response to critics pointing out the possibility that the costs of the new deal are likely to exceed the original and already high costs of the project, the Trust responded with vague reassurances that the deal was 'commercially confidential' and will 'deliver multi-million pound gains' for the local users (Settle, 2006). The most recent Report of the Committee of Public Accounts (HM Treasury, 2007) indicates that, while the Treasury had been aware for some time of the potential problems with excessive private sector gains, it was reluctant to intervene with measures which 'might affect the private sector's interest in bidding for the early PFI contracts' (HM Treasury, 2007:7).

Overall the UK government's approaches to the regulation of refinancing have been characterised by a continued unwillingness to recognise the failure of voluntary approaches. In the long run there is every possibility that this failure to regulate refinancing and excessive private sector profit making in PFI projects in general is likely to undermine the credibility of PFI as a means for ensuring VFM in the provision of infrastructure and services to the public.

4.4   PFI Profits and Refinancing

Focusing on the post-refinancing profitability of UK PFI projects, this section will investigate a number of issues associated with PFI refinancing in the UK, including the level of returns from refinancing transactions, the relationship between levels of return and levels of disclosure provided by private sector contractors or investors, sectoral differences in terms of refinancing returns and PFI returns in general and, lastly, changes in the levels of gross returns on PFI refinancing over the past 7 years.

The most recent 'Update on PFI Debt Refinancing and the PFI Equity Market' by the Comptroller and Auditor General, which was published by the NAO in April 2006 (NAO, 2006), provides the most extensive set of financial details on refinancing deals in England. Overall the report provides data on 46 PFI projects which had been completed in England. Unfortunately no comparably detailed information is available for Northern Ireland, Scotland and Wales.[1] This data includes project-based information on contractors' shares of refinancing gains, their equity investment as well as a disclosure rating with regard to the detail of financial information provided. In the following sections this data serves as the basis of an analysis which seeks to identify some of the key financial and sectoral characteristics of refinancing deals.

The data is presented in Table 4.1, analysed by sector and contractor in panels a and b respectively. Average returns and disclosure scores are shown in both cases in Table 4.1a(i) and Table 4.1b respectively and comparative

Table 4.1   a) (i) Rates of return on PFI refinancing transactions and disclosure scores by sector.

Sector	Average return	Disclosure score
Health	25.24%	4.22
MOD	8.81%	4.00
Devolved (Scotland, Wales, N. Ireland)	7.81%	n/a[1]
Education	6.38%	2.33
Home Office	5.56%	8.17
Transport	5.18%	0.00
Other (DEFRA, HMRC)	4.64%	7.33
Community & local government	1.17%	5.67
Overall	**8.04%**	**6.00**
Public sector share as % of total contract value	**3.13%**	

(ii) Internal rates of return before and after refinancing.

Sector	Project IRR		Ratio
	Before refinancing	After refinancing	
Health	17.21%	48.64%	2.83
Education	12.43%	41.15%	3.31
MOD	16.45%	33.33%	2.03
Transport	16.48%	26.12%	1.59
Community & local government	13.96%	16.10%[2]	1.15
Other (DEFRA, HMRC)	12.28%	15.39%	1.25
Home Office	14.02%	14.45%	1.03
Overall	**14.88%**	**30.96%**	**2.08**

b) Rates of return on PFI refinancing transactions and disclosure ratings by contractor/investor.
Calculated from appendix 9 (NAO, 2006).

Contractor/ Investor	No. of transactions	Average return	Sector/activity	Disclosure score
Taylor Woodrow	2	149.63%	Road, hospital	5.50
Laing	3	75.89%	Road, hospital, MOD	3.33
Carillon	3	68.72%	Road, hospital, prison	4.33
Innisfree PFI	5	62.94%	Hospitals, schools	7.80
Barclays Capital	7	43.08%	Hospitals, schools	3.57
Jarvis[3]	10	36.75%	Schools, local authority, rail	5.27
Wakenhut	6	19.77%	Prisons, school	6.67
Serco	9	19.12%	Prisons, school, MOD, hospitals	3.56
WS Atkins	3	18.38%	Roads, prison	3.33
Balfour Beatty	3	11.07%	Roads	0.00
28 others	28	41.00%		4.20
Total	**79[4]**	**43.51%**		**4.58**

Notes:
1. Devolved authorities were not required to respond to the C&AG survey.
2. In line with Figure 2, note1, evidence 49 (National Audit Office, 2006), North Wiltshire is assumed to have a post refinancing IRR of 13%.
3. Excludes Tube Lines Ltd, an atypical investment with a shared gain of £4.45m on an investment of £1.2bn. If added to its other 10 investments, Jarvis's average return falls to 0.7%.
4. The discrepancy between the number of transactions (79) and the number of PFI contracts from which they originated (46) is due to several contracts having more than one company involved and the need to define company specific transactions in determining profit shares. Differences in overall average disclosure scores occur for the same reason.

Part One

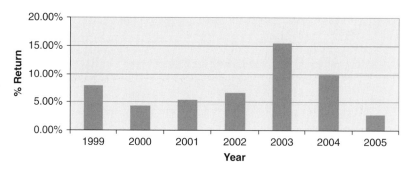

Figure 4.1 Gross returns on PFI refinancing contracts, 1999–2005. Calculated from appendix 9 (NAO, 2006).

internal rates of return (IRRs) before and after refinancing in Table 4.1a(ii). Figure 4.1 shows the trend in refinancing returns through time. Table 4.1a returns are gross average returns (i.e. the combined returns received by the private and pubic sector shares) on all refinancing transactions in a specific sector of public sector activity together with the average disclosure score for transaction in that sector. Average return has been calculated as the simple average gross return of all refinanced projects in that sector, with the gross return on an individual project being computed as the gain available to all parties to the contract (including the public sector) divided by total capital value (i.e. the total of debt and equity finance). The average return to the public sector (public sector share of gains) has been calculated by totalling the public sector share in all contracts and dividing the result by the capital value of all contracts. The average disclosure score by sector has been calculated by taking the number of items voluntarily disclosed to the NAO (up to a maximum of 10) in the first survey for each contract in that sector and averaging the disclosure score across all contracts in that sector. IRRs in Table 4.1a(ii) are taken directly from the transaction specific IRRs reported in the NAO report and averaged by sector. Average rates of return by contractor/investor in Table 4.1b have been computed by taking the annualised weighted average contractor's share of the refinancing gains from the projects invested divided by the weighted average contractor's share of the equity investment in the same projects. The equity share was calculated by subtracting the debt finance from each project's capital value. Where no data existed for the financing mix, the ratio of debt to project capital value was assumed to be 90%, which is the typical mix (see NAO, 2006:1).

Subsequent sections below examine each of the five areas for investigation outlined in the chapter introduction.

4.4.1   Refinancing returns and disclosure levels by sector

A refinancing transaction represents an opportunity for the private and public sector to benefit from better financing terms once a project has become operational and some of the initial procurement risks have been eliminated. However, consistently high refinancing gains in a particular sector can also be indicative of a pattern of mispricing of risk in a particular sector and as such should be of concern to those who seek to ensure VFM in PFI projects.

One of the outstanding results of the analysis in Table 4.1a is the high level of average returns on refinancing in the health sector (25.24%). Similarly, health has the highest IRRs both before and after refinancing. Although the ratio between IRRs before and after refinancing in health is lower than in education it appears to be attributable primarily to the considerably lower before refinancing IRR of the education sector.

It is possible that the underlying risk in the construction phase is higher than in other sectors. It seems unlikely that the risk is three times higher, as suggested by the differential between health and the next highest sector in Table 4.1a(i), whatever the special features of hospital construction as opposed to schools, roads or MOD buildings. It seems even less likely when evidence is considered that suggests procedures in schools tend to overestimate the risks at this stage (Ball *et al.*, 2001). The disproportionately high levels of return on refinancing in the health sector should therefore be worrying for a number of reasons, including, above all, the fact that these returns are indicative of comparatively high initial costs of PFI projects in this sector. Additionally, the large differential in refinancing returns between the health sector compared to the next highest sector (defence with 8.81%) suggests that this sector suffers from particularly poor risk pricing and financing practices which are likely to result in problematic and potentially costly procurement and management practices.[1] Lastly, given that the private sector has in the past been able to secure the lion's share of refinancing gains, and continues to do so under the voluntary 30% agreement (with 70% of gains going to the private sector and only 30% to the public sector), this analysis would also indicate that health is probably the most difficult sector for public clients to secure VFM.

4.4.2   Refinancing returns and disclosure levels by contractor/investor

Table 4.1b presents data on the average rates of return on PFI refinancing transactions by company. In Table 4.1 these average returns and disclosure rates are reported for individual contractors/investors which have been involved in anything from two (Taylor Woodrow) to up to ten (Jarvis, which has 11 if Tube Lines is also included) refinancing deals. Data for contracts which have been involved in only one refinancing deal are not separately reported but have been included in the '28 others' category).

Overall the average return figures indicate that refinancing deals have been highly profitable for most contractors/investors with an average return of 43.51% for all transactions. However, as regards individual contractors/investors, there has been a significant spread in the level of these average returns ranging from 11.07% (for the three transactions concluded by Balfour Beatty) to 149.63% (for the two transactions completed by Taylor Woodrow).

Typically the rate of return accruing to an investing firm is much higher than the rates reported according to sector averages. The principal reason is that, as pointed out above, debt is left out of the denominator for firm level calculations. These rates of return correspond to the equity or risk-bearing stakeholders in the private sector firms. In other words in addition to the underlying risk associated with the project, the return is also compounded by the leveraging effect of debt finance. Such returns are governed by the prior claims of third-party debt holders in a standard capital market. In PFI transactions debt and equity holders are related through consortium membership and board representation of senior debt holders.

Bearing these points in mind, the rates of return being earned by many firms are impressive and in some cases excessive. The top five firms by profitability are all involved in hospitals, so some of the reasons for risk mispricing in health discussed above seem to be translating themselves into large profits for these five firms in particular.

It is interesting to note that most contractors/investors appear to specialise in two sectors. With the exception of Serco, all other contractors/investors who were involved in multiple transactions, conducted business in no more than three areas of activity with the majority working in no more than two. It is possible that this provides the firm with some benefits of risk diversification whilst retaining the advantages of sector-specific knowledge. If this is the case, it is likely that policy makers will need to encourage more firms to enter the PFI market for reasons of capacity as well as competitive bidding processes.

4.4.3   Refinancing returns over time

Figure 4.1 shows the time trend for annual gross returns on refinancing transactions for the period from 1999–2005. Since opportunities for refinancing arise in part from an initial mispricing of projects risks, it should be expected that, notwithstanding significant fluctuations in the volume of PFI project completions, these returns should show a significant decrease over time as a consequence of improved pricing and management of PFI risks[2]. Figure 4.1 indicates that has not been the case. In fact 2003 was a peak year following a rising trend suggesting there is no obvious learning curve effect, since most of the profits analysed by firm and by sector above seem to have been earned in 2003 and 2004. There is no evidence, therefore, of increasing adeptness of public sector officials at dealing with risk pricing and commercial contract negotiation, unless 2005 data suggest the beginning of a new trend. Very large refinancing profits on hospital contracts at Dartford and Gravesham (£122m, 2003), Norwich and Norfolk (£229m, 2003) and Bromley (£150m, 2004) help explain the pattern in Figure 4.1. These deals attracted especially critical attention from the Public Accounts Committee (HM Treasury, 2007), and it is possible therefore that political accountability may exert downward pressure on profit levels in future transactions.

4.4.4   The public sector share

The public sector share of the profits totals £137.5m, representing a return of 3.13% on the total assets in the scheme. If the unrepresentatively large Tube

Lines project is excluded, these figures are £95.7m and 1.61% respectively. The public sector share of the total profits on all projects is 34% including higher shares (60%) on the more recent Tube Lines project and also a 60% share of a significant profit from Newcastle Estates. Excluding Tube Lines the figure is 28%. The public sector also took 30–35% of the profit on the three hospital contracts referred to in the previous section. The evidence therefore suggests that although returns to the private sector are usually significant and indeed often excessive, there is some clawback for the public sector which mitigates the effect. It is nonetheless interesting to note that the public sector rate of return is considerably less by comparison.

4.4.5   The pattern of disclosure

The evidence in Table 4.1a suggests that there is an inverse relationship between profitability and the level of disclosure. For the transactions where both returns and disclosure scores are available (n = 35), there is a clear and statistically significant negative relationship between returns and level of the disclosure.[3] Health, which is associated with the excessive profits discussed above, has an average disclosure score of 4.22 compared to an average of 6.00 across all sectors. As far as the companies are concerned, there is no clear relationship between profitability and secrecy. The obvious interpretation is that secrecy is driven by sector participation rather than board room imperative. In other words, although firms benefit from risk mispricing in certain sectors, especially health, and reduced disclosure occurs partly as a result of excessive profits on those transactions, they are likely to have investments in other sectors in their portfolios where they are less bound by the imperatives of secrecy. Again it is worth noting that in the cases of the three hospitals referred to above (Dartford & Gravesham, Norwich & Norfolk, and Bromley), there was an absence of any disclosure in the first call for information. Although the information is now in the public domain, these transactions attracted zero disclosure scores in the analysis as a result of their default behaviour.

4.5   Conclusion

The striking result of the cross-sector comparison is the excessive returns obtained in health. There are two possible reasons for this. Either the construction phase risk for health projects is three times higher than other public sector activities, or the level of risk has been mispriced. Further investigation is required into the dynamics of health sector contract negotiating and financing, particularly on the three contracts where very large profits were earned. However the analysis also revealed that the health sector was below average in disclosure practice, suggesting that further investigation may be problematic in an atmosphere of secrecy.

Another interesting aspect of the analysis, whether conducted by sector or by contractor, is that the pay-off in profit terms is always positive. In financial markets, the positive net value of the risk premium is a function of some combination of negative and positive outcomes with the latter on average prevailing. In the PFI case, the average risk premium on refinancing lies in the middle of a distribution of returns which are always positive. If there are no negative outcomes *ex post*, this begs the question of whether there is any genuine downside risk in the contract terms and if the 'risk' transferred to the private sector is merely upside risk. That the public sector feels the need to make substantial *ex post* clawback of refinancing gains suggests that this is the case. It is also consistent with the view that such contract terms are necessary to induce the private sector to enter these markets, in view of their higher opportunity cost of capital, which is a function of government borrowing rates plus the average risk premium for private sector investments. All the private sector participants have made healthy profits from PFI refinancing deals, suggesting that they are likely to extend or deepen their relatively narrow portfolio exposure engaged in hitherto. Much of the profit has been the result of using levels of leverage that would be unusual in private sector settings. Where the operating risk is high, it makes little sense to layer financial risk on top, which is why private sector venture capital is usually equity based. Again, this suggests that the private company's perception of the risk is lower than they would have their negotiators make out.

From the point of view of government and public officials involved in PFI, there is little evidence of a learning curve in contract specification, at least as far as the trend in negotiated profit levels is concerned. The government has succeeded in clawing back around 30% of profits on refinancing. Again, this begs the question of why the contracts are not priced more aggressively from a public sector point of view in the first place, thereby avoiding the necessity of a claw back mechanism. If this economises on the transaction cost of policing the deals, so much the better, since the levels of disclosure, particularly on the most profitable contracts, especially in the health sector, are rather low.

In summary there is a need for greater adeptness on the public sector's part in negotiating PFI deals. Insofar as the scheme substitutes for public expenditure it can never do so totally since the public sector will always need some investment for training its staff to negotiate good deals for the taxpayer. To deliver VFM, on the basis of the evidence presented above, more such investment is probably needed than has been made hitherto.

Notes

1. Unfortunately no comparably detailed information is available for Northern Ireland, Scotland and Wales.
2. This is even more problematic when considering that the health sector was amongst the pioneers in PFI procurement, and by 2005 accounted for the largest share of PFI contracts of any government department with 236 of a total of 780 contracts for all UK government departments (International Financial Services, 2006).
3. Such a decrease would be likely to reflect a concomitant increase in the VFM of PFI projects to the pubic sector.

4. The correlation coefficient is -0.361 which is statistically significant at the 95% confidence level (p-value $= 0.033$). The negative sign confirms that firms with high returns have lower disclosure and vice versa.

References

Akintoye, A., Beck, M., Hardcastle, C. *et al.* (2001) *Standardised Framework for Risk Analysis and Management in PFI Projects.* Glasgow Caledonian University, Glasgow.

Allen, G. (2001) *The Private Finance Initiative (PFI)*, Research Paper 01/117, prepared by the Economics Policy and Statistics Section. House of Commons Library, London.

Anon (2006) The Issues Explained, PFI Refinancing. *The Times*, May 16.

Ball, R., Heafey, M. and King, D. (2001) Private Finance Initiative – a good deal for the public purse or a drain on future generations? *Policy and Politics*, 29(1), 95–108.

Bates. M. (1997) *First Review of The Private Finance Initiative by Sir Malcolm Bates.* HM Stationery Office, London.

Bates, M. (1999) *Second Review of the Private Finance Initiative by Sir Malcolm Bates.* HM Stationery Office, London.

Beck, M. and Hunter, C. (2002) PFI uptake in UK local government. In: Akintoye A., Beck M., Hardcastle C. (eds.) *Public Private Partnerships: Managing Risks and Opportunities.* Blackwell Science, Oxford.

Dawson, D. (2001) The Private Finance Initiative: a public finance illusion? *Health Economics*, 10, 479–486.

Edwards, P. and Shaoul, J. (2003) Partnership: for better or worse? *Accounting, Auditing and Accountability Journal*, 16(3), 371–385.

Froud, J. and Shaoul, J. (2001) Appraising and evaluating PFI for NHS hospitals. *Financial Accountability and Management*, 17(3), 247–270.

Gaffney, D. and Pollock, A. M. (1999) Pump-priming the PFI: why are privately financed hospital schemes being subsidized? *Public Money and Management*, January–March: 55–62.

Gaffney, D. and Shaoul, J. (1999) Pump-priming the PFI: why are privately financed hospital schemes being subsidised? *Public Money and Management*, 17(3), 11–16.

Glaister, S. (1999) Past abuses and future uses of private finance and public private partnerships in transport. *Public Money & Management*, 19(3), 29–36.

Grout, P. (1997) The economics of the Private Finance Initiative. *Oxford Review of Economic Policy*, 13(4), 53–66.

Hall, J. (1998) Private opportunity, public benefit? *Fiscal Studies*, 19(2), 121–140.

Hencke, D. (2007) Sharp business people outwitting Whitehall over PFI refinancing deals. *The Guardian*, May 15.

HM Stationery Office (1998) *Better Quality Services, A Handbook on Creating Public/Private Partnerships through Market Testing and Contracting Out.* HM Stationery Office, London.

HM Stationery Office (2000) *Public Private Partnership: The Government Approach.* HM Treasury, London.

HM Treasury/Private Finance Panel (1995) *Private Opportunity, Public Benefit: Progressing the Private Finance Initiative.* HM Stationery Office, London.

HM Treasury (1999) *Modern Government Modern Procurement.* HM Treasury, London.

HM Treasury (2000) *Management of Risks – A Strategic Overview*. HM Treasury, London.

HM Treasury (2003) *House of Commons, Committee of Public Accounts, PFI Refinancing Update*, Twenty-second Report of Session 2002–03. HM Treasury, London.

HM Treasury (2004) *Standardisation of PFI Contracts (Version 3)*. HM Treasury, London.

HM Treasury (2006) *House of Commons, Committee of Public Accounts, Refinancing of the Norfolk and Norwich PFI Hospital*, Thirty-fifth Report of Session 2005–06. HM Treasury, London.

HM Treasury (2007) *House of Commons, Committee of Public Accounts, Update on PFI Debt Refinancing*, Twenty-fifth Report of Session 2006–07. HM Treasury, London.

House of Commons Committee on Public Accounts (2007) *Tendering and Benchmarking in PFI*. House of Commons, London.

International Financial Services (2006) PFI in the UK: Update. International Financial Services, London. Hompage: http//:www.ifsl.org.uk.

Jenkinson, J. (2003) Private finance. *Oxford Review of Economic Policy*, 19(2), 323–334.

Minutes of Evidence – Public Accounts (2007) *Update on PFI Debt Refinancing and the PFI Equity Market*, HC 1040. House of Commons, London.

National Audit Office (1999) *Examining The Value for Money of Deals under The Private Finance Initiative*, Report prepared by the Controller and the Auditor General for the House of Commons. HM Stationery Office, London.

National Audit Office (2000) *The Refinancing of the Fazakerley PFI Prison Contract*, Ordered by the House of Commons, Prepared by the Controller and the Auditor General. HM Stationery Office, London.

National Audit Office (2002) *NAO Focus, Making the Project a Good Deal*. National Audit Office, Homepage: http://www.nao.gov.uk.

National Audit Office (2006) *Update on PFI Debt Refinancing and the PFI Equity Market*, Ordered by the House of Commons, Prepared by the Controller and the Auditor General. HM Stationery Office, London.

Office of Government Commerce (1999) *Office of Government Commerce Guidance for Government Departments: Refinancing of PFI Projects*. Homepage: http://www.ogc.gov.uk/procurement.

Office of Government Commerce (2001) *Office of Government Commerce Guidance for Government Departments: Refinancing of PFI Projects*. Homepage: http://www.ogc.gov.uk/procurement.

Pollock, A., Shaoul, J. and Vickers, N. (2002) Private finance and 'value for money' in NHS hospitals: a policy in search of a rationale? *British Medical Journal*, 324, 1205–1209.

Private Finance Panel (1995) *Private Opportunity, Public Benefit: Progressing the Private Finance Initiative*. HM Stationery Office, London.

Robinson, P. (2000) The Private Finance Initiative: The Real Story. *Consumer Policy Review*, 10(3), 83–85.

Rozenberg, G. (2006) Taxpayers punished by lack of deal-broking skills at councils. *The Times*, May 15.

Russell, B. (2007) City runs rings round taxpayers in PFI refinancing. *The Independent*, May 15.

Settle, M. (2006) PFI contractors may be compelled to reveal their 'obscene' profits. *The Herald*, December 29.

Spackman, M. (2002) Public-private partnerships: lessons from the British approach. *Economic Systems*, 26, 283–301.

Timmins, N. (2001) Labour's private determination to deliver. *The Financial Times*, Homepage: http://globalarchive.ft.com/globalarchive/articles.htm.

Timmins, N. (2006) Contractors pressed to reveal PFI profits. *Financial Times*, December 28.

Treasury Task Force (1997) *Treasury Taskforce Guidance, Partnerships for Prosperity. The Private Finance Initiative*. HM Stationery Office, London.

Treasury Task Force (1998) *Treasury Taskforce Policy Statement No.2, Public Sector Comparators and Value for Money*. HM Stationery Office, London.

Treasury Task Force (1999) *Treasury Taskforce Guidance, Standardisation of PFI Contracts*. HM Stationery Office, London.

Treasury Task Force (2000) Value for money drivers in the Private Finance Initiative. Arthur Andersen and Enterprise LSE.

Part One

5

The Dedicated PPP Unit of the South African National Treasury

Philippe Burger

After the first democratic election in South Africa in 1994, the South African government set about reforming the approach of government towards the management of state assets. It did this in a manner that can best be described by what Flinders (2005: 216) calls the increasing use of institutional hybridity and a move from 'government to governance'. This approach towards state assets is broader than just privatisation (Department of Public Enterprises, 2005a). It includes (Department of Public Enterprises, 2005b):

- Concessions
- Strategic equity as well as management partnering
- PPPs
- Privatisation (partial and full)
- Flotation of State Owned Enterprises (SOEs) (initial and secondary)
- Securitisation

Thus, the restructuring and management of state assets also includes the use of PPPs. At the heart of the South African PPP structure is the National Treasury's PPP Unit constituted in 2000. This dedicated PPP unit plays a key role particularly in the creation of PPPs where it has the final authority in the approval of PPP agreements. It has this authority even though the initiative and ultimate management of PPP agreements originates, and rests with, individual government departments and provinces. This chapter explores the role of this unit in the South African context. It commences with a discussion of the theoretical rationale for PPPs and, in particular, for having a dedicated PPP unit. This is followed by a brief history of PPPs and the dedicated PPP unit in South Africa, whereafter the discussion turns to the role and operation of the unit itself as well as its future challenges.

5.2   The Rationale for PPPs

Though the PPP concept is often confused with privatisation proper, it shares a commonality with privatisation in that PPPs also entail the introduction of private sector management and/or ownership of what traditionally has been the sole preserve of government. A PPP is an institutional and contractual partnership arrangement between government and a private sector operator to deliver a good or service to the public, with the following distinctive elements (Fourie and Burger, 2000):

- A true partnership relationship (i.e. alignment of objectives through the alignment of the incentive structures facing the public and private partners).
- A sufficient amount of risk transfer to the private operator to ensure that there are sufficient incentives for the private operator to operate efficiently. This entails that risk is allocated to the party best suited to carry it.

The main rationale to use PPPs is the perceived efficiency of the private sector and inefficiency of the public sector. In terms of economic literature three kinds of efficiency can be distinguished: allocative efficiency, i.e., the use of resources so as to maximise profit and utility; technical efficiency; and X-efficiency, i.e., the prevention of a wasteful use of inputs (Fourie and Burger, 2000). The perceived efficiency that the private sector brings to a PPP agreement refers especially to technical and X-efficiency. Companies are driven to be technically and X-efficient by the technical, operational and financial risk that they carry. These are mostly supply-side risks. The perception that private sector participation brings improved efficiency seems to be vindicated by experience in, for instance, the UK where Hodge (2004) cites studies that indicate that government departments that implemented PPPs registered between 10% and 20% in cost savings. In addition, Gosling (2004) notes that, according to the UK's National Audit Office, 76% of PFI deals are constructed on time, while in the case of projects completed under conventional procurement, it is only 30%. In terms of projects constructed to budget the figures are respectively 78% and 27%. In South Africa PPP projects, in general, appear to be completed in time and early indications are that these projects yield the expected cost-saving and VFM benefits (Dachs, 2006).

Instead of fully privatising the delivery of a good or service, government could enter into a PPP agreement, if the good or service to be delivered is a public good or a good characterised by an externality. Public goods or goods characterised by externalities suffer from the free-rider problem. This means that demand is not fully revealed, causing private companies not to be able to estimate the future demand for the good. As such, government may need to estimate the full social demand, so as to either supply the demand itself, or to reveal it to a private producer who then supplies to government. Through this action government is supposed to improve the allocative efficiency of the goods or services delivered. If government uses a private producer to deliver the good or service, it usually pays the private operator who delivers the service fully or augments the user fee that the private operator levies by an additional amount. Note that in the absence of a free-rider problem, when the good is a private good, demand is fully revealed, enabling a private company to estimate demand and, subsequently, to carry the demand risk involved. In such a case privatisation, instead of a PPP, may be the best mode of delivery.

In the case where a good is a public good, or a good characterised by an externality so that demand-side risk is present, the choice between delivery through a PPP or by government itself, depends first on the ability of government to transfer sufficient supply-side risk to the private operator, and secondly on the level of competition or contestability facing a private operator (Fourie and Burger, 2000; Hodge, 2004; Grimsey and Lewis, 2005). These two conditions ensure that the private operator behaves with technical efficiency and X-efficiency. In the absence of these two conditions, private sector delivery may not necessarily be more efficient, whereas its costs, such as interest cost and the profit that it has to pay to its shareholders, may cause the cost of delivery through a PPP to exceed that of government delivery (Fourie and Burger, 2001; Grimsey and Lewis, 2005). Indeed, Hodge cites the UK study of Anderson and LSE Enterprise (2000) that indicates that 60% of cost saving in the PFI projects it examined took place as a result of risk transfer, while for six of the 17 cases examined VFM depended completely on risk transfer.

However, efficiency is not necessarily the only reason for using a PPP. A PPP can be preferred to both pure public provision and full-blown privatisation when effectiveness, in addition to efficiency, is also an aim of government policy. A policy is effective if the level of service that government planned to deliver is delivered, irrespective of whether or not this has been done in an efficient manner. Effectiveness becomes linked to issues of equity where, for instance, poverty levels prevent the poor from making an effective demand for a good or service, even when the need is large. Through the PPP contract, and the per unit amount it pays the private operator, government can ensure that the right level of services is delivered (hence the decision not to privatise, since a privatised entity can decide to deliver less on grounds of profitability), while also improving efficiency through private sector involvement (hence the decision not to rely on pure public production and delivery).

One exception where government may decide not to use a PPP, noted by Flinders (2005) and Fourie and Burger (2000), is those services that government considers to be so important to the public interest that it does not want the private sector to deliver them. These are services that may be said to have an 'inelastic social demand'; both the public and government consider their delivery so essential that government does not want to run the risk of a private operator failing in their delivery.

5.3   The Rationale for a Dedicated PPP Unit

Several reasons exist for the creation of a dedicated PPP unit. First, the danger exists that departments do not appreciate fully the budgetary implications of PPPs due to the off-budget nature of PPPs. In particular, a department or province may reason fallaciously that because in most cases a private operator is responsible for the initial capital outlay, government spending is reduced, thereby allowing government to spend more on other categories of expenditure (Fourie and Burger, 2001). The existence of this type of fallacious reasoning generates the fear that lack of knowledge about the financial intricacies of PPPs may lead government departments to over-commit financially. That such fear still exists is also clear from Gosling (2004), who notes that in resource-constrained departments the off-balance sheet nature of the capital acquisition component of a PFI/PPP creates a clear advantage in favour of going the PPP route. As such, Gosling (2004) states that the off-balance sheet nature of PPP creates a potential bias in the policy environment. This bias highlights the importance of ascertaining the affordability of a project in terms of the current and the expected future budgets of a department *prior* to exploring whether to use either the conventional procurement route or a PPP. A dedicated PPP unit is the ideal instrument to monitor and judge the affordability of a project, in particular since it acts as a regulatory body within government, but at an arm's length from the department that wants to implement the PPP.

Secondly, where departments do fully appreciate the budgetary implications of PPPs, there may nevertheless be the further danger of a principle-agent and free-rider problem. This problem may exist between an individual department, only responsible for its own budget, and the national treasury that is responsible for the overall budget. More specifically, an individual department knows that government as a whole is ultimately responsible for any agreement that the department may conclude, including the payment obligations emanating from such agreement. Therefore, since it knows that central government will have to make good on the agreement, a department may commit to an agreement even though it cannot afford to do so in terms of its allocated budget. A dedicated PPP unit could eliminate such a free-rider problem by still leaving the initiative to initiate a PPP, as well as the ultimate day-to-day management of the contract, to the individual government department, while the unit, situated in the treasury, has the authority to judge and approve the ability of an individual department to afford the PPP agreement. Such approval will then constitute a precondition for the final conclusion of the PPP agreement.

Thirdly, a dedicated PPP unit may be established to create a centre of knowledge and expertise that can provide individual departments with technical assistance during the creation process of a PPP and keep a watchful eye on departments through its regulatory approval mechanism. This is the main reason for its creation in South Africa. A dedicated PPP unit that serves as a centre of expertise also increases the confidence of potential private sector partners. In this respect Ahadzi and Bowles note:

> . . . it is not surprising that the private sector is more concerned to see an established PPP unit within the client organization. A PPP unit suggests an experienced and able client team that has the power and authority necessary for an effective negotiation process. The absence of such a team may raise concerns about the public sector's project management strengths. This will be particularly pertinent where the functions of the public sector client are fragmented across a number of departments. (Ahadzi and Bowles, 2004: 976)

5.4   A Brief History of PPPs and the PPP Unit in South Africa

PPPs have a relatively short history in South Africa. In April 1997 the cabinet approved the appointment of an interdepartmental task team to develop policy, legislation and institutional reforms to enable the use of PPPs. From 1997–2000 the government operated six pilot projects. These are (PPP Unit, 2005):

- SA National Roads Agency: N3 and N4 toll roads
- Department of Public Works and Correctional Services: two maximum security prisons
- Two municipalities: water services
- SA National Parks: tourism concessions

The Strategic Framework for PPPs was endorsed in December 1999, while the National Treasury issued regulations for PPPs in April 2000. By mid 2000 a PPP Unit was established in the National Treasury. In terms of the legislation, PPPs on national and provincial level are regulated in terms of Treasury Regulation 16, issued in 2004 to the Public Finance Management Act (1999). Government has also, in terms of the Public Finance Management Act, issued a series of National Treasury PPP Practice Notes. These notes constitute a PPP manual that government departments and provinces use to guide them through the project lifecycle of a PPP. Municipal PPPs operate under the Municipal Public-Private Partnership Regulations, issued in 2005 in terms of the Municipal Finance Management Act of 2003.

Since 1997, the creation of PPPs in South Africa on national and provincial level occurs at roughly two per annum, though in the 2006–07 fiscal year the pace increased with the approval of six (however, none were approved in 2005–06). The main reason for this rather slow roll out is the lack of skilled staff capacity in individual departments and provinces to develop a PPP and take it through its project lifecycle. Between March 2000, i.e. since the acceptance of the Strategic Framework and Treasury regulations, and December 2006, 16 project agreements were signed (another one was signed in the first quarter of 2007), with roughly 50 projects still in the pipeline. Table 5.1 contains the details regarding the 16 PPPs that were approved between March 2000 and December 2006. It also shows the duration of the individual PPP agreements, as well as the dates on which they were concluded. Lastly, Table 5.1 also indicates the nature of the project and the government institution responsible for their enactment. What is also notable from this list is that 12 of the 16 projects are provincial projects, with only four on national government level. (Details on projects in the pipeline, as well as information on the private parties involved in the concluded agreements, can be found in the PPP Quarterly (PPP Unit, 2007).

By the end of the first half of 2007–08 the National Treasury expects a further two projects to be signed (National Treasury, 2007). In addition, there are also six municipal PPPs covering services such as solid waste management, commercial property development and water services (National Treasury, 2007).

Table 5.1   PPP projects agreements concluded as of September 2006 (PPP Unit, various issues).

Project	Government institution	PPP type	Contract duration and date closed
Fleet Management	Northern Cape Dept of Transport, Roads and Public Works	DFO	5 years; Nov 2001
Inkosi Albert Luthuli Hospital	KwaZulu-Natal Dept of Health	DFBOT	15 years; Dec 2001
Eco-tourism	Manyeleti three sites Limpopo Dept of Finance, Economic Affairs, Tourism	DFBOT	30 years; Dec 2001
Universitas and Pelonomi Hospitals	Co-location Free State Dept of Health	DFBOT	16.5 years; Nov 2002
Information Systems	Dept of Labour	DFBOT	10 years; Dec 2002
Chapman's Peak Drive	Western Cape Dept of Transport	DF(part)BOT	30 years; May 2003
State Vaccine Institute	Dept of Health	Equity partnership	4 years; April 2003
Humansdorp District Hospital	Eastern Cape Dept of Health	DFBOT	20 years; Jun 2003
Fleet Management	Eastern Cape Dept of Transport	DFO	5 years; Aug 2003
Head Office Accommodation	Dept of Trade and Industry	DFBOT	25 years; Aug 2003
Cradle of Humankind Interpretation Centre Complex	Gauteng Dept of Agriculture, Conservation, Environment and Land Affairs	DBOT	10 years; Aug 2003
Social Grant Payment System	Free State Dept of Social Development	DFO	3 years; Apr 2004
Gautrain Rapid Rail Link	Gauteng Dept of Public Transport, Roads and Works	DBFOT	20 years; Sept 2006
Fleet Management	Dept of Transport	DFO	5 years; Sept 2006
Western Cape Rehabilitation Centre & Lentegeur Hospital	Western Cape Dept of Health	Facilities management	No info on length of contract; Nov 2006
Polokwane Hospital Renal Dialysis	Limpopo Dept of Health	DBOT	10 years; DBOT

PPP type indicated by combination of private party risk for: D: design; F: finance; B: build; O: operate; T: transfer of assets back to government.

Part One

Of the 45 projects that were in the pipeline in December 2002, eight were concluded successfully and form part of the 14 agreements that were signed by the end of September 2006, while a further 11 were still in the pipeline at the end of September 2006 (almost 4 years later). The remaining projects never reached the contract signing stage and were deregistered. In addition, though the services of these projects are now not provided through PPPs, many are also not provided through the conventional procurement process. In short, many of these projects disappeared altogether. Again, the main reason for the deregistration of these projects (as well as their non-delivery altogether) is not so much that these proposed projects failed the tests of affordability, VFM or sufficient risk transfer, but rather the absence of capacity in departments and provinces.

Although the legal and regulatory framework for PPPs in South Africa is quite advanced, the country has a long way to go in the rolling out of PPPs. Though one should be careful to compare like with like, this becomes particularly clear when its record is compared with that of the UK, where PPP legislation enabled the creation of private finance initiatives (PFIs) since 1992. In April 2007, the number of PFI/PPP projects signed in the UK stood at 590 projects with a total capital value of £53.4bn (HM Treasury, 2007), or £35.8bn if the three London Underground projects to the value of £17.6bn are excluded. Sixty-four of these projects were in education and a further 69 in health. Hodge (2004) notes that in 2004 the Blair government had some £100bn committed to 400 PFI contracts for the following 5 years. In Australia the amount of private finance that could flow into public assets was AUS$20bn, also for the 5 years following 2004. In South Africa the net present value (NPV) of benefits to government for six of the eight projects for which this data is available, is lower than R100m (roughly £1 = R13). The other three are the Chapmans Peak Drive toll road where the NPV equals R450m, the Gautrain project with a capital value of R23.09bn and the latest fleet management project of the Department of Transport where the NPV equals R919m. Eight of the projects have a unitary charge. The NPV to government of these eight projects range between R18.9m and R4.5bn (only two have a value that exceeds R1bn). The Gautrain project to be completed in 2010–11 will be the largest PPP to date (with government contributing about 87% of its capital).

Although the roll out of PPPs in the UK has been significantly more exten-sive than in South Africa, even in the UK it remains a small proportion of total public investment. Gosling (2004) notes that PFI constitutes no more than 11% of total public service investment in any given year. While the South African government has still a long way to go before reaching it, the view is held in the PPP Unit that investment through PPPs in South Africa should not exceed 20% of the total public service investment in any given year (Dachs, 2006). The revised estimate by the National Treasury (2007) indicates that, as a percentage of infrastructure expenditure by general government, PPP infrastructure expenditure constitutes 5.5% in the 2006–07 fiscal year and is budgeted to remain approximately at that level until the 2009–10 fiscal year (see Table 5.2).

Part One

Table 5.2 PPP and other infrastructure expenditure (2007) (R million) (National Treasury, 2007 (Budget Review 2007/8)Table 3.2, p. 45).

	2003/4	2004/5	2005/6	2006/7 Revised estimate	2007/8	2008/9	2009/10
					Medium-term estimates		
National departments	4005	4566	4936	4923	5783	6908	7766
Provincial departments	18 729	19 955	22 535	26 591	35 383	41 561	42 203
Municipalities	16 529	16 865	21 084	23 441	28 214	32 413	33 537
Public-private partnerships	**1552**	**1106**	**728**	**3444**	**3458**	**5197**	**4160**
Extra-budgetary public entities	3053	3470	3144	4262	5298	5608	6385
General government	**43 868**	**45 962**	**52 427**	**62 661**	**78 136**	**91 687**	**94 051**
Non-financial enterprises	21 375	22 145	26 424	38 322	44 681	50 324	56 929
Total	**65 243**	**68 107**	**78 851**	**100 983**	**122 817**	**142 011**	**150 980**
Total as % of GDP	5.1	4.8	5	5.8	6.3	6.6	6.4
GDP	1 288 952	1 430 673	1 580 119	1 755 340	1 938 934	2 141 747	2 379 299
PPP as % of general government	**3.5**	**2.4**	**1.4**	**5.5**	**4.4**	**5.7**	**4.4**

5.5   The Role of the South African Dedicated PPP Unit

The main function of the South African PPP Unit is to ensure that all PPP agreements comply with the legal requirements of affordability, VFM and sufficient risk transfer. In seeking to meet these objectives, the PPP Unit must guide government departments and provinces to follow international best practice that will ensure the successful creation of PPPs. Several authors (Fourie and Burger, 2000, 2001; Gosling, 2004; Hodge, 2004; Grimsey and Lewis, 2005) have indicated that a successful PPP is characterised by affordability, VFM and sufficient risk transfer. Grimsey and Lewis (2005) and Fourie and Burger (2000) argue that the main drivers of VFM and efficiency are risk transfer and competition. In addition, risks must be allocated between the public and private partners in such a manner that the VFM is maximised. Lastly, Grimsey and Lewis emphasise that the comparison between publicly and privately funded options should be fair, realistic as well as comprehensive. This implies the use of a public sector comparator (PSC).

A further prerequisite to ensure VFM is affordability. Gosling (2004) questions whether a proper appraisal of VFM can take place if a department knows that, due to budget constraints, the PPP route is the only route to obtain the finance needed for the project. This refers to the balance sheet bias discussed above. In addition, Grimsey and Lewis (2005) note that one of the assumptions made when using a PSC – the instrument used to ascertain VFM – is that the capital funds needed for the up-front investment are available. Thus, not only could the balance sheet fallacy cause departments to engage in PPP agreements that they cannot afford, but it could also affect the level of seriousness with which they approach the VFM assessment. Therefore, a government department should only consider the use of a PPP when it has a real choice in terms of financial capacity between the PPP route and the conventional procurement route.

To fulfil the above-mentioned function the PPP Unit in the National Treasury has two broad tasks:

- To provide technical assistance to government departments, provinces and municipalities who want to set up and manage PPPs.
- To provide National Treasury approvals during the pre-contract phases of a PPP agreement.

Though focusing primarily on the pre-contract period, the PPP Unit provides technical assistance throughout all the phases of the PPP project lifecycle. The lifecycle comprises six phases:

- Inception
- Feasibility study
- Procurement
- Development
- Delivery
- Exit

The first three phases represent the pre-contract or project preparation period, while the last three phases represent the contract or project term.[1] During the inception phase departments and provinces must inform the PPP Unit of their intent to set up a PPP. They also need to inform the PPP unit of their available expertise and appoint a project officer and team. The availability within a department or province of capacity and skills to create and manage a PPP is of fundamental concern to the PPP Unit. The Unit registered many PPP projects in its early years, but many of these projects were later deregistered due to departmental or provincial capacity and skill shortages. To prevent a repeat of such large-scale deregistration and the accompanying waste of resources, the PPP Unit is currently busy developing what could be seen as a checklist that departments will need to complete in the inception phase (Dachs, 2006). This checklist will serve to weed out early on projects that are not feasible, thereby saving time and cost.

The inception phase is followed by a feasibility study. This study must clarify the function that the private party will perform and include an analysis of the needs that will be addressed and the options available to government. The feasibility study must pass the three regulatory tests of affordability, VFM and risk transfer. The PPP Unit applies these tests in what is called Treasury Approval:I, which takes place after the feasibility study has been completed. This approval is needed before the department or province may proceed with the procurement phase.

The feasibility study entails several stages (see National Treasury PPP Manual, Module 4). First the department or province must ascertain the need for the service they contemplate delivering. This is done prior to the decision as to whether the conventional method or a PPP will be used to deliver the service. Subsequent to the needs analysis the department or province must consider the various options through which the service can be delivered. These options may include a PPP, but also the conventional procurement method. Affordability constitutes a key aspect of this stage. Subsequent to ascertaining the various options a project due diligence and value assessment must be made.

The value assessment is a very rigorous process that includes the compilation of a PSC. First a base PSC and then a risk-adjusted PSC are compiled, followed by the compilation of a PPP reference model and a risk-adjusted PPP reference model. The PPP Unit is not prescriptive with respect to the discount rate that a department or province must use in compiling the PSC and PPP reference models. However, it recommends that a department or province uses the rate of a government bond of which the term corresponds with that of the PPP agreement. Furthermore, all values are nominal, including the discount rate. In addition, the risk-adjusted PSC and PPP reference models do not adjust the discount rate to cater for risk, but rather prefer to cater for it in the expected (probability-weighted) cash flows. After the construction of these models a sensitivity analysis is performed.

Following these stages a budget must already exist for the project. This budget is then analysed to ascertain affordability and VFM. In addition, those projects that are either greenfield or capital projects, or projects with externalities must also submit to an economic valuation. The department or province must furthermore submit a procurement plan as part of the feasibility study. The feasibility study is then submitted for approval by Treasury Approval:I.

During the procurement phase two more treasury approvals take place. The procurement phase starts with the government department or province preparing the procurement documentation. The documentation also includes a draft contract. In what is called Treasury Approval:IIA the PPP Unit approves this documentation, after which the department can proceed with the procurement process. Procurement takes the form of a bidding process, which has as key elements accountability, responsiveness and openness in the decision-making process of the department or province. Throughout the bidding process all bidders must have an equal chance.

After the bidding process, the department or province needs to evaluate the bids. Before the department or province can appoint the preferred bidder it needs to submit a report to the PPP Unit that demonstrates that in its evaluation of all the bids it applied the criteria of affordability, VFM and substantial risk transfer. It must also demonstrate how the preferred bidder fulfils these criteria. This report forms the basis for Treasury Approval:IIB.

Competition in the bidding process forms a key element of this phase given its importance as a driver of VFM. Should only one bidder emerge, the PPP Unit considers the possibility that the low turn out of bidders is the result of a contract design that fails to attract bidders. However, given the small size of some markets in South Africa, only a small number of companies may possess the capacity and skill to undertake a project. In such cases the PPP Unit follows a second-best strategy where the bidder competes against the PSC to ensure VFM.

Following Treasury Approval:IIB the department or province finalises the detail of the contract, draws up a management plan to manage its part in the PPP and completes a due diligence on all the parties concerned to establish their competence and capacity to enter the agreement. However, before the contract can be signed, the PPP Unit needs to issue Treasury Approval:III, in which it approves that the contract meets the requirements of affordability, VFM and substantial risk transfer. Treasury Approval:III also must approve the capacity, mechanisms and procedures of the department or province to manage the contract successfully. After the contract is signed no further approvals need to be obtained from the PPP Unit. However, should any party contemplate any significant changes to the agreement after it has been concluded, the PPP Unit must approve the changes. The management of the agreement, once it is signed and the pre-contract period is over, rests with the individual department or province and is not the responsibility of the PPP Unit. Nevertheless, the PPP Unit still provides technical assistance where needed.

For the projects for which contracts have been concluded, the length of the pre-contract period in South Africa is roughly 8–18 months (Dachs, 2006). One exception is the Gautrain project that took 54 months to finalise because of the complexity and scale of the contract (National Treasury, 2007). The 8–18 months compares well with the UK. Ahadzi and Bowles (2004) note that in the UK there are excessive time overruns in the pre-contract stages, resulting in large advisory cost overruns. They reviewed 42 UK projects spanning health, education and civil engineering projects (Ahadzi and Bowles, 2004). Of these, 98% had time overruns of between 11 and 166%. The overruns for the schools were the highest, while those for the civil engineering projects were the lowest. Total negotiation time scales were also considered high, with some close to 50 months.[2] Therefore, though the scale of PPPs in South Africa is much smaller than in the UK, those that were concluded (with the exception of the Gautrain) were finalised within a year and a half. Notwithstanding these relative successes, the discussion above also indicated that there are several projects that were in the pipeline in 2002 that are still in the pipeline in 2007.

In the UK the pre-contract time and cost overruns are largely due to the different perceptions of the public and private sector about the relative importance of public and private party attributes such as the importance of communication and the ability and willingness to accept risk. For instance, Ahadzi and Bowles (2004) argue that in the UK, compared to the private sector, the public sector attaches more importance to open and frank communication, the willingness of the private party to accept risk and to commit to earlier negotiated terms. The public sector also attaches more importance to the ability of the private party to commit equity for a long period of time. In addition, relative to the private sector, the public sector attaches less importance to the private party's previous experience. The private sector, in turn, is more concerned about the previous experience and the capacity of the government department that deals with PPP procurement. This also explains why the private sector attaches more importance to the existence of a dedicated PPP unit.

The situation is not much different in South Africa. When the public sector wants to transfer risk in a PPP agreement in South Africa, private contractors tend to be less willing to accept risks that they are not familiar with (Dachs, 2006). In addition, the pre-contract period in South Africa sometimes lasts longer than expected if the parties involved need to obtain environmental approvals as part of the project.

5.6   Future Challenges

With an average of two PPP contracts concluded per annum since 1997, the PPP Unit does expect an increase in the pace at which contracts are concluded. But it does not expect it to increase dramatically in the foreseeable future (Dachs, 2006). This is largely due to capacity constraints within departments and provinces. One of these constraints results from the phenomenon that contract managers and staff of departments and provinces involved in the creation of a PPP contract tend to continue working on the contract after it has been concluded. Thus, valuable skills obtained during the creation and development of a PPP contract are not transferred to other contracts, implying that departments need to create capacity anew with each new contract. Thus, one way departments and provinces can deal with capacity constraints, and one that the PPP Unit might be considering, is to transfer skilled staff from project to project (Dachs, 2006). In addition, government and the PPP Unit are also busy creating capacity within government departments and provinces to deal with PPPs.

Three areas that possess significant potential for the increased use of PPPs are health, education and infrastructure development, and in particular the building and maintenance of clinics, schools and roads. However, the initiative to set up such projects rests with the relevant government departments and provinces and not with the PPP Unit. Therefore, these departments and provinces need to consider seriously the potential that PPPs hold. Moreover, they should consider approaching the issue in a structured and systematic manner where they first ascertain and prioritise the needs that they must address. This must then be followed by a clear analysis of what would in terms of VFM constitute the best method for delivering these services: the conventional procurement path or a PPP. Once this is done, a department or province has to compile a portfolio of projects that are structured in terms of policy priorities and that can be procured using PPPs. Such a strategy will undercut the rather *ad hoc* manner in which departments and provinces currently undertake PPPs. In addition, in the case for schools and clinics there is scope for the creation of standardised contracts that will shorten the pre-contract period significantly.

A further development that might increase the pace at which PPPs are created is the implementation of provincial dedicated PPP units that Finance Minister Trevor Manuel (2006) announced on 5 June 2006. As mentioned above, 12 of the 16 PPPs approved are provincial PPPs, with many other in the pipeline. Currently officials of some of the provinces are trained to take up positions in such units. These units will be rolled out in provinces as they develop the necessary capacity to run such units. This also implies that not all units will be rolled out simultaneously, while some provinces might even opt to not have such units. Again the difficulty is the shortage of capacity on provincial level that might limit the ability of provinces to even implement a unit successfully (not to mention the need for skilled PPP managers in provincial departments such as health and education that ultimately need to initiate and manage PPP contracts). Hence, given that it requires less skilled people power it is also foreseen that provincial units will mostly be dealing with issuing Treasury Approvals. The national treasury PPP Unit (in cooperation with the provincial units) will then still be the predominant centre of technical assistance, even in the case of provincial PPP agreements.

Municipal PPPs are a case apart. Not only do they fall under a separate legislative framework, but unlike provinces that are dependent on central government transfers for more than 90% of their revenue, municipal authorities raise most of their own revenue (through the sale of water and electricity and the levying of municipal rates and taxes). This relative financial independence also leaves municipalities more scope to approve their own PPPs. However, both the national and provincial PPP units can provide technical assistance to municipal authorities given that the skills shortage is even starker on local government level.

An issue that the national PPP Unit will need to deal with concerns the maintenance of competitive pressure on private operators, particularly in long-term contracts. Currently, the PPP Unit considers competition as a crucial element in ensuring VFM. Bidders compete against each other, thereby minimising the cost to government or, as mentioned above, in the absence of multiple bidders a single bidder competes with the PSC, also to minimise the cost to government. However, competitiveness becomes more of a problematic issue during the contract or project period. Often the service rendered through the PPP is not available on an open and well-developed market. This means that once a contract is awarded to a bidder, the unsuccessful bidders disappear altogether or conduct business in markets for services other than the ones delivered through the PPP. Thus, the competition of the successful bidder disappears and in the worst-case scenario the market becomes uncontested (i.e. there are not even any potential entrants to the market). Therefore, the private operator becomes a monopolist supplier to government. Particularly during long-term contracts such operators can place undue pressure on government to renegotiate terms of the contract to ensure more favourable terms to the private operator. This will undermine the VFM aspect of the PPP arrangement.

5.7   Conclusion

From the above it can be concluded that the role of the dedicated PPP unit comprises the authority to approve PPP agreements (and changes to concluded agreements) and the rendering of technical assistance in the creation and maintenance of PPPs. However, the initiative, ultimate management and accountability regarding PPP agreements originates and rests with individual government departments and provinces.

Currently capacity and skills shortages in government departments and provinces tend to constrain the pace at which the South African government is able to roll out PPPs. The intended creation of provincial PPP units might alleviate some of this pressure. Unfortunately, the ability of provincial governments to operate provincial PPP units might be constrained even more than the ability of national government by the shortage of skills and capacity. This means that government will need to pay special attention to the creation of skills within government to deal with PPPs, not only within PPP units, but also within government departments.

Note

An earlier version of this paper was presented at the workshop on PPPs for Infrastructure Financing in the MENA region, organised by the OECD, held in Istanbul, Turkey, 8 November 2006, as well as at the Symposium on Agencies and PPPs, organised by the OECD and IGEA, held in Madrid, Spain, 5–7 July 2006

Notes

1. These phases correspond broadly with the four main stages of the PPP procurement process identified by Ahadzi and Bowles (2004). The stages are 1) the planning and

feasibility stage, 2) the bidding and negotiation stage, 3) the construction stage and
4) the possible transfer/renegotiation stage.

2. In addition to the pre-contract time overruns, there were also substantial pre-
contract cost overruns ranging from 25–200%. These were due to the continued
retention of advisors by both the government and the private party during the
negotiations. Ahadzi and Bowles (2004) also note that both the cost and time
overruns were lowest in the civil engineering projects, most probably because of
the central procurement of these projects.

References

Ahadzi, M and Bowles, G. (2004) Public–private partnerships and contract negotia-
tions: an empirical study. *Construction Management and Economics*, 22, 967–978.
Dachs, W. (2006) Interview conducted by author. 21 June.
Department of Public Enterprises (2005a) *Overview and History.* Online: www.dpe.
gov.za. (Accessed 4 November 2005)
Department of Public Enterprises (2005b) *Corporate Structure and Strategy Overview.*
Online: www.dpe.gov.za. (Accessed 4 November 2005)
Flinders, M. (2005) The politics of public-private partnerships. *British Journal of
Politics and International Relations*, 7, 215–239.
Fourie, FCvN and Burger, P. (2000) An economic analysis and assessment of public–
private partnerships (PPPs). *South African Journal of Economics*, 68(4), 693–725.
Fourie, FCvN and Burger, P (2001) Fiscal implications of public–private partnerships
(PPPs). *South African Journal of Economics*, 69(1), 147–167.
Gosling, T. (2004) Three steps forward, two steps back: reforming PPPs. *New Econ-
omy*, 229–235.
Grimsey, D. and Lewis, M.K. (2005) Are public–private partnerships value for money?
Evaluating alternative approaches and comparing academic and practitioner views.
Accounting Forum, 29, 345–378.
HM Treasury (2007) *PFI Signed Projects List*, April. www.hmtreasury.gov.uk/
documents/public_private_partnerships/ppp_index.cfm.
Hodge, G.A. (2004) The risky business of public–private partnerships. *Australian
Journal of Public Administration*, 63(4), 37–49.
Manuel, T.A. (2006) Address to parliament on the budget votes of the Ministry of
Finance. Online: www.treasury.gov.za.
National Treasury (2004) *National Treasury PPP Manual. National Treasury.* Online:
www.treasury.gov.za.
National Treasury (2007) *Budget Review.* Online: www.treasury.gov.za.
PPP Unit (Various issues) PPP Quarterly. National Treasury. Online: www.treasury.
gov.za.

6

PPP in Greenfield Airport Development: A Case Study of Cochin International Airport Limited

Thillai A. Rajan, Sheetal Sharad and Sidharth Sinha

6.1   Introduction

It is generally believed that the economy of India is on the threshold of achieving significant growth in the coming years. Since the turn of the millennium, fuelled by the growth in information technology, construction, automobile and telecommunication industries, India's GDP has registered an yearly growth of around 8%. With the liberalisation of the Indian economy in 1991, many sectors that were the prerogative of the public sector were thrown open for private sector investment. Attracting private sector investment in the infrastructure sector was an important component of the above economic reforms and liberalisation policy.

The government of India has recognised the importance of creating adequate infrastructure facility for increasing, if not sustaining the economic growth. For the 11th 5-year plan covering the years 2007–11, the government estimates to make an investment of Rs.16 trillion in the infrastructure sector. To accelerate the process of creating infrastructure capacity, the government of India has opened up many infrastructure sectors for private sector investment. Several project development models are being experimented in an attempt to accelerate the creation of infrastructure capacity. PPP is one such mechanism that is being increasingly used in several infrastructure sectors, where the private sector is sought to be involved in many ways other than just an investor.

The Indian civil aviation industry has benefited from the strong overall economic growth over the past decade. During the years 1995–2005, the Indian aviation sector has witnessed consistent growth in passenger traffic, cargo and overall aircraft movement (Table 6.1). The growth in passenger traffic was estimated to be around 50% in 2006, with major airports in Mumbai and Delhi registering a growth of over 20% and 35% respectively. Smaller airports have registered much higher growth rates in traffic, as in the case of Pune airport, where traffic grew by 80% in 2006 (Business Standard,

Table 6.1   Indian air traffic trends (Airports Authority of India website).

Year	Aircraft movement (number of flights)		Passenger traffic (million passengers)		Cargo (thousand tonnes)	
	International	Domestic	International	Domestic	International	Domestic
1995–1996	92515	314727	11.45	25.56	458.21	222.04
1996–1997	94884	324462	12.22	24.28	484.34	225.99
1997–1998	98226	317531	12.78	23.85	493.84	252.16
1998–1999	99563	325392	12.92	24.07	481.00	258.71
1999–2000	99701	368015	13.29	25.74	538.64	303.35
2000–2001	103211	386575	14.01	28.02	565.16	327.51
2001–2002	107823	402108	13.62	26.36	568.23	333.46
2002–2003	116442	444208	14.83	28.90	654.83	372.55
2003–2004	136193	505196	16.64	32.14	700.81	413.54
2004–2005	163274	554323	19.42	39.86	830.99	492.23

2007).[1] Reports indicate that India has the fastest growing passenger figures in the world, which is projected to increase more than fivefold by 2020, to 200 million a year (The Hindu, 2007a).[2]

The growth in the aviation sector has increased interest among private players who are eager to invest in all segments including airport development, maintenance and operation of carriers, and allied services. Expectedly, private sector participation has been the highest in the airlines sector thus far. Given the current low penetration of air travel in India,[3] many new operators, both full-service and low-cost carriers, have entered the market expecting an increase in air travel penetration in the coming years. From only a couple of private airlines about 5 years ago, the industry has now significantly expanded. Kingfisher Airlines, Spice Jet, Go Air, Indigo airways, Air Deccan, Paramount airways, East-West Airlines, Indus Airways and Jagson Airlines are some of the private players that began operations recently. In addition to new entrants, the incumbents have also added more aircraft to their fleets with an objective to increase the network and connectivity both within and outside the country.

To capitalise on this momentum the industry needs good infrastructure and creation of international standard airport facilities along with associated facilities is an important component of such new infrastructure creation. The government allocated Rs.14.73bn and Rs.237.89bn for civil aviation industry in the 2004 and 2005 annual budget respectively. During the 11th 5-year plan, the government of India is planning to invest Rs.400bn in airports. To start with, existing airports in the metropolitan cities of Delhi, Mumbai, Kolkata and Chennai are being expanded. As a part of the expansion plan, the government of India has restructured existing airports in Delhi and Mumbai and has invited private sector participation through long-term lease of these airports. For example, the Delhi International Airport Limited is a PPP initiative between GMR Group, Airports Authority of India (AAI), Fraport Eraman Malaysia and India Development Fund. This consortium has been given the mandate to expand the existing airport in Delhi to make it a world-class facility, with a capacity to cater to 37 million passengers per annum by 2010.

Similarly, the state government of Tamil Nadu recently decided not only to expand the existing international airport in Chennai but also create a new greenfield airport to meet the increased traffic in the future (The Hindu, 2007b).[4] Greenfield airports are also being planned at Hyderabad and Bangalore. The new international airport at Shamshabad, in Hyderabad, would involve an investment of around Rs.23bn and is expected to become operational by 2008 (The Economic Times, 2007).[5] The greenfield airport for Bangalore, being constructed at Devanahalli, would involve an investment of Rs.35bn and would also become operational by 2008. Both the greenfield airports in Hyderabad and Bangalore are being implemented in a PPP format, with significant investment from the private sector.

This study analyses the experiences of Cochin International Airport Limited (CIAL), the first commercial airport in India to come up under the PPP format. The rest of this chapter is structured as follows: section 6.2 provides an overview of private participation in new airport development; section 6.3 describes the Indian aviation sector; section 6.4 provides details of CIAL; section 6.5 analyses the performance of CIAL; and section 6.6 provides a summary of learning's based on CIAL experience.

6.2   Private Participation in New Airport Development

6.2.1   Features of infrastructure projects

Infrastructure projects are characterised by large upfront investments, long asset lives and asset specificity. Because of these characteristics, infrastructure investments are subject to post-investment opportunistic behaviour between stakeholders (Williamson, 1988). Therefore, to attract private sector investment in the infrastructure sector necessary institutional support structures (regulation, legal environment to enforce contracts, transparent political regime, etc.) need to be in place. Gramlich (1994) provides a good overview of characteristics of infrastructure investments. Infrastructure projects may also have public good characteristics and monopoly features. For example, unless there is adequate demand or there is a potential for traffic growth, it would be wasteful to have more than one airport in a specified geographical area. The monopoly nature of airports may be questioned by the fact that London has five airports in its vicinity. However, it needs to be understood that the presence of multiple airports in this case has been necessitated because of the increased demand for airport capacity as Heathrow airport would not have been able to meet the requirements.

Many infrastructure projects also have strategic importance to the economy. For example, bridges, airports, power plants and transmission lines play an important role in the day-to-day life of society at large. Any disruption in these services has the potential to affect many segments and thereby the economic activity of society.

In emerging economies such as India, provision of infrastructure services has been traditionally in the domain of public sector. The liberalisation of the sector, which allows for private sector participation started only in the 1990s. Therefore, the experience of privatisation in infrastructure sectors is limited. For smoother and faster implementation of projects, several countries adopt a PPP approach in many infrastructure sectors, rather than full and complete private sector participation. To enable the success of PPP programmes in infrastructure, governments need to choose the form of private participation with great care.

6.2.2   Private financing of infrastructure projects

Privately funded infrastructure projects often have a BOT (build, operate, transfer) type of arrangement, which is a form of project financing designed to attract private participation in financing, constructing, and operating infrastructure projects. BOT type arrangements can provide an effective mechanism for optimal allocation of risks involved with such projects. Various contracts that exist among multiple parties in a BOT arrangement can be seen as risk-management devices, which have been designed to shift a variety of project risks to those parties best able to appraise and control them (Brealey *et al.*, 1996).

There are many variations of the BOT scheme. But the widely used schemes apart from BOT are BOO (build, own, operate), and BOOT (build, own, operate, transfer) arrangements. While the underlying characteristics do not differ significantly between these variations, several differences exist between these arrangements. For example, there is a difference in the degree of privatisation that can be said to occur with different project structures. On one end of the continuum is a totally government owned project and on the other extreme is the fully private owned entity. Between these two is a continuum of various project formats which involve varying levels of public sector–private sector participation. BOO structure is very similar to any other conventional private investment. In between is the whole range of other structures like BOT, BOLT, BOOT, in which the permanent ownership of assets exists with the state or reverts to the state after the concession period. Thus, it could be said that the degree of privatisation increases correspondingly as we go from BOT to BOOT to BOO format.

Successful project completion would require identifying an appropriate project structure which can match the project characteristics. In an earlier paper (Rajan, 2004), the author has highlighted the variations between the different arrangements as well as suggested suitability of appropriate project structures for different projects. Based on this study, it emerged that:

- Projects whose output has a public good characteristic or represent strategic interests of the government need to be structured on BOT format.
- Projects which can support significant privatisation can be structured on BOO format. Inviting full private ownership in 'socially sensitive' projects may lead to a political backlash. BOO projects need to be structured only in

those sectors that do not involve large-scale people opposition for private sector participation.

- Projects characterised by a far higher degree of uncertainty need to be structured on BOT format.

6.2.3   Features of airport projects

Infrastructure projects can be broadly classified into two types (Esty, 2002): 'stock type' projects that involve a fixed resource that is depleted over time such as mines, oil wells etc., and 'flow type' projects that require use to generate value such as toll roads. Additionally, projects can also be classified as having either retail (where output is sold to individual end users) or wholesale (where output is sold to producers or distributors) customers. Seen in this framework, airports are wholesale, flow type projects, whose main customers would be the airlines, and other service providers at the airport such as duty-free shops, hotels etc. However, the passenger experience can play an important role in achieving the traffic projections at the airport. As is today seen in most airports, the services and airport facilities need to be marketed to the retail passengers as well, given their role as influencers.

Airports can be considered as strategic assets to the economy. While there has been private sector participation in terms of airport maintenance and renovation, most of the airports worldwide are owned by the government or local authorities. Compared to other infrastructure sectors, such as a thermal power project, or an oil refinery, the revenue uncertainties are higher in an airport project and are a function of traffic flow, extent of non-aeronautical revenues etc.

Services provided by an airport can be divided into two types: airside and landside. The airside services include the runways, taxiways, aprons and terminals, and are usually funded in part from landing fees, passenger fees and profits from fuel, ground handling and in-flight catering. The landside services include passenger check-in, retail and duty-free shops, food and beverages, car parking, hotels etc.

6.2.4   Private participation in airports

Private involvement in airports varies greatly, but it is common in landside services as compared to airside services. Private involvement in airside facilities are complicated because of associated externalities:

- Airports are considered as catalysts of local economic growth and the airside services provide the essential foundation for all airport activities.
- Aircraft movement can lead to increase in noise levels and a potential source of controversy with neighbouring communities.
- Airside investments are lumpy, resulting in periods of low utilisation followed by periods of congestion.
- Airside services are considered as natural monopolies as well as strategic economic assets. It is for this reason that most of the new airports in Asia

Table 6.2 Potential for private participation in airports: economically advantageous and politically acceptable.

Criteria	Airport airside	Airport landside
Prospects for competition	Often good	Good
Prospects for financial self-sufficiency	Often poor	Good
Impact of externalities	Poor	Good
Prospects for real efficiency gains	Modest	Good
Overall prospects for privatisation	Poor	Good

have significant government presence (Hooper and Walder, 2002). For example, the new airports in Osaka and Hong Kong (costing in the range of $15–$20bn each), Bangkok ($4bn), Kuala Lumpur ($4bn), Seoul ($3.5bn) and Shanghai ($1.6bn) were financed primarily by governments. Some new airports in Seoul, Macau and Osaka are structured as private companies, but the regional or national governments own most of the shares.

Table 6.2 indicates the potential for privatisation among airside and landside activities (Gomez-Ibanez, 2002). Given the poor overall prospects of privatisation for airside activities, airports are largely funded by the public sector. Private participation has been widely seen on the landside services. While there have been many instances of privatisation of airports, only existing airports are being privatised. Table 6.3 lists some international experiences in airport privatisation. Private funding of greenfield airports has been more of an exception. The largest privately funded greenfield airport has been the Athens International Airport which opened in 2001. Going by the rationale provided earlier, private sector airports need to have either a BOT or BOOT arrangement. Athens International Airport has been structured under a BOOT arrangement with a 30-year concession period.

Table 6.3 International experiences in airport privatisation.

Country	Privatisation
UK	Full privatisation: privatisation of British Airport Authority responsible for the operation of seven airports
Australia	Full privatisation: 17 airports sold on long-term leases of 50 years, with an option for additional 49 years
Colombia	BOT/concession schemes: BOT contract to build second runway and operate both runways at El Dorado International Airport, Bogota
Canada	BOT/concession schemes: private entity invited to build and operate a third terminal at Pearson International airport in Toronto on long-term lease
Thailand	Strategic partner was being sought to participate in development and operation of a second international airport at Bangkok
Hong Kong	Management contract: private company awarded management contract of Kai Tak new airport

Financing of airport infrastructure has some inherent problems. These projects have a large element of sunk cost, a very long gestation period and highly uncertain returns on investment based on several assumptions of traffic growth that may fail to materialise. Flyvbjerg *et al.* (2003) have indicated most large infrastructure projects are frequently characterised by cost overruns and lower than predicted revenues. The cost overrun for Denver's $5bn new international airport, which opened in 1995, was close to 200% and passenger traffic in the opening year was only half of that projected. Similarly operating problems with Hong Kong's new $20bn Chek Lap Kok airport, which opened in 1998, were said to have cost the Hong Kong economy $600m (Economist, 1999). While these could have been start-up problems, this type of expense is very rarely taken into account when planning mega projects. Even the Cochin airport experienced a cost overrun of about 50%. The initial cost estimates were Rs.2045m in 1995, but the final costs in 2000 worked out to Rs.3050m.[6] Such large extant downside risks make it difficult to attract private sector investment for new airports unless they are sufficiently compensated for undertaking the risks.

6.3   Indian Aviation Sector

The political governance in India is based on a federal structure, with clearly identified areas of responsibility for the central and state governments. The Constitution of India refers to civil aviation as a subject under the responsibility of the central government. Accordingly, the responsibilities of central government towards airports include:

- Investment in airport infrastructure
- Clearance of greenfield airport projects
- Airspace management, safety and security of airports
- Bilateral air services agreements, including those involving international cooperation for modernisation and upgradation of airports
- Licensing of airports and air traffic control personnel
- Environmental aspects and removal of obstructions around airports
- Approval of aeronautical charges

The state governments are expected to support any new airport project in the following manner:

- Acquisition of private land and allotment of government land
- Supply of water and power, and provision of sanitation and sewage services
- Provision of surface access through multi-modal linkages
- Prevention of environmental pollution
- Maintenance of law and order
- Protection of airports from encroachments and vandalism

Part One

6.3.1   Airports in India

In 2006, there were about 449 airports and airstrips in India that can be classified in the following categories[7]:

- International airports: available for scheduled international operations by Indian and foreign carriers. There were 14 international airports in the country in 2006.
- Domestic airports: available for scheduled domestic operations by Indian carriers. There were 94 domestic airports in the country. Domestic airports were classified in 4 categories:
  - Custom airports: these have customs and immigration facilities for limited international operations by national carriers and for foreign tourist and cargo charter flights.
  - Model airports: these are domestic airports which have minimum runway length of 7500 feet and adequate terminal capacity to handle Airbus 320 type of aircraft. These can cater to limited international traffic, if required.
  - Other domestic airports: all other airports are covered in this category.
  - Civil enclaves in defence airport: there are 14 civil enclaves in defence airfields. At these airports, air traffic control was managed by the military and Airports Authority of India (AAI) used the facilities on payment basis.

The AAI, a public sector organisation under the Ministry of Civil Aviation is the nodal organisation that handles all matters relating to the international and the domestic airports in the country. AAI's main functions include Airport Development and Construction Services, Air Traffic Management Services (ATM), Communication, Navigation and Surveillance (CNS) Services and Ground Support & Safety Services. Most of AAI's revenue is generated from landing/parking fees and fees collected by providing air traffic control services to aircraft over the Indian airspace.

6.3.2   Airport financing in India

The existing pattern of financing has been predominantly based on internally generated resources of the AAI. Funding through external assistance, external commercial borrowings, loans and equity have been negligible. However, given the magnitude of investment requirement for modernisation and upgrading of existing airports as well as for new airports, the government has been actively seeking investment from the private sector. After exploring various alternatives for financing airports, the Ministry of Civil Aviation has indicated that:

> In the final analysis, looking at the quantum of investment required, the answer to all the problems lies in the infusion of private (including foreign) investment in this sector. This needs to be encouraged by adopting a flexible and positive attitude towards such proposed ventures. The possibility of international aid and

cooperation for building of new airports or for modernization and upgradation of existing ones will be seriously explored.

The truth of the matter is that public funds for development of airports are getting more and more scarce and private sector involvement has, therefore, got to grow. There is a definite worldwide movement from monopoly state ownership of airports to corporatization, in the first phase, with the final aim of privatization of ownership and management. India has to be a part of this global transition. (Ministry of Civil Aviation website)[8]

To facilitate private sector investment, the government has announced the following incentives:

- Foreign equity participation in such ventures may be permitted up to 74% with automatic approvals, and up to 100% with special permission. Such participation could also be from foreign airport authorities.
- An Airport Restructuring Committee in the Ministry of Civil Aviation would identify existing airports, in respect of which private sector involvement for development and upgrading of infrastructure is desired. It would also prepare a shelf of projects in respect of greenfield airports. The pre-feasibility reports of such projects would be made available to private investors.
- If existing airport operators desire private participation in their airports, no government approval is required.
- For faster decisions on greenfield airports, the central government planned to set up an independent statutory body called the Airport Approval Commission, having adequate technical and financial expertise to examine proposals for new airports quickly.
- Fiscal incentives that were provided to investors in infrastructure projects would be applicable to those investing in airports. Currently, the following incentives are available:
  □ 100% deduction in profits for purposes of income tax for the first 5 years.
  □ 30% deduction in profits for the same purpose for the next 5 years.
  □ Full deduction to run for continuous 10 out of 20 fiscal years of the assessee's choice.
  □ 40% of the profit from infrastructure is also deductible for financial institutions providing long-term finance for infrastructure projects.
  □ The above incentives were made available not only to new companies investing in airport infrastructure but also to agencies investing in upgradation of existing airport infrastructure.

6.4   The Cochin International Airport Project

Cochin International Airport was the first airport to have been built in India with private sector investment. The key stakeholders in the project comprised private participants such as high net worth individuals and industrialists, mainly the NRIs (non-resident Indians) from over 30 countries and public sector participation from the government of Kerala, which was also the single largest shareholder in the project. The new Cochin airport project was an alternative to the existing civil enclave in the naval airport, which was not equipped to handle large-bodied civil aviation aircraft due to lack of appropriate facilities and other technological limitations. The airport can be credited to have helped in evolving a policy on airport infrastructure in India. This section provides the context and an overview of the airport project.

6.4.1   Location

Cochin, now known as Kochi, is the largest city in Kerala with a population of more than 1.5 million. The city is also known as the commercial and spice capital of the state of Kerala. Owing to its proximity to the Arabian Sea, Kochi is a natural seaport and lies at the northern end of a narrow neck of land, about 19 km long and less than 1.6 km wide in many places. Kochi is also considered as the safest natural harbour in India. It is one of the most visited Kerala Backwater destinations. The economy of the city is largely dependent on the service sector. Major business areas are gold and textile retail, seafood and spices export, information technology, tourism and allied services, health services, banking, ship building, fishing and allied activities. Cochin is also home to the International Pepper Exchange, where pepper is globally traded.

Kerala is one of the smallest Indian states located in the south western coast of the Indian peninsula. With a population of close to about 30 million, it is considered as India's most prosperous state in terms of education, literacy and health. Kerala is basically an agrarian economy. Kerala's per capita income and production lags behind many of the Indian states but in terms of Human Development Index and life standard of the people it is much ahead of most other states in India, and in fact, on certain development indices it is on a par with some of the developed countries across the world.

Being a land of great natural beauty it has been named 'God's own country' and attracts many tourists from around the world. Kerala is famous for its backwaters and lagoons. The state was nicknamed as one of the '10 paradises of the world' by the *National Geographic Traveler*. Tourist traffic to Kerala has been constantly increasing over the years. The total number of tourists (domestic and foreign) visiting Kerala in 2000 was about 5.2 million and it has steadily increased to about 6.3 million tourists in 2004 (Department of Tourism, 2004).[9]

Significant percentage of Keralites work abroad, mainly in the Gulf region. In 2000, 15% of Kerala's workforce, amounting to approximately 1.5 million, were working abroad. The state's migration prevalence ratio (MPR) and household migration rate (HMR) were 59 and 38.5 respectively. Such high numbers of people working abroad resulted in significant inward remittances from those working abroad. In 2000, inward remittances contributed to 21% of the state's GDP (Centre for Development Studies, 2000).[10]

Politically, Kerala has always been the stronghold of the leftist and communist parties. A highly politicised region, Kerala hosts two major political alliances: the United Democratic Front (UDF – led by the Indian National Congress) and the Left Democratic Front (LDF – led by the Communist Party of India (Marxist)). The presence of an active left front was considered to be one of the reasons why the state's economy has been largely operated under welfare-based communist principles. Nevertheless, the state was increasingly liberalising its economy with a greater role for the free market and a facilitative environment for foreign direct investment. Since 2000, the government of Kerala has been giving priority to the establishment of information technology and business process outsourcing enterprises by initiating projects like the Kochi Info Park, and the SEPZ (Special Export Processing Zone).

6.4.2   Need for a new airport at Cochin

The main source of air traffic to Kerala is a combination of tourists and expatriate Keralites and their families, mainly from the Gulf region. For many years, there had been a demand from NRIs from Kerala to build an international airport in Cochin. Earlier, many of the NRIs visiting the state had to take a detour via Mumbai to reach Kerala for their vacations. During the 1990s, three international airports, Cochin, Tiruvananthapuram and Calicut, served Kerala. Despite capacity and technological constraints, air traffic at Cochin was much higher when compared to the other two airports in the state. Cochin airport was the busiest airport in Kerala, with air traffic growth rates much higher than those of Tiruvananthapuram and Calicut. There was no doubt that Cochin was in need of a bigger airport to meet the increasing demand.

Before the new international airport, Cochin was served by an airport at nearby Willington Island, which belonged to the Indian Navy. In the early 1990s, Indian Airlines started phasing out the 737 series aircrafts and started incorporating wide-bodied and more fuel-efficient aircraft (such as Airbus A320 and A300) within its fleet. However, the existing facility at the naval airport was unsuitable for handling the new generation aircraft. Cochin airport was a civil enclave at the naval airbase and suffered from capacity constraints such as a smaller apron area that could only serve a limited number of aircraft at the airport at any point of time. The runway could only handle the smaller Boeing 737 aircraft and that too with limited passengers and fuel.

The AAI had been examining the feasibility of expanding the existing airport for two decades. However studies indicated that the cost of expanding the airport would almost equal the cost of constructing a new airport. Therefore, AAI came up with the final recommendation to construct a new airport at a new location. In October 1991, the government decided to do away with the idea of expanding the naval airport and instead build a new airport.

Although the AAI suggested the construction of a new airport, both the Director General of Civil Aviation and the AAI expressed their inability to invest the kind of funds that were needed to build a new international airport and stated it would take a very long time to get the funding and clearance from the government. As indicated by the Business Head (Airports), Larsen & Toubro during an interview for this study:

> The government's focus has been on social engineering projects that cater to the basic needs of the population. The main beneficiaries of airports are the upper middle and upper class. It becomes difficult to get funding from the government in such projects as priorities may not match. This is also a reason why such projects are taking the PPP route.

When the demand for a new airport was being made for Cochin, the district administration was headed by V.J. Kurien, a bureaucrat from the Indian Administrative Service. Since Kurien found the idea of an international airport in Cochin appealing, he wasted no time in meeting the then Kerala Chief Minister, K. Karunakaran, with a project report recommending different ways of financing the project.

6.4.3   Financing of Cochin international airport

Kurien had several interesting ideas for financing the airport from private sources. He pointed out that there were more than 2 million native Keralites working abroad, particularly in the Gulf countries. Kurien proposed seeking the help of these NRIs for financing the construction of the airport.

To implement this idea he incorporated the Kochi International Airport Society (KIAS) as a charitable society in July 1993. Kurien believed that collecting the required funds would not be a difficult task considering the huge NRI Keralite population and their much felt need for good airport infrastructure in Cochin. Therefore, KIAS appealed to the NRI population from Kerala to invest in the new airport project.

As a first step towards knowing the minds of the NRIs, KIAS issued an advertisement in the local newspapers telling the public about the proposed airport project at Cochin and asking those interested to fill in a coupon and mail it back. The advertisement got a positive response from the public.

Following the encouraging response, the KIAS offered an interest-free deposit scheme to the public. As per the scheme, individual investors would have to provide interest free deposits of Rs.5000 for a period of 6 years. For every individual with a minimum investment of Rs.5000, the society would purchase Kisan Vikas Patra (KVP)[11] worth Rs.2500 in the person's name which would double itself in 5.5 years. All investors would also be entitled to facilities like waiver of entry fee, special lounge in the airport, a separate check-in counter etc. The original plan for financing the airport is given in Table 6.4.

Going by calculation, if 400 000 people extended an interest-free loan of Rs.5000 each, the society would be able to raise Rs.2000m in cash which was the estimated project cost. 50% of the money raised (Rs.1000m) would be used for the purchase of KVPs. This would enable the project to repay the loan to the investors at the end of loan period. During that period, government of India had a scheme of lending back 75% of the investment in KVPs to the state government for developmental purposes as a low-interest, long-term loan. This loan would be serviced by accrued income from the airport and

Table 6.4   CIAL: Initial financing plan (1992) (Varkkey and Raghuram, 2001).

	Funding source	Rs. Million
A	Interest free deposits from 400 000 overseas Keralites @ Rs.5000 per person	2000
B	Money invested in KVPs of Rs.2500 each for repayment in 6 years when the amount doubles	1000
C	Cash in hand (A − B)	1000
D	Loans against KVPs (75% of investment)	750
E	Donations	250
	TOTAL (C+D+E)	**2000**

from the sale of any excess land that had been already acquired. An amount of Rs.250m was expected to be mobilised as donations from industrial houses and an interest-free loan from the airport service providers would be used to service the remaining debt. Thus as per the initial financing plan, the required amount of Rs.2000m would be raised.

However, in reality the public deposit scheme was not as successful as was initially expected. It could only collect Rs.40m, i.e. only 2% of the initial target amount of Rs.2000m. This necessitated the need to look at other financing options. Therefore as an alternative, in March 1994, Kurien incorporated a new public limited company called the Cochin International Airport Limited (CIAL) to build, operate and maintain an international standard airport at Cochin. To focus more on development of CIAL, Kurien relinquished his role as head of district administration and took charge as the managing director of CIAL. The company had planned to raise equity capital of Rs.700m (authorised capital of Rs.900m) and debt of Rs.1300m. Once again, the initial investment in equity was below expectations.

In March 1995, Housing and Urban Development Corporation (HUDCO), a premier government institution involved in extending loans to housing and other infrastructure projects in the country, sanctioned a short-term loan of Rs.250m at 16.5%. Simultaneously, Federal Bank, one of the leading private sector banks in South India, granted a bridge loan (short-term debt) of Rs.100m for 6 months for land acquisition for the airport. These short-term loans were guaranteed by the government of Kerala as CIAL did not have anything to offer as collateral or security for the debt.

In addition to the short-term loan, HUDCO also provided a long-term (10 year) loan of Rs.980m to CIAL. The loan was guaranteed by the government. This loan was used to start construction of airport and runway. CIAL also used a part of this long-term loan to clear off the bridge loan from Federal Bank and other short-term loans. The HUDCO loan had to be serviced starting from the year 2000, the year when CIAL was expected to start operations, for a period of 10 years at an interest rate of 18% per annum.

After CIAL was able to raise the initial finance in the form of debt, it wanted the state government to contribute equity capital for the company. During the period 1994–96, the government agreed to take up equity in the company, but no money was disbursed for the same. After repeated requests, the government released Rs.10m towards equity in CIAL. However, in 1996 the state witnessed a change of power and the new government was quick to release Rs.292m towards equity. The government also appointed the Chief Minister as the Chairman of the Board of Directors of CIAL.

After the government's investment in CIAL, it became easier to get greater participation from Indians living overseas, airport service providers and other entities. The majority of the NRI as well as domestic investors were attracted to the project through word of mouth and news about CIAL. The public relations drive, directly handled by the Managing Director, prompted many service providers as well as small investors to consider investing. CIAL issued shares at Rs.10/- each, but insisted that an individual shareholder should apply for at least 250 shares worth Rs.2500/-. About 10 000 NRIs invested in the airport. The inflow of funds to the new airport showed that the project had credibility. In the same light, it was decided that the government of Kerala and state government undertakings would have a majority shareholding in the company, with at least 51% of the equity investment, whereas the balance would be invested by the private sector along with the NRIs and individual investors. After many efforts, CIAL was successfully able to complete its financing. Table 6.5 provides the capital structure of CIAL.

6.4.4   Inauguration of the new airport

During the 5 years of construction of the airport, CIAL officials had to deal with three Civil Aviation Ministers, four Civil Aviation Secretaries, four Chairmen of AAI at Central Government, and three Chief Ministers, four Transport Ministers and four Transport Secretaries at the State Government. The former President of India, Mr. K.R. Narayanan, formally inaugurated the airport on 25 May 1999. Basic details of the airport are given in Table 6.6. Soon after the inauguration, the domestic operation from the old naval airport was shifted to the new airport. Cargo operations also commenced at the new airport by September 1999. Duty-free shop operations commenced from May 2002. A chronology of key events in the development of CIAL is given in Table 6.7.

6.4.5   Revenue model and initial performance

When CIAL became operational, regulatory provisions did not allow for private sector participation in airport operations. A public sector entity, AAI, was the sole entity vested with management and operations of commercial airports in India. As CIAL was the first airport with the PPP structure, the DGCA issued a temporary licence valid for 3 months. The licence was renewable every 3 months based on regular inspections. Awarding landing rights to airline companies to operate from a specific airport was also regulated by the government of India. To avoid potential confrontation with AAI because of private participation, CIAL followed the same tariff structure adopted by AAI for landing charges.

Table 6.5 Capital structure of CIAL (as on 31 March 2001) (various sources including newspaper articles and online papers).

Equity	in Rs. million	% of total equity
Public sources		
Government of Kerala	324.50	41%
BPCL	52.30	7%
SBI	50.00	6%
IOB & Dhanlakshmi Bank	7.50	1%
Air India	50.00	6%
Total public		*62%*
Private sources		
NRI directors	141.40	18%
Indian residents	44.80	6%
NRIs	54.00	7%
Federal Bank	30.00	4%
Alpha Retail	30.00	4%
Total private		*38%*
TOTAL EQUITY	**784.50**	**100%**
LOANS (including accrued interest)		
HUDCO Term loans	1527.20	
Federal Bank	246.90	
State Bank of Travancore	275.10	
District Cooperative Bank	120.00	
TOTAL LOANS	**2169.20**	
INTEREST-FREE DEPOSITS		
Air India	110.00	
Thomas Cook	5.00	
Indian Oil Corporation	7.50	
Alpha Retail – Duty Free Shop	100.00	
Retail Outlets	27.50	
TOTAL DEPOSITS	**250.00**	
TOTAL	**3203.70**	

Part One

The revenue model for airports consists of two components – aeronautical revenue and non-aeronautical revenue. The proportions of these two revenue streams for different airports are given in Table 6.8. Though non-aeronautical revenues can account for significant proportion of revenues in many cases, CIAL decided to go in for a 50:50 spilt between the two revenue sources. The revenue streams for CIAL from both the sources are given in Table 6.9.

However, CIAL could not achieve the anticipated aeronautical revenues because of the shortfall in achieving projected passenger traffic flight estimates. Further, delay in commencement of duty-free shops and lower rental incomes from these shops resulted in aeronautical revenues accounting for about 80% of its total revenues. Revenues from visitors' entry into the airport also fell short of expectations, because of the increased security, and restriction on entry of non-passengers inside the airport building following the 9/11 terrorist

Table 6.6   CIAL project details.

Location	Nedumbassery, 25 km northeast of Cochin
Area	1400 acres
Project cost	Rs. 3.15 billion
Passenger capacity	2.8 million per year (in 2008)
Cargo capacity	32 000 tonnes per year (in 2008)
Operation start date	10 June 1999
Project conception	October 1991
Construction begins	1993
Runway length	3400 m
Runway width	45 m width (additional 7.5 meter shoulder on either side)
Apron area	61 500 m^2
Terminal building	23 550 m^2
No of aircraft stands	9
Check in counters	12 (international) and 10 (domestic)
Aero bridges	2 + 3 (to be constructed in expansion plan)
Sponsors	Government of Kerala, NRIs, Indian residents, Air India, BPCL, SBI, Alpha Retail, Federal Bank, IOB, Dhanlakshmi Bank
Lead contractors, designers, architects and engineers	E & M Associates, New Delhi; KMC constructions Ltd, Hyderabad; NATPAC; Hellmuth, Obata & Kassabaum (HOK) Inc, USA
Financing	State Bank of Travancore, Federal Bank, HUDCO, District Co-operative Bank

attacks on the World Trade Center. Cargo estimates too fell well short of the projections. The projected cargo for the first year of operations was about 15 000 tonnes, whereas what was achieved was only about 5000 tonnes. This was attributed to CIAL's lack of focus on the marketing of its cargo services.

6.5   Performance of CIAL

Initial lower revenues affected the cash flow position of CIAL, which resulted in difficulties for CIAL to service its debt. To find a solution to this problem, CIAL appointed a consultant to suggest means to increase the revenues and reduce the increasing interest burden. In order to raise the required money, CIAL offered a 1:1 rights issue in 2001 to increase the capital base of the company from the existing Rs.900m to Rs.2000m. The rights issue was undersubscribed as the government of Kerala was unable to contribute its share and other investors insisted on government subscription. CIAL also sought to impose a surcharge fee of Rs.500 for every passenger using the airport but the move came under severe criticism and eventually it had to be withdrawn.

In 2001–02, CIAL registered a net loss of Rs.188m mainly because of the shortfall in expected revenues and problems with high-cost debts as part of its balance sheet. This led to difficulties for CIAL with its investors. However, the difficulties proved temporary, as the scenario changed in the very next year on account of stronger revenues and CIAL achieved a net income of Rs.125m for the year 2002–03. The opening up of the duty-free shops and other related

Table 6.7 CIAL: Chronology of key events.

Before incorporation Oct 1991 to Feb 1994	Company incorporation Zero Date March 1994	<12 months April 1994 to May 1995	<5 years May 1995 to April 1999	<8 years May 1999 to April 2002	<12 years May 2002 to April 2006
• Govt of Kerala decides to construct a new airport at Cochin • Ministry of civil aviation gives clearance • June 1991: UDF (United Democratic Front) comes to power in the state • July 1993: Kochi International Airport Society (KIAS) is incorporated to raise funds	• 30 March 1994: Cochin International Airport Limited (CIAL) is incorporated • Tentative date of inauguration is fixed as 15 August 1997 (construction to take 3 years)	• CIAL issues shares to NRIs and public to raise funds for the airport construction • Federal Bank sanctions loan. • Runway construction commences • 21 August 1994: Foundation stone is laid for CIAL • March 1995: HUDCO sanctions term loan and contributes Rs.10m towards equity	• Design, construction, equipment supply, installation, testing and commissioning start on the airport • Private players invest about Rs.150m towards equity in CIAL • BPCL is given exclusive refuelling rights for contributing Rs.50m towards equity • March 1996: Govt of Kerala (UDF Govt) contributes Rs.10m towards equity in CIAL • May 1996: Change in state government. LDF (Left Democratic Front – CPI-M) comes into power. Issues Rs.292m towards equity of CIAL	• 25 May 1999: Honourable President of India inaugurates the airport • First inaugural flight by Air India on 10 June 1999. • 1 July 1999: Domestic operations from old naval airport shift to CIAL • September 1999: Cargo operations commence • Nov 1999: Kurien transferred from CIAL • 2001: Rights issue to raise additional capital for expansion of airport facilities • April 2001: Change in state government. UDF (United Democratic Front – Congress) comes to power • CIAL lands into debt servicing problems with banks, FIs and HUDCO	• May 2002: Duty-free shop operations commence • CIAL approaches HUDCO for converting its debt into equity • March 2003: CIAL registers profit for the first time since operations commenced in 1999 • June 2004: Declaration of 8% dividend • August 2005: 10% dividend proposed • December 2005: CIAL proposes to divest 26% stake through IPO • February 2006: BODs for CIAL clear another 1:1 rights issue to raise Rs.1500m to fund expansion plans • 12 May 2006: DRT gives verdict in favour of HUDCO asking CIAL to allocate 26% equity to HUDCO before the public offer is made

Part One

Table 6.8   Revenue distribution for various airports (source: Annual Statements of Airports).

Airports	Aeronautical revenues	Non-aeronautical revenues
British Airports Authority	28%	72%
Toronto	38%	62%
Sydney	29%	71%
Houston	19%	81%
Heathrow	47%	53%
Kuala Lumpur	46%	54%
Los Angeles	57%	43%
Changi, Singapore	42%	58%
Paris	51%	49%
Zurich	49%	51%

infrastructure development in the nearby areas boosted the non-aeronautical revenues. On the other hand, a surge in the number of airlines operating from the airport added to the aeronautical revenues of CIAL. From then on, CIAL has been consistently generating profits. The financial statements of CIAL are given in Tables 6.10 and 6.11. For the financial year 2005–06, CIAL registered a turnover of Rs.1.1bn and a net profit of Rs.313.8m, up from Rs.287.9m in the previous year (The Hindu, 2006).[12]

In mid 2004 CIAL declared a maiden dividend of 8% to its shareholders. With increasing profitability, CIAL declared a 10% dividend to all its shareholders in August 2005. The strong performance of CIAL resulted in several investors expressing an interest to invest in CIAL. During 2004–05, the Reliance group of industries (India's largest private sector group), Hyderabad-based GMR group and the consortium led by the GVK industries & airports company of South Africa expressed their interest in buying the stake in the airport held by NRIs.

6.5.1   Servicing debt investors

CIAL was generating positive net operating income right from the first year of its operations but this was not sufficient to service its debt obligations. In the year 2001, CIAL had debt amounting to Rs.1720m from HUDCO. After having registered a loss during 2001–02, CIAL had difficulties in servicing the loans obtained from FIs, HUDCO and other banks. To tide over the situation,

Table 6.9   Main revenue sources for CIAL (Varkkey and Raghuram, 2001).

Aeronautical revenues	Non-aeronautical revenues
Landing, parking, X-ray and TNLC	Fuelling operations (land lease and royalty)
Passenger service charges	Star Hotel/Flight Kitchen
Housing charges	Entry and parking charges (visitors and vehicles)
Cargo handling charges	Rental income
Ground services royalty from IA	Duty-free shops

Table **6.10**   CIAL balance sheet (in Rs. Million) (Prowess Database, 2007).

	31 March 03	31 March 04	31 March 05
LIABILITIES			
Authorised capital	2000.00	2000.00	2000.00
Paid-up equity capital	1436.50	1479.10	1488.60
Total reserves and surplus	−347.40	−268.60	−149.70
Net worth	**1089.10**	**1210.50**	**1338.90**
Short-term bank borrowings	0.00	0.00	0.00
Long-term bank borrowings	1000.00	501.70	202.80
Financial institutional borrowings	550.70	550.70	550.70
Total borrowings	**1550.70**	**1052.40**	**753.50**
Deferred tax liabilities	**231.90**	**240.50**	**223.00**
Sundry creditors	201.40	220.90	252.20
Interest accrued/due	0.00	57.20	84.20
Other current liabilities	280.60	281.50	306.40
Total current liabilities	**482.00**	**559.60**	**642.80**
Provisions	**12.20**	**169.60**	**230.50**
Total liabilities	**3365.90**	**3232.60**	**3188.70**
ASSETS			
Land and building	2571.10	2612.30	2664.90
Other assets	601.80	613.60	748.00
Gross fixed assets	3172.90	3225.90	3412.90
Less: cumulative depreciation	499.10	631.40	764.90
Net fixed assets	**2673.80**	**2594.50**	**2648.00**
Deferred tax assets	**418.10**	**301.80**	**114.20**
Raw materials and stores	2.50	2.80	4.00
Finished and semi-finished goods	10.40	36.40	48.30
Total inventories	**12.90**	**39.20**	**52.30**
Sundry debtors	158.70	190.70	235.00
Other receivables	31.90	49.50	88.90
Total receivables	**190.60**	**240.20**	**323.90**
Cash and bank balance	**68.90**	**56.90**	**50.30**
Other misc. expenses not written off	**1.60**	**0.00**	**0.00**
Total assets	**3365.90**	**3232.60**	**3188.70**

CIAL approached HUDCO during 2002 with a suggestion to convert the outstanding debt into equity. CIAL expected HUDCO to convert at least 20% of the outstanding debt into equity and both parties also ended up signing a contract for the same. However, CIAL renegotiated the interest rates with HUDCO from 16–17% down to 11–12% per annum.

By the year 2004, CIAL generated sufficient profits and it paid off Rs.1200m of the debt to HUDCO, with another Rs.520m left to be paid. Soon after, the interest rates further declined and CIAL decided to refinance the remaining debt of Rs.520m at 11% by taking a loan from Federal Bank and Punjab National Bank at 6.25% per annum. Given the increasing profitability of the company, CIAL withdrew the offer it had made to HUDCO for converting the Rs.520m debt into equity and instead sent them a cheque for

Table 6.11   CIAL income statement (in Rs. Million) (Prowess Database, 2007).

	2002–03	2003–04	2004–05
Operating income	559.80	778.20	938.00
Other income	43.90	60.50	64.60
Change in stocks	−12.60	26.00	8.30
Non-recurring income	152.20	13.90	10.80
Total Income	**743.30**	**878.60**	**1021.70**
Operating expenses	25.80	29.40	37.30
Purchase of finished goods	22.60	85.50	111.70
Energy (power and fuel)	28.70	34.90	37.50
Salaries and wages	34.10	52.00	59.80
Other expenses	94.90	63.90	95.60
Total Expenditure	**206.10**	**265.70**	**341.90**
PBDIT	**537.20**	**612.90**	**679.80**
Less: Financial charges	207.70	119.60	54.50
PBDT	**329.50**	**493.30**	**625.30**
Less: Depreciation	131.20	132.80	134.30
PBT	**198.30**	**360.50**	**491.00**
Less: Tax provision	72.80	149.30	203.10
PAT	**125.50**	**211.20**	**287.90**
Appropriation of profits			
Dividends	0.00	132.40	169.00
Retained earnings	**125.50**	**78.80**	**118.90**

the same amount to settle the loan. HUDCO then returned the cheque to CIAL saying that it wanted equity in the company as per the contract signed between the two parties. CIAL refused to give any equity to HUDCO and eventually the matter finally landed up with the Debt Recovery Tribunal (DRT). On 12 May 2006 the Debt Recovery Tribunal (DRT) gave its verdict in favour of HUDCO asking CIAL to allocate 26% of the equity to HUDCO (in lieu of the Rs.520m outstanding debt) before it went ahead with the public issue.

6.5.2   Expansion plans and developments

With the aviation sector growing fast, CIAL felt the need to expand its existing airport facilities to meet the higher expected traffic demand in the near future. In 2004, CIAL appointed Ernst & Young as consultants for the proposed commercial expansion of the airport on 400 acres of land at an approximate cost of Rs.35bn. Some of the proposed developments for the expansion plan were to:

- Increase the international passenger arrival terminal area from 11 600 m^2 to 18 580 m^2 in order to increase the peak hour passenger capacity from 400 to 700 passengers.
- Construct a full-length rapid parallel taxiway (3400 m) that can serve as a runway during emergency situations.

- Increase the apron area (100 m wide and 137 m long) in order to handle the wide-bodied Airbus A-380 aircraft.
- Increase the number of aerobridges from two to five including one for Airbus A-380 aircraft.
- Install visual docking guidance system and automated fuel hydrant system in the parking bay.
- Launch a private low-cost airline. Consultancy agency IL&FS conducted a feasibility study for the project and suggested that a budget airline, flying to national and international destinations from CIAL had immense scope. But later developments indicate that the proposed air line, Air Kerala, would not be able to offer international services to start with (The Economic Times, 2006a).[13]
- Construct five-star hotels, a shopping mall, a golf course and an airport township on the land which CIAL has under possession around the airport.
- Set up India's largest aircraft maintenance company by creating a large aircraft maintenance hub at the airport. A sharp increase in the number of airlines flying the Indian skies, with only Indian Airlines and Air India having full-fledged aircraft maintenance facilities, created a huge gap in the demand and supply of aircraft maintenance engineers in the country. On 1 December 2005 CIAL announced the setting up of a new company called the Kochi International Aviation services, a joint venture between CIAL (51% stake) and licensed aircraft maintenance engineers (49% stake) working for different airlines across the globe. Apart from aviation maintenance services, the new company would also extend cabin crew and flying training services.

In December 2005, the Board of Directors for CIAL decided to tap the capital markets to raise money for its future expansion plans. They decided to go for public listing and would divest 26% of their stake through an initial public offer (IPO). In February 2006, the board also gave the green signal to CIAL for increasing its equity capital base from the existing Rs.2000m to Rs.4000m by allowing a 1:1 rights issue to its existing shareholders. While CIAL had delayed its IPO process, the expansion and modernisation plan of the airport were on track. At the 12th Annual General body Meeting of the company held in December 2006, the Chief Minister of Kerala, V.S. Achutanandan, who is also the Chairman of CIAL, announced a Rs.50bn plan for expansion and modernisation. The first phase of investment involving Rs.10bn is expected to be completed by 2009 (The Economic Times, 2006b).[14]

6.6   Summary and Lessons

CIAL began its operations when there was neither a history of private sector involvement in airports in India nor a policy that allowed for private sector investment. Notwithstanding this lack of clarity on policies, issues and regulations concerning the construction and operation of greenfield airport projects, the performance of CIAL had been encouraging. The success of CIAL provided the necessary impetus to the centre and state governments to adopt the PPP route for the construction of new international standard airports in the future. The key lessons from CIAL can be summarised as follows:

- *Evidence of success in smaller projects can be helpful to attract subsequent investment in larger projects.* The initial capacity of CIAL is much less than the capacity of greenfield airports that are being currently constructed in Bangalore and Hyderabad. For example, CIAL is spread over 1400 acres and with passenger capacity 2.7 million per year, whereas the upcoming Bangalore airport is spread over 3886 acres and has an initial passenger handling capacity of 6.7 million per year. The investment in CIAL was Rs.3.15bn whereas the proposed project cost of Bangalore airport is Rs.18.11bn. However, the success of CIAL could be considered to be an important factor for private sector committing large investments in the airport sector. The policy changes that happened in the Indian civil aviation sector, which initially did not allow for private sector investment in airports, were based on the successful experience of CIAL. Without the experience of CIAL, the policy changes in the sector would have followed international trends, without giving much importance to the ground realities that exist in India. This, in turn would have been questioned by many stakeholders, making the process of privatisation difficult in the sector. Following the successful operation of CIAL, the obstacles have been less for private sector investment in airports in India. Though the private sector does not have a majority shareholding in CIAL, the experience provided the comfort factor to the government to increase the level of private sector investment in airports. In the case of the upcoming Bangalore airport, the private sector has the majority shareholding with 74% of equity. The initial success also seems to have benefited CIAL, since it found it easier to raise resources for future expansion and modernisation.
- *Political risk management is very important for successful implementation of large infrastructure projects.* Infrastructure projects such as airports take many years to develop and very often involve acquisition of land from private players for the project. In a democratic country like India, a change in the government after the elections can be an important risk factor. Large projects that are initiated by the incumbent government and those that are still in the development phase are usually reviewed by the new government. Given the political element in such reviews, project risks increase considerably during changes in government. It therefore becomes important to have a structure that facilitates political commitment at the highest level. In the case of CIAL, the Chief Minister of Kerala is the Chairman of the company. Since the project is directly overseen by the Chief Minister, the magnitude of political risk is reduced as no Chief Minister would like to see the projects directly under their oversight to fail, even when there is a change in government. The land market in India does not operate in a transparent manner. The market value of land transactions is not recorded for a variety of reasons. Most of the delays in infrastructure projects in India could be attributed to problems in acquiring land for the project. The government finds it difficult to purchase land from private

players because of unavailability of information on the market value of land. In many cases, affected people go to court against the government decision to acquire their land, thereby delaying the project. Strong political leadership would help in overcoming such hurdles and help in smooth implementation of the project.

- *Ensuring management continuity during the project conceptualisation and construction phase provides focus and demonstrates commitment.* Continuity in project management structure is very important for faster implementation of project. While airport development is an initiative from the state government, it still has to depend on the central government for new landing rights and increase of flights from the national carriers. The political differences between the state and central governments could increase the project risks. Projects which on an average take 5–8 years in development and construction could witness a change in government midway through the project. Any change in government is also likely to result in cascading changes in the bureaucracy. In the case of CIAL, there was a change in government in May 1996 when the Left Democratic Front (LDF) came to power, defeating the previous government formed by the United Democratic Front. As mentioned earlier, there were also several changes in the political and bureaucratic set-up during project construction. Management leadership in the project organisation can play an important role in reducing the risks during such transitions. For example, Mr V.J. Kurien, Managing Director of CIAL, was responsible for the project from 1992 to November 1999. Not only was there a continuation in leadership, but Kurien lobbied effectively for political support for the project during regime changes. There have been various examples (Orissa power sector reform programme for instance) where the bureaucrats played a key role in programme/project continuation and implementation. Because of the familiarity with the Indian political system the bureaucrats are better equipped to manoeuvre around regime changes. The commitment of the government to the project was also evident when Kurien relinquished his position as the head of district administration and took charge as Managing Director of CIAL with sole responsibility for developing the airport.
- *CIAL provides further evidence that allocating various risks to those participants who are better equipped to manage them can reduce the overall risk of the project.* Infrastructure projects are generally characterised by strong operating cash flows (CIAL had an operating margin of 66% and 70% in FY05 and FY04 respectively), and different parties could try to appropriate the free cash flows thereby increasing the project risk. A risk management framework, which allocates different project risks to those who are best equipped to manage those risks, can reduce post-investment opportunistic behaviour. For example, inclusion of Air India and BPCL as sponsors in CIAL is a good strategy for reducing post-investment opportunistic behaviour. An investment from Air India, the national carrier and a key flight carrier from Cochin, would ensure that there are adequate interests on the part of the airline to operate and increase its frequency of services from the airport. The trend of airlines investing in new airports

Part One

Table 6.12   CIAL agreements with different companies (Varkkey and Raghuram, 2001).

Work	Entity	Contribution
Ground handling of aircraft	Air India	Contribution of Rs.50m in the equity capital, Rs.110m as interest-free deposits and 15% royalty on gross revenue
Aircraft refuelling	BPCL	Contribution of Rs.52.3m to equity, royalty charges and lease rent
Forex counters	SBI	Contribution of Rs.50m in the equity and Rs.250m as term loan
Forex counters	Thomas Cook	Rs.5m interest-free deposit
Forex counters	Federal Bank	Contribution of Rs.30m in the equity capital of Rs.250m as term loan
Restaurants	Oberoi	Interest free deposit of Rs.25m and payment of royalty
Retail and duty-free shops in terminal buildings	Alpha Retail	Contribution of Rs.30m towards equity share capital and Rs.100m as interest-free deposit
Petrol outlets	IOC	Interest-free loan of Rs.7.5m

could be seen elsewhere too. In Shanghai airport for example, Lufthansa had taken a 29% share in a joint venture to build and operate the cargo terminal. Similarly, China Southern Airlines was a significant investor in Guangzhou's $1.3bn new airport (Hooper and Walder, 2002). On the other hand, the investment from BPCL in CIAL would ensure adequate and uninterrupted availability of fuel for flight operations. In addition to the above sponsors, CIAL had tied up with different contractors as a part of its risk management strategy. Table 6.12 lists the agreements that CIAL had made with other companies.

- *Given the widespread impact on large infrastructure projects, it helps to create an environment of trust and involvement with the public and society at large.* The project was based on a felt need from the NRIs to have an airport in Cochin. Investment opportunities in CIAL were given to the public and the NRIs to obtain their support for the project. To create a favourable impression among different stakeholders, publicity and communication regarding the airport project were directly handled by the Managing Director, which helped in quickly responding to the queries and concerns. The presence of the Chief Minister as the Chairman of CIAL, ensured that there would be adequate political support for the project at the grassroots level. Such mass support is necessary to prevent the project being delayed from land acquisition related issues and other public interest litigations.

Notes

1. 'Smaller airports see passenger traffic surge,' *Business Standard*, 15 May 2007.
2. 'India's airlines: losing money, buying planes,' *The Hindu*, 09 May 2007.

3. It is estimated that just three persons per 100 travel by air each year in India, compared with about 10 per 100 in China. 'India's airlines: losing money, buying planes,' *The Hindu*, 09 May 2007.
4. 'Lands for airport expansion, new airport identified,' *The Hindu*, 23 May 2007.
5. 'Shamshabad: Schiphol in the making,' *The Economic Times*, 2 May 2007.
6. Report on CIAL by PricewaterhouseCoopers Ltd.
7. Airports Authority of India website: http://www.airportsindia.org.in/AAI/main.jsp (accessed 2007).
8. Ministry of Civil Aviation website: http://civilaviation.nic.in/ (accessed 2007).
9. Tourist Statistics 2004, Department of Tourism, Govt of Kerala.
10. Dynamics of Migration in Kerala: Dimensions, Differentials and Consequences, Centre for Development Studies, October 2000.
11. Kisan Vikas Patra (KVP) is a government of India bond which doubles in value over a certain period of time. In 1992, money invested in a KVP doubled in 5.5 years.
12. 'CIAL net up to Rs. 31.38 crores,' *The Hindu*, 19 May 2006.
13. 'Air Kerala project crash-lands,' *The Economic Times*, 25 May 2006.
14. 'CIAL air traffic surges by 38%,' *The Economic Times*, 29 December 2006.

References

Brealey, R.A., Cooper, J.A. and Habib, M.A. (1996) Using project finance to fund infrastructure investments. *Journal of Applied Corporate Finance*, 9(3), 25–38.

Business Standard (2007) Smaller airports see passenger traffic surge. *Business Standard*, 15 May 2007.

Centre for Development Studies (2000) *Dynamics of Migration in Kerala: Dimensions, Differentials and Consequences*. October 2000.

Centre for Monitoring Indian Economy (2007) Prowess database of large and medium Indian Firms. Mumbai, India.

Department of Tourism (2004) *Tourist Statistics 2004*. Department of Tourism, Govt. of Kerala.

Esty, B. (2002) Harvard Business School Document No. 5-202-031 on Poland's A2 Motorway, March 6.

Economist (1999) Bad Landing. *Economist*, 352(8134), 28 August 1999.

Flyvbjerg, B., Bruzelius, N. and Rothengatter, W. (2003) *Megaprojects and Risk – An Anatomy of Ambition*. Cambridge University Press, Cambridge.

Gomez-Ibanez, J.A. (2002) *Case Notes on Cochin International Airport*. Kennedy School of Government Harvard University, Document No. CR14-02-1650.2

Gramlich, E.M. (1994) Infrastructure investment: a review essay. *Journal of Economic Literature*, 32, 1176–1196.

Hooper, P. and Walder, J. (2002) *Cochin International Airport: Gateway to God's Own Country*. Case Study CR14-02-1650.0, Kennedy School of Government, Harvard University.

Rajan, T. (2004) Observations on project structures for privately funded infrastructure projects. *Journal of Structured and Project Finance*, 10(1), 39–45.

The Economic Times (2006a) Air Kerala project crash-lands. *The Economic Times*, 25 May 2006.

The Economic Times (2006b) CIAL air traffic surges by 38%. *The Economic Times*, 29 December 2006.

The Economic Times (2007) Shamshabad: Schiphol in the making. *The Economic Times*, 2 May 2007.

The Hindu (2006) CIAL net up to Rs. 31.38 crores. *The Hindu*, 19 May 2006.

The Hindu (2007a) India's airlines: losing money, buying planes. *The Hindu*, 9 May 2007.

The Hindu (2007b) Lands for airport expansion, new airport identified. *The Hindu*, 23 May 2007.

Varkkey, B. and Raghuram, G. (2001) *Public Private Partnership In Airport Development – Governance And Risk Management Implications From Cochin International Airport Ltd*. IIM Ahmedabad Working Paper.

Williamson, O.E. (1988) Corporate finance and corporate governance. *Journal of Finance*, 43, 567–591.

PPPs for Physical Infrastructure in Developing Countries

Akintola Akintoye

7.1   Introduction

PPPs are now commonly used in both developed and developing countries to accelerate economic growth, development and infrastructure delivery and to achieve quality service delivery and good governance. Given the changing economic, social and political environment, coupled with globalisation and budgetary constraints, PPP has become unavoidable and indeed is considered desirable by many countries. For many developing countries that are facing major challenges in the provision of infrastructure, PPP has become about the only show in town in order to control public sector borrowing. The need for PPP in developing countries has been intensified by the public sector realisation of the vital role of modern infrastructure in economic growth and poverty alleviation, which cannot be supported by the existing level of public sector income.

In essence, PPPs have been recognised as an important avenue for funding major public sector infrastructure projects. PPPs are joint ventures in which business and government cooperate, each applying its strengths to develop a project more quickly and more efficiently than government could accomplish on its own. The private sector may be responsible for designing, financing, constructing, owning and/or operating the entire project. The private sector may want to be assured that the PPP structure is designed to provide competitive rates of return commensurate with a financial rate of return similar to alternative projects of comparable risk.

This chapter is divided into five sections. The first part presents an overview of PPP and the second discusses PPP in developed countries to contextualise the use of PPP in developing countries. The third section presents general information on the use of PPP in developing countries and various initiatives that have been developed to encourage developing countries. The fourth part provides an analysis of the extent to which PPP for infrastructure development has emerged in developing countries. The final section discusses how to create enabling environments for the use of PPPs in developing countries.

7.2   An Overview of Public-Private Partnerships

PPP can be described as a contractual agreement of shared ownership between a public agency and a private company, whereby, as partners, they pool resources together and share risks and rewards, to create efficiency in the production and provision of public or private goods.

A PPP implies that there is some shared responsibility between the public sector and private sector for outcomes or activities (Collin, 1998). This differs from other relationships between the public and the private sectors in which the public sector retains control over policy decisions after receiving the advice of organisations in the private sector. PPPs often are separate organisational structures, rather than bargaining relationships which have been established among otherwise autonomous organisations. Grant (1996) argues that shared authority and responsibility, joint investment, sharing liability/risk-taking and mutual benefit stand at the core of a partnership. According to Plummer (2000), PPP specifically refers to those forms of partnership in which government establishes an arrangement with the private sector in which the private sector provides some form of investment. As such, the terminology of PPP tends to exclude service and management contract arrangements, but includes leases and concessions.

PPP can take different forms, and the UK government has identified about seven PPP models (HM Treasury, 2000). In the British models context, PPP includes privatisation which essentially involves the outright sale of assets to a private company. French models of privatisation take a variety of forms of private sector participation, not necessarily involving the sale of assets at all but include management contracts, leases, sub-contracting, management or employee buyout, and outsourcing or contracting specific activities to private actors. The World Bank takes a holistic view which describes as PPP all investment (public and private) in projects with private participation in provision of public sector infrastructure. Specifically it has identified four categories of PPP (World Bank, 2005):

1. Management and lease contracts. These are contracts where a private entity takes over the management of a state-owned enterprise for a fixed period while ownership and investment decisions remain with the state. In a management contract the government pays a private operator to manage the facility and assumes the operational risk, whilst in lease contracts government leases to the private operator who takes on the operational risks.
2. Concessions. A private entity takes over the management of a state-owned enterprise for a given period during which it also assumes significant investment risk. This includes: rehabilitate, operate and transfer; rehabilitate, lease or rent, and transfer; and build, rehabilitate, operate and transfer projects.

3. Greenfield projects. This has four categories: build, lease and own; build, own, transfer, or build, own, operate, transfer; build, own and operate; and merchant project, where a private entity or a public-private joint venture builds and operates a new facility for the period specified in the project contract.

4. Divestitures. Full (100%) or partial government transfers of the equity where a private entity buys an equity stake in a state-owned enterprise through an asset sale, public offering, or mass privatisation programme.

In essence, the level of private sector involvement in public sector service delivery might range from a purely service provision, without recourse to public facilities, through service provision based on public facilities usage, up to 'public facilities' ownership. Gentry and Fernandez (1998) noted that the form of ownership adopted depends on such issues as: the degree of control desired by the government; the government's capacity to provide the desired services; the capacity of private parties to provide the services; the legal framework for monitoring and regulation; and the availability of financial resources from public and private sources. For example, South African PPPs exclude an agreement between an institution and a private party, where the latter performs an institutional function without accepting the significant risks (South Africa Government Gazette, 2000).

The Institute of International Project Financing (IIPF) has produced a list of how PPP project finance has been used internationally (IIPF, 2005). According to IIPF, whether termed 'international project finance', 'global project finance' or 'transitional project finance', the financing technique of bringing together development, construction, operation, financing and investment capabilities from throughout the world to develop a project and deliver public sector service in a particular country has been relatively successful. The technique is now being used throughout the world, in emerging and industrialised societies. Examples of facilities developed through PPP project financing include, energy generation, pipelines developments, mining development, toll roads, waste disposal and telecommunications (IIPF, 2005).

7.3   PPP: Developed Economies

The use of PPP in developed economies has been the subject of rigorous research investigation. For example, Jones (1998) analysed the INFRAFIN project which was funded by the European Commission under the RTD programme of the 4th Framework to examine issues in the planning, financing and operation of major transport infrastructure projects, whether undertaken as PPPs or as traditional publicly financed schemes. Poole (1995) reported the need for PPP in America to empower cities and states to tap into private capital and rebuild America. Poole (1995) noted that the case for PPP included new sources of capital, time saving, capital saving, risk reduction and new tax revenues. In addition, he suggested that 'an added benefit of encouraging investor-owned infrastructure in America would be the development of world-class US infrastructure firms'.

Li and Akintoye (2003) analysed the pattern of PPP use across continents. What is noticeable in their study is that, while PPP is used predominantly in public sector infrastructure developments in developing economies, it is used in the developed economies to deliver various government public services, goods and facilities. In addition, the extent of dominance of either the public sector or private sector in PPP in developed economies is often dependent on the ideological positioning of the country (Savitch, 1998). For example, the PPP arrangements in social/political driven economies or unitary forms of governance like France and Sweden are often characterised by public sector dominance, whereas countries like Canada, the USA and Hong Kong with profit-driven, private economy and associated loose governance tend to have private sector dominance. The UK falls in between these two extremes.

The UK presents an example of a developed country where PPP has been widely used since 1992. The most popular form of PPP in the UK is the Private Finance Initiative (PFI) in which the private partner builds a facility to the output specifications agreed with the public agency, operates the facility for a specified time period under a contract or franchise with the public sector client, and then transfers the facility to the latter party when the contract expires. The PFI in facilities development can involve private bodies in the design, financing, construction, ownership and/or operation of a public sector utility or service. In the majority of PPP/PFI projects, building and infrastructure may be required in the development of the solution to the client's requirements. However, PPP/PFI is about the delivery of public sector services that are communicated to the private sector in form through the output specification.

In the UK system of PFI procurement, the public sector is expected to produce a business case, or proposal. This specifies both the functional and performance or output requirements for the scheme. The private sector then transforms this to a service design which should meet the performance requirements specified by the client. The consortium is expected to operate, repair and maintain the asset throughout the contract period to an agreed quality standard and ensure continuity and quality of service of the asset. Because the private sector consortium takes control of the design, construction, operating and financing of the scheme, there is opportunity to introduce innovation that will ensure sustainability of the service provision. At the design stage, for example, the PPP consortium has the opportunity to assess the environmental quality of the scheme and the consequent potential for reducing environmental damage by improving the design. In essence, this PPP procurement method provides the consortium with a greater opportunity to assess the environmental, economic and social issues associated with a scheme at the design stage. The overall implication of the environmental, economic and social assessments at this stage is that the lifecycle of the scheme can be considered to achieve a sustainable solution.

Four inter-related principles at the heart of the UK PPP/PFI approach are: genuine risk transfer; that the contracts should specify the service output required by the public sector client from the private sector (output specification); whole-life asset performance; and performance-related reward to the contractor under a PFI contract. The payment mechanisms are characterised

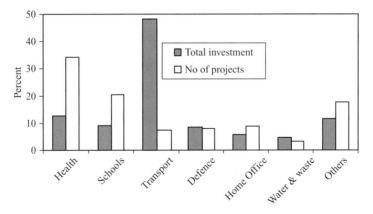

Figure 7.1   PPP/PFI usage in the UK across sectors.

as a regular 'unitary' fee for services which must be subject to performance appraisal in relation to specific and quantified criteria in the contract. Where the users pay directly for the service rendered, this will be in the form of a toll payment.

PPP is being used in the UK to deliver different types of projects, and in different sectors: education (schools, education facilities, leisure), healthcare (hospitals and equipments), transport (car parks, airports, rail, tram, roads, bridges), custodial infrastructure (prisons, court house), public buildings (non-housing accommodation), housing, utilities (water, waste water etc.), defence and IT facilities. Despite this, private sector investments have mostly been in three sectors: transport, health and defence (Figure 7.1). Combined projects in these three sectors represent 51.12% of the total number of the signed PFI projects and 78.16% of the total capital value. The transport sector has the highest share of the PFI schemes undertaken in the UK in terms of value of schemes. Although transport PPP projects are responsible for less than 6% of the signed projects, they account for about 50% of the capital value. The average capital value of projects in this sector is £368m, with 74% of the schemes over £50m capital value.

Locke (1998) argues that the interest in transport schemes is attributable to a large backlog of road and bridge projects held 'on-the-shelf' plus interest in light rail or guided bus schemes. Most road and bridge PFI schemes in England are sponsored centrally by the Highways Agency under the Department of Transport. The total PFI investment in transport PFI schemes between 1989 and 2003 was £15bn. Since then many new transport PFI schemes have been signed. It is expected that the investment in road PFI schemes will increase now that the local authorities are venturing into road maintenance PFI. Akintoye *et al.* (2005) have produced a comprehensive analysis of the PPP trends in the UK.

7.4   PPP: Developing Economies

PPP in developing countries has not advanced to an extent that is comparable with developed countries. For example, Jütting (1999) has shown how the implementation of PPP in the health sector infrastructure, although theoretically appealing, is still not very common in developing countries. Many developing countries depend on extraction, exportation and exporting of raw natural resources to support the economy; these activities are areas where foreign investment in infrastructure development is predominant.

The infrastructures mainly needed by the developing countries to support their economic activities are those related to transportation, energy and potable water and, more recently, telecommunication. Although these are needed, many developing countries cannot afford them without affecting other economic activities because of cost considerations (initial capital outlay and cost of operation and maintenance) and lack of appropriate technology to support them. Existing levels of productivity have also been identified as factors militating against infrastructure performance in developed economies. All this opens avenues for PPPs to play a role in the design, construction, operation, maintainance and finance infrastructure of developing countries.

Funding major infrastructure development is a major problem for many developing countries that often rely on government annual capital investment budget or foreign aid. Lifset and Fernandez (1998) reported a summary of their internet conference on 'The Search for Best Practices in Urban Solid Waste Management Services in Developing Countries', which showed that the financing of solid waste management in many developing countries is conducted through traditional sources such as municipal government. Finance through collection by private entrepreneurship and own-income communities that pay for this is low, while community financing is not universal. Given the current position in many developing countries they suggested a need for further institutional development and more effective regulatory frameworks to facilitate greater involvement of private capital.

The UNDP (2006) Memorandum to the UK Select Committee on International Development identified major barriers to the implementation of PPPs; these included an absence of efficient, transparent and participatory policies, mechanisms, and institutions in the developing countries which has consequential effects on an increase in the transaction costs of PPP projects. Other barriers identified for private sector development and investment were 'lack of adequate capacity and the absence of innovative partnerships and business models, of a policy environment to facilitate cooperation and partnerships between public and private actors and access to financing, of safety net mechanisms and basic services'.

However, UNDP is of a firm belief that 'it is through PPP that the developing countries can create employment and income growth as well as improve the quality of life for the poor'. Bennett *et al.* articulated the need for PPP:

> It is becoming increasingly clear that governments cannot meet the continually growing demand for water, waste and energy services acting alone. Governments are finding that their tax revenues are not providing sufficient resources to meet these needs, and official development assistance has not been able to fill the gap. New approaches to addressing these problems that involve collaboration among an increasing number of stakeholders are urgently needed. Public-private partnerships are one of the most promising forms of such collaboration. (Bennett *et al.*, 1999)

Sader (2000, cited by Thomsen 2005) identified the main obstacles with PPP in the water sector in developing countries as those associated with conflicting aims of development policy objectives, lack of transparency and objective evaluation criteria in award procedures, weak legal environment/ regulatory frameworks, public governance, preferential treatment for existing service providers and lack of political commitment where governments renege on contractually agreed terms. On the other hand, Estache (2004) summarised a list of promises, in order of importance, that have been made in the context of reforms leading to PPP. These are its contribution to fiscal stabilisation, increased investments, improved efficiency from a more competitive environment, contribution to growth, better access and affordability for residential users and improved governance.

There have been many initiatives launched to encourage PPP in the developing countries. The Commonwealth Initiative on Public-Private Partnerships is an example which seeks to promote PPPs, mainly in infrastructural facilities, in Commonwealth developing countries, in general, and post-conflict ones in particular. The activities in this area are aimed at: bringing would-be private investors into contact with officials in the potential host developing countries; raising awareness of member countries of opportunities for, and benefits of, PPP; enabling member countries to share experiences of PPP; and building the capacity of government institutions and officials on PPP matters (Smith, 2007). One consensus from the Monterrey, Mexico International Conference on Financing for Development is to:

'... support new public/private sector financing mechanisms, both debt and equity, for developing countries and countries with economies in transition, to benefit in particular small entrepreneurs and small and medium-size enterprises and infrastructure. Those public/private initiatives could include the development of consultation mechanisms between international and regional financial organizations and national Governments with the private sector in both source and recipient countries as a means of creating business-enabling environments.' (United Nations, 2003).

Follow-up processes to the international conference were World Economic Forum (2005) multi-stakeholder consultations on PPPs for improving the effectiveness of development assistance in relation to water, education and health.

7.5   PPP: Analysis of Private Sector Participation

This is based on a combination of desk top review of PPP trends and development and analyses of secondary data from the UK and World Bank: Private Participation in Infrastructure Database (http://ppi.worldbank.org/ reports/customQueryAggregate.asp). The World Bank database of PPP infrastructure projects covers PPP projects in developing economies by income that have reached financial closure. The World Bank PPI Project database tracks all investment (public and private) in projects with private participation. The database covers four infrastructure sectors:

- Energy (electricity generation, transmission and distribution; natural gas transmission and distribution)
- Telecommunications (fixed or mobile local telephony, domestic long-distance telephony and international long-distance telephony)
- Transport (airports runways and terminals; railways fixed assets, freight, intercity passenger, and local passenger; toll roads, bridges, highways, and tunnels; seaports, channel dredging and terminals)
- Water (potable water generation and distribution; sewerage collection and treatment)

The database considers projects to have private participation if a private company or investor bears a share of the project's operating risk. The analysis contained in this section is based on private sector participation in infrastructure development from 1990–2005; a period which coincides with rapid growth in the use of PPP in both the developed and developing countries.

7.5.1   PPP investment: regional analysis and trends

Table 7.1 shows that most private sector participation in the delivery of public sector infrastructure, in terms of number and value of projects, has been in the Latin America and Caribbean (LAC) region (36% and 44.4% respectively). The figures show that the Middle East and North Africa (MENA), South Asia (SA) and sub-Sahara Africa (SSA) countries, although representing 50% of the developing countries, have not benefited significantly from PPP, compared with the remaining three regions that have continuously used PPP to deliver public sector infrastructure. The two regions, LAC and East Asia and Pacific (EAP), are responsible for 62.1% of the total number of PPI projects and 67.8% of the total PPI investment of the developing economies.

Figure 7.2 shows an overall trend in private investment in infrastructure since 1990 with the associated number of projects while Figure 7.3 shows the regional trends. Estache (2004) and Thomsen (2005) have shown how private

Table 7.1 Private participation in infrastructure investment: regional analysis (1990–2004).

Region	Number of projects Total	%	Investment US$m	%
East Asia and Pacific	764	26.1	197 282	23.4
Europe and Central Asia	550	18.8	136 911	16.2
Latin America and Caribbean	1051	36.0	374 622	44.4
Middle East and North Africa	87	3.0	42 041	5.0
South Asia	224	7.7	52 844	6.3
Sub-Saharan Africa	246	8.4	39 291	4.7
Total	2922	100.0	842 991	100.0

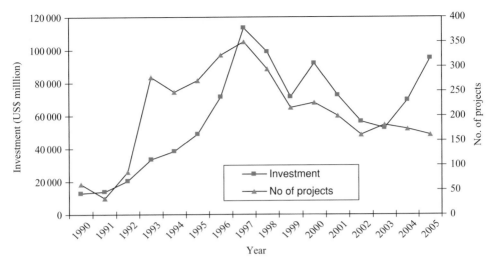

Figure 7.2   Trends in private participation in infrastructure (1990–2005).

participation in infrastructure in developing countries peaked in 1997; this was followed by a steady drop, reaching less than US$50bn in 2003. Reasons for the decline include the Asian crisis of 1997. Another reason is the failure of PPP to deliver acclaimed promises which is said to have led to a decline in its use in many Latin American countries.

Figure 7.2 shows that PPP investment is now picking up following the trough in 2003, while at the same time project size is becoming bigger. Figure 7.3 shows that the significant drop in PPI investment from its peak in 1997 resulted from a reduction in private investment experienced by LAC and EAP regions compared with insignificant but upward growth experienced by SSA and SA regions. It is now generally accepted that private investment in infrastructure in developing countries will grow in all the regions

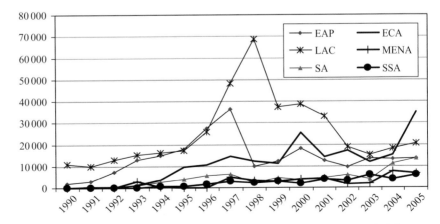

Figure 7.3   Regional investment (US$ million) trends in private participation in infrastructure (1990–2005).

Part One

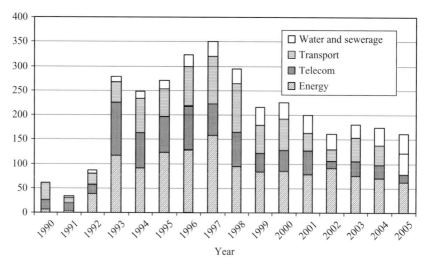

Figure 7.4   Number of PPI projects by sector (1990–2005).

given budgetary constraints that many developing countries are experiencing. There is an upward trend across the regions in 2004 and 2005. Evidence for growth in private investment in infrastructure could be gleaned from infrastructure development needed in Latin America in the next five years, which has been estimated to be in the range of US$70bn. Many countries in the region are now considering PPPs as the only realistic option for them to meet their needs. For example Brazil has responded to this by passing a PPP Act that came into force in December 2004 to promote private sector investment in public infrastructure delivery.

7.5.2   PPP investment: sector analysis and trends

Figure 7.4 shows the trends in the number of projects in the energy, telecommunication, transport and water and sewerage sectors between 1990 and 2005. Again the figure shows that the number of projects peaked across all the sectors in 1997. An exception to this is the water and sewerage sector that remained at a very similar level throughout the period.

Figure 7.5 shows the percentage of investment in each sector with the associated percentage of number of projects. Although the energy sector has the largest number of PPI projects (41.4%), almost half of the total investment within this period is telecommunication (47.3%) suggesting that private capital was a significant source of financing for these two sectors.

While energy and transport projects dominate PPI in the EAP (79.6%), LAC (74.1%) and SA (74.6%) regions, telecommunication dominates the SAA region projects (51.2%). The number of projects (10.5%) and the amount of investment (4.5%) in water and sewerage infrastructure is generally low across the regions. Telecommunication projects (average of US$628.74m/project) are generally larger compared with energy

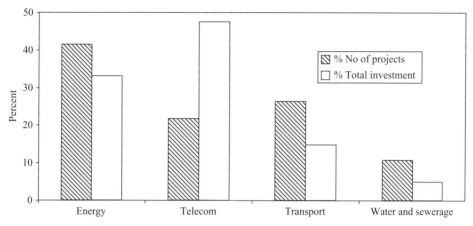

Figure 7.5   Sector analysis of private participation in infrastructure.

(US$229.85m), transport (US$161.35m) and water and sewerage PPI projects (US$134.55m/project). The overall average is US$288.50m. Figure 7.6 shows that telecommunication accounted for a growing share of investment in recent years while the shares of energy and transport have reduced significantly.

7.5.3   PPP: types of PPI investment

Table 7.2 shows the types of private participation in infrastructure development across the regions in terms of number of projects and amount of investment. The table shows that EAP, SA and SSA regions are mainly involved in greenfield projects while LAC region is engaged in all four types of PPP (concession, divestiture, greenfield, and management and lease

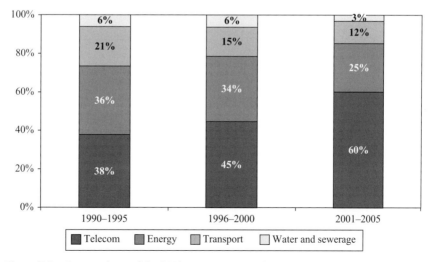

Figure 7.6   Sectors share of the PPI investment commitment.

Table 7.2   Regional and type of PPP investment.

Region	Concession		Divestiture		Greenfield		Management and lease contract	
	Number of projects							
East Asia and Pacific	176	23.0%	124	16.2%	448	58.6%	16	2.1%
Europe and Central Asia	35	6.4%	250	45.5%	222	40.4%	43	7.8%
Latin America and Caribbean	329	31.3%	221	21.0%	471	44.8%	30	2.9%
Middle East and North Africa	17	19.5%	4	4.6%	54	62.1%	12	13.8%
South Asia	19	8.4%	15	6.6%	189	83.6%	3	1.3%
Sub-Saharan Africa	40	16.1%	23	9.3%	146	58.9%	39	15.7%
		20.4%		23.7%		50.0%		5.9%
	Investment (US$m)							
East Asia and Pacific	30 634	15.5%	43 945	22.3%	122 644	62.2%	60	0.0%
Europe and Central Asia	5911	4.3%	71 495	52.2%	59 220	43.3%	283	0.2%
Latin America and Caribbean	64 451	17.2%	201 496	53.8%	108 383	28.9%	293	0.1%
Middle East and North Africa	7817	18.6%	11 427	27.2%	22 794	54.2%	3	0.0%
South Asia	1478	2.8%	5771	10.9%	45 595	86.3%	0	0.0%
Sub-Saharan Africa	3962	10.1%	13 114	33.4%	22 217	56.5%	28	0.1%
		13.6%		41.2%		45.2%		0.1%

contracts). Europe and Central Asia (ECA) region has the largest number of management and lease contracts and tends to favour divestiture and greenfield PPI. Although most of the projects are greenfield PPP projects to build and operate a new facility for the period specified in the project contract, the divestiture projects, where private entities buy equity stakes in state-owned enterprises through asset sale, public offering, or a mass privatisation programme, are generally larger (see Figure 7.7) with an average project size of US$676.90m. Figure 7.7 shows that greenfield PPI projects represent 50% of the project number and 45.2% of the total investment compared with 23.7% and 41.2% respectively for divestiture PPI projects. Management and lease contracts are comparatively insignificant in terms of number of projects, value of investments and size of projects.

Figure 7.8 shows the trends in investment across the PPP types from 1990–2005. The figure shows that greenfield projects accounted for an increasing share of the investment. Divestiture projects have however reduced significantly from the peak in 1993. The figure shows a diminishing investment in divestiture projects which might suggest either that it is no longer fashionable for private entities to buy equity stakes in state-owned enterprises through asset sale, public offering or a mass privatisation programme, or that there are diminishing state-owned enterprises for private enterprises to buy.

Figure 7.9 further shows a growing share of investment of greenfield projects from 45% in the early 1990s to 58% in 2001–05 compared with a significant drop in divestiture projects from 40% to 34%. Based on number of projects, Table 7.3 shows the investment index by the type of PPI against the four infrastructure sectors. Using a nominal value of 100 to represent

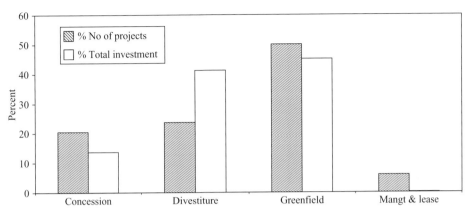

Figure 7.7 PPP type analysis of private participation in infrastructure.

all the total number of PPI projects in developing economies from 1994–2005, the indices show that greenfield PPI in the energy sector (24.49) has the largest number of the projects. This is followed by telecommunication using greenfield (16.74) and energy sector using divestiture (14.96). The table shows that concession is most popular with transport projects and greenfield is most popular with energy and telecommunication projects. Management and lease contracts are mostly used in the water and sewerage projects while divestiture is mainly used for energy projects.

7.5.4   Comparative analysis of top two PPI investment countries in each region

Table 7.4 shows analysis of the top two countries for PPI investment in the six developing economy regions in terms of the income level, total investment,

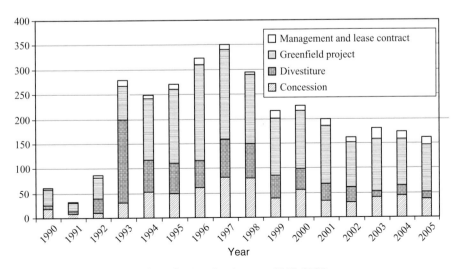

Figure 7.8 Number of PPI project by type: 1990–2005.

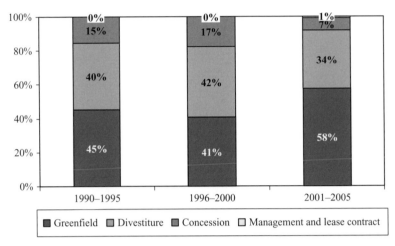

Figure 7.9   PPP type share of investment commitments.

type of PPI and sectors where they are located. With the exception of ECA region (33%), PPI investments in all the top two countries represent more than 50% of the total PPI investment in each region. For SA region, PPI investments in India and Pakistan represent 92% of the region's total PPI investment. The PPI in the ten countries represents about 56% of the total PPI investment in the 139 developing countries listed in the World Bank database of developing countries that had private participation in infrastructure. Overall the analyses show PPI investment is dominated by a small group of developing countries with relatively fast-growing markets; for example, three Latin American countries (Brazil, Argentina and Mexico) accounted for more than a third of total PPP investment in developing countries.

Table 7.4 shows that most PPI investments are in the lower middle and upper middle income developing economies. While most PPI investments in the lower middle and upper middle income countries are in the energy and transport sectors, the telecommunication sector dominates the low income countries. PPI investment in Brazil, Argentina and Malaysia, which are upper middle income countries, is significantly higher than PPI investment in low income countries like India, Pakistan and Nigeria, which tends to suggest that PPP in developing countries is determined by the income level. While the

Table 7.3   Private participation in infrastructure investment: PPP type and sector cross tabulation analysis.

Sector	Concession	Divestiture	Greenfield	Management and lease contract
Energy	1.86	14.96	24.49	0.58
Telecom	0.29	3.28	16.74	0.15
Transport	14.77	2.18	8.52	1.46
Water and sewerage	4.48	0.73	3.42	2.11

Table 7.4 Top two countries for PPI investment in the developing economy regions (based on an analysis of the World Bank database, 2005).

Regions	No of countries with PPI	Top two countries	Total investment	Sector / PPI type	Energy Concession	Telecom Divestiture	Transport Greenfield project	Water & sewerage Managt & lease	% of total investment Regional	Overall Total	Income level
EAP	18	China	66 955	Sector	40.9%	21.6%	32.8%	4.6%	34	8	Lower middle
				PPI Type	12.2%	40.0%	47.7%	0.1%			
		Malaysia	37 845	Sector	32.6%	16.3%	37.7%	13.5%	19	4	Upper middle
				PPI Type	15.0%	13.8%	71.2%	0.0%			
ECA	26	Poland	23 480	Sector	12.4%	84.0%	3.5%	0.1%	17	3	Upper middle
				PPI Type	3.1%	66.6%	30.2%	0.0%			
		Russia Federation	22 485	Sector	15.1%	80.0%	2.3%	2.6%	16	3	Low
				PPI Type	0.0%	49.2%	50.7%	0.1%			
LAC	28	Brazil	150 395	Sector	39.1%	44.7%	13.7%	2.5%	40	18	Upper middle
				PPI Type	14.8%	62.8%	22.4%	0.0%			
		Argentina	72 575	Sector	36.2%	34.3%	18.1%	11.4%	19	9	Upper middle
				PPI Type	29.2%	50.8%	20.0%	0.0%			
MENA	14	Morocco	15 642	Sector	58.6%	41.4%	0.0%	0.0%	37	2	Lower middle
				PPI Type	41.2%	30.0%	28.8%	0.0%			
		Saudi Arabia	8834	Sector	0.0%	96.6%	2.8%	0.6%	37	1	Upper middle
				PPI Type	2.8%	46.2%	51.0%	0.0%			
SA	6	India	39 571	Sector	42.2%	49.0%	8.2%	0.6%	75	5	Low
				PPI Type	2.4%	11.2%	86.4%	0.0%			
		Pakistan	8940	Sector	66.4%	28.4%	5.2%	0.0%	17	1	Low
				PPI Type	3.2%	11.2%	85.6%	0.0%			
SSA	47	South Africa	19 015	Sector	6.6%	82.4%	9.9%	1.1%	48	2	Lower middle
				PPI Type	7.7%	47.5%	44.8%	0.0%			
		Nigeria	5441	Sector	13.0%	86.5%	0.4%	0.0%	14	1	Low
				PPI Type	0.4%	0.0%	99.6%	0.0%			
Total	139		471 178	Sector	35.0%	44.1%	16.4%	4.5%		55.89	
				PPI Type	14.3%	45.3%	40.4%	0.0%			

Source: EAP = East Asia and Pacific; ECA = Europe and Central Asia; LAC = Latin America and Caribbean; MENA = Middle East and North Africa; SA = South Asia; and SSA = Sub-Saharan Africa.

Part One

Part One

Table 7.5 Private sector participation in infrastructure: low, lower middle and upper middle incomes (based on an analysis of the World Bank database).

	Low income	Lower middle income	Upper middle income
Total number of countries	59	49	30
Project investment (US$m)			
Energy	11.2%	40.5%	48.3%
Telecom	9.2%	32.8%	58.0%
Transport	4.3%	34.2%	61.5%
Water and sewerage	1.6%	40.5%	57.9%
Grand total	8.7%	36.0%	55.3%
Percent	8.71%	35.97%	55.33%
Project count			
Energy	14.2%	49.5%	36.3%
Telecom	31.5%	42.2%	26.3%
Transport	13.2%	40.6%	46.3%
Water and sewerage	5.0%	41.4%	53.6%
Grand total	16.8%	44.7%	38.5%
Percent	16.84%	44.69%	38.47%
Investment/scheme (US$m)	150.38	234.13	418.37

lower middle and upper middle income countries have wider spread of the use of concession, divestiture and greenfield PPP procurement, PPP in the low income countries is dominated by greenfield (India, 86.4%; Pakistan, 85.6%; and Nigeria, 99.6%).

Table 7.5 again shows that the middle income (lower middle and upper middle) developing economies are associated with most of the private sector participation (91% total project value and 83% total project count). Overall, the table shows that low income developing economies have insignificant private sector involvement in public sector facilities (in terms of project investment and project count) compared with lower middle income and upper middle income developing economies. In the telecommunication sector where the low income countries accounted for a significant number of PPI projects (31.5%) these were smaller in terms of size (these only accounted for 9.2% of the telecommunication investments). Overall, the size of PPP projects in low income countries is relatively smaller than in lower middle and upper middle income countries.

7.6   Discussion

The need for private sector involvement in the provision of infrastructure is now common across developing countries. For many of these countries, private sector provision of public sector services has become one of the government's main policies to tackle public sector borrowing requirements and public sector waste. Despite the higher need for infrastructure development in low income developing countries, the level of private sector participation in infrastructure investment is significantly low, particularly in the low income countries that are mostly located in sub-Saharan Africa, compared with other lower and middle income developing countries like Indonesia, Thailand, Malaysia etc. The sectors where investments are urgently needed in many low income developing countries are energy (mainly electricity generation, transmission and distribution), water (potable water) and transport (roads, bridges and highways). Nonetheless PPP investments in these developing countries are mainly in the telecommunication sector, and infrastructure investment in the other sectors is very low. The main reason for low investment in some sectors could be attributable to what is reported about India that 'it is hardest to attract private sector investment in sectors such as water and sanitation, where prices have been very heavily subsidised and there is often political opposition to the concept of private provision of what are regarded as essential services'. This has not being the case in sectors that are already, or close to, commercially viable, such as telecom, ports and airports where private investment is easier to attract (Sharma, 2006).

That the middle income (lower and middle) developing economies are associated with most of the private sector participation might suggest that they are uniquely positioned compared with low income countries. Although some regions are associated with many PPI projects compared with others, this may not translate to the success of the projects after financial close and operation phase. World Bank (2005) reports the percentage of investments that are either cancelled or under distress by 2004: EAP, 12%; ECA, 3%; LAC, 13%; MENR, 2%; SA, 6%; and SSA, 2%. This suggests that the regions with the majority of PPI investments (EAP and LAC) have the largest percentage of PPI projects suffering from cancellation or distress.

Despite that many of the projects may be cancelled or under distress, Malhotra (1997) has attributed the growth in private participation in infrastructural investment in the power sector in Asia (mainly middle income developing countries) to a rapidly changing environment and attractive opportunities available to private investors. These include government commitment, increased private interest, move to competitive processes, greater availability of information, acceptable prices, high developer returns and the large size of projects.

Sharma (2006) listed some bottlenecks identified by stakeholders in PPP in India (low income) which are responsible for low adoption of PPP compared with countries like Brazil, Chile and Colombia (private investment in infrastructure in India, in the last decade, has averaged only around 1% of GDP compared with Chile (averaged 3–4%); Brazil (1.5%), Colombia (2–3%)). The bottlenecks include: problems of land acquisition which has an impact on project completion within a stipulated time period; few bankable projects in the infrastructure sector; lack of sufficient capacity in the public system for a critical mass of bankable projects under the PPP format; and inability of government to develop innovative and attractive financial models to encourage financial institutions and the private sector to participate.

To encourage private sector participation in infrastructure development in developing countries, the private sector sponsor and, in particular, foreign investors would want to be assured that the project is technically and

Part One

economically feasible, financially viable and socially and politically acceptable (Akintoye *et al.*, 2006). Where an international agency is involved in the
project funding, socio-political acceptability in terms of the extent to which
the project meets government objectives and goals of job creation, society
transformation and creation of opportunities for the local enterprises might
be important.

Malhotra (1997) has identified six ways in which the governments of Asian
developing countries can create an enabling environment to facilitate greater
private sector participation in infrastructure projects: transparency of process; competitiveness of bids; appropriate allocation of risk; developer returns
commensurate with risks; stable policy regime; and government guarantees
and credit enhancements. He has not included stable political and financial
environments because most Asian developing countries have fairly stable government and foreign exchange profiles (until the Asian financial crisis of late
1990s) compared with the sub-Sahara African countries. Qiao *et al.* (2001)
have shown how a stable political and economic situation is one of the major
success factors for BOT projects in China. The factors identified by Malhotra
in addition to stable political and financial environments are important for
many developing economies, particularly the low income countries, to engage
private sector participation in infrastructure projects development.

It is generally recognised that developing countries' governments would
have major roles to play in encouraging private investment in infrastructure
development. For example, stakeholders in PPP in India have argued the roles
for the government (Sharma, 2006): they want the government to build up
capacity and develop innovative and alternative financing models to encourage private investments. This could be in the form of financial relief like
exit options for initial lenders which incorporate limited recourse financing
under which creditors are re-paid primarily from the revenues generated by
the project itself. They would want PPP policy to clarify the contribution of
the central government, state governments and multilateral agencies. Others
are (1) deepening of the long-term debt market to enable banks and finance
institutions to participate more in infrastructure financing; (2) creation of
tradable debt-based securities and development of a corporate bond market;
(3) government to increase the viability gap funding (VGF)[1] and arrange low
cost financing through reduction in duties and procedures for importing ma-
chinery; and (4) one window clearance for all licences and permissions and
assistance in form of guarantees and assurances to financial institutions to
enable raising funds at low costs.

PPP thrives on effective procurement, project implementability, government guarantee, favourable economic conditions and an available financial
market. Jütting (1999) identified macro level conditions in favour of setting
up of a PPP: these include a political environment supporting the involvement
of the private sector; an economic and financial crisis leading to pressure for
the public sector to think of new ways of service provision; and a legal framework which guarantees a transparent and credible relationship between the
different actors. At the micro level, the capacities of the actors, e.g. their
personal interest, skills and organisational and management structure are
identified as being important (Jütting, 1999).

In spite of the dire need for private participation in infrastructure in developing countries, even where the enabling environment is conducive, it is worth mentioning that PPP is unsuitable if: (1) the project is not affordable; (2) public sector needs to own the asset; (3) public sector disapproves of innovation; (4) public sector is the expert for the specific facility provision rather that the private sector; (5) public sector does not want a long-term relationship; (6) public sector does not want to use private sector employees; (7) output-based specification is not possible; and (8) performance-related payment is not possible.

Some enabling factors for PPP development include: creation of contractual and legal frameworks to expedite PPP projects; development of guidelines that promote PPP contracts; partnering role in procurement process; and PPP strategy that focuses investment in optimum areas. The strategic implementation framework for PPP in developing countries is needed to create a favourable environment that will facilitate PPP project implementation, provide comfort to potential investors in PPP projects, and guidance and direction to implementing government agencies. The key elements of the framework should be, amongst others, a clear guiding policy, appropriate legislation, an institutional set-up capable of efficient implementation and facilitation of PPP projects, as well as standard procedures and guidelines for setting out the process to be followed in implementing PPP projects.

7.7   Conclusions

Infrastructure development under PPP can involve the private bodies in the design, financing, construction, ownership and/or operation of a public sector utility or service. From the results of the analysis of the World Bank data it is evident that private sector investment in infrastructure is a major source of investment for delivery of public sector services in the developing economies. Private sector investment has been mainly in telecommunication in low income developing countries rather than energy, transport, water and sewerage. Although the low income countries constitute the bulk of developing economies, private sector investment in infrastructure in this category of countries is very insignificant compared with middle income developing countries. The amount of private sector participation in infrastructure in sub-Sahara Africa is comparatively low.

However, private sector investment in public sector infrastructure development has some benefits that the developing countries need to tap. This will enable the governments in developing countries to develop capacity for integrated solutions for infrastructural development, reduce time and cost to deliver projects, reduce risk associated with infrastructure projects, attract larger and potentially more sophisticated project sponsors and achieve technology and knowledge transfer. To achieve these sustainable results, an enabling environment needs to be created in the form of appropriate guidelines, contractual and legal frameworks to promote PPP, government guarantees and stable economic, social and political environment, and a PPP strategy that focuses investment in optimum areas.

Part One

Part One

In addition, it is important for the governments of developing countries to take on board the key lessons identified by Malhotra for private sector participation in infrastructure: government commitment, increased private interest, move to competitive processes, greater availability of information, acceptable prices and high developer returns as incentive to the private investors and large size of projects. To make PPP projects attractive to key private sector companies in the UK, small projects in the healthcare sector and education have been bundled to achieve large projects. This has an advantage to reduce overheads to both public sector and private sector associated with such projects.

Note

1. VGF is a special facility by the Indian government to support PPP where an infrastructure project is economically justifiable but not viable commercially, at least in the initial years, due to long gestation periods and economic externalities. The VGF scheme provides funding for state or central PPP projects implemented by the private sector developer on a BOT basis. Funding is available for 20% of the project cost; an additional 20% can be made available by the public sector sponsor if required.

References

Akintoye, A., Bowen, P.A. and Evans, A. (2005) Analysis of development in the UK public private partnership. *Proceedings of the CIB W92/T23/W107, International Symposium on Procurement Systems – The Impact of Cultural Differences and Systems on Construction Performance*, University of Nevada, Las Vegas 8–10 February, 1, 113–124.

Akintoye A., Kyaw, T., Ngowi, A. and Bowen P.A. (2006) Development in public private partnerships for construction-based projects in the developing countries. *Proceeding of CIB W107 International Symposium on Construction in Developing Economies: New Issues and Challenges*, Santiago, Chile, January 18–20.

Bennett, E., Grohmann, P. and Gentry, B. (1999) *Public-Private Cooperation in the Delivery of Urban Infrastructure Services (Options & Issues)*, Public-Private Partnership for Urban Environment (PPPUE) Working Paper Volume 1. http://www.undp.org/pppue/library/publications/working1.pdf. (Accessed August 2007).

Collin, S. (1998) In the twilight zone: a survey of public-private partnerships in Sweden. *Public Productivity and Management Review*, 21(3), 272–283.

Estache, A. (2004) Countries. *Proceedings Of International Workshop On Public-Private Partnerships: Theoretical Issues & Empirical Evidences*, European School on New Institutional Economics, Paris, France, October.

Gentry, B. and Fernandez, L. (1998) Evolving public private partnerships: general themes and examples from the urban water sector. *OECD Proceedings, Globalisation and the Environment*, Perspectives from OECD and Dynamic Non-Members Economies, pp. 99–125.

Grant, T. (1996) Keys to successful public–private partnerships. *Canadian Business Review*, 23(3), 27–28.

HM Treasury (2000) *Public Private Partnerships – The Government's Approach*. The Stationery Office, London, http://www.hm-treasury.gov.uk/docs/2000/ppp.html.

IIPF (2005) International project finance. http://members.aol.com/projectfin/project_finance_links.htm.

Jones, I. (1998) *INFRAFIN*, Final Report for Publication Contract No: ST-97-SC.1218, project funded by the European Commission under the transport RTD programme of the 4th Framework Programme.

Jütting, J. (1999) Public private partnerships and social protection in developing countries: the case of the health sector. Paper presented at the ILO workshop on '*The extension of social protection*' Geneva, 13–14 December, Centre for Development Research, University of Bonn, Germany.

Li, B. and Akintoye, A. (2003) An overview of public private partnership. In: Akintoye, A., Beck, M., and Hardcastle, C. (eds) *Public Private Partnership: Managing Risks and Opportunities*. Blackwell, Oxford.

Locke, D. (1998) On the move. *Private Finance Initiative Journal*, 3(4), 82–85.

Lifset, R., and Fernandez, L. (1998) *The search for best practices in urban solid waste management services in developing countries*. Internet conference 1997/98, http://www.undp.org/pppue/pppueold/global/waste.html.

Malhotra, A.K. (1997) Private participation in infrastructure: lessons from Asia's power sector. *Finance and Development*, December, 33–35.

Plummer, J. (2000) *Private Sector Participation In Water And Sanitation Services In Stutterheim, South Africa*. Building Municipal Capacity for Private Sector Participation, Working paper 44201, GHK International, 526 Fullham Rd, SW6 5NR, London, http://www.undp.org/pppue/library/publications/stutterheim.pdf.

Poole, Jr. R.W. (1995) *Revitalizing State And Local Infrastructure: Empowering Cities And States To Tap Private Capital And Rebuild America*. Policy Study No. 190, May, http://www.rppi.org/privatization/ps190.html#11.

Qiao, L., Wang, S.Q., Tiong, R.L.K., and Chan, T. (2001) Framework for critical success factors of BOT projects in China. *The Journal of Project Finance*, Spring, 53–61.

Sader, F. (2000) *Attracting foreign direct investment into infrastructure: Why is it so difficult?* Foreign Investment Advisory Service, Washington, DC.

Savitch, H. (1998) The Ecology of Public-Private Partnerships: Europe. In: Pierre, J. (ed) *Partnerships in Urban Governance: European and American Experience*. MacMillan, London, pp. 175–186.

Sharma, S.N. (2006) Funding the infrastructure blue-print. *Times News Network*, October, 15.

Smith, R. (2007) *Challenges of International Finance for Development and Attempts by the Commonwealth Secretariat to address them*, by Commonwealth Deputy Secretary General at the Workshop on Debt, Finance and Emerging Issues in Financial Integration 6 March 2007, Marlborough House, London. http://www.un.org/esa/ffd/Multi-StakeholderConsultations/FFDO/SovereignDebt/ransford%20smith-speech.pdf. (accessed August 2007).

South Africa Government Gazette (2000) *Pubic Finance Management Act, 1999*. Regulation Gazette No. 6780, http://www.gov.za/gazette/regulation/2000/21082.pdf.

Thomsen, S. (2005) Encouraging public-private partnerships in the utilities sector: the role of development assistance. Proceedings of Roundtable *Discussion on Investment for African Development: Making it Happen, NEPAD/OECD Investment Initiative*, Uganda, 25–27 May.

UNDP (2006) Memorandum to the UK http://www.publications.parliament.uk/pa/cm200506/cmselect/cmintdev/921/921m01.htm   http://www.publications.parliament.uk/pa/cm200506/cmselect/cmintdev/921/921m34.htm#n160,

United Nations (2003) *Financing For Development. The Final Text Of Agreements And Commitments Adopted At The International Conference On Financing*

Part One

For Development Monterrey, Mexico, 18–22 March 2002, United Nations Department Of Economic And Social Affairs, Financing For Development Office, Www.Un.Org/Esa/Ffd

World Economic Forum (2005) *Multi-Stakeholder Consultations On Financing For Development, 2004–2005.* Http://Www.Un.Org/Esa/Ffd/Indexmulti-Stakeholderconsultations.Htm

World Bank (2005) Private participation in infrastructure database. http://ppi.worldbank.org/reports/customQueryAggregate.asp. (Accessed August 2005).

Part One

8

Team Building for PPPs

Mohan M. Kumaraswamy, Florence Y.Y. Ling and Aaron M. Anvuur

8.1   Introduction

Many developing countries, and indeed others with limited financial resources, still look to PPPs as an alternative source for financing much needed
infrastructure such as roads and bridges (Akintoye *et al.*, 2003; Branco *et al.*,
2006; Anvuur and Kumaraswamy, 2006). However, many developed countries have transitioned to a new wave of PPPs that focus heavily on achieving
VFM by mobilising private sector efficiencies, innovations and flexibilities
in delivering both infrastructure and services to a more discerning public
(Akintoye *et al.*, 2003; Akbiyikli and Eaton, 2006; Clifton and Duffield,
2006).

However, drivers for mobilising alternative finance are being overtaken in
many countries by the focus on better value services, and the consequential
need for a shift in cultures, mind sets and capacities of the teams to be entrusted with the invariably wide-ranging and far-reaching PPP projects. This
chapter compares a new wave of PPPs in Hong Kong and Singapore with
developments in some other regions. Specifically we focus on how teams
were assembled and on lessons learned therefrom. In this context we have
developed a general framework to demonstrate how suitable teams may be
selected and developed in order to meet the special requirements of PPPs.
Compared to non-PPP projects, such requirements are inherent in the usually longer-term multi-functional and inter-disciplinary demands of PPPs. For
example, special needs emerge for better and longer-term team building and
relationship management during design and construction, as well as during
operation and usage of the built facilities; and smoother interactions at the
many interfaces.

Recent definitions and descriptions of PPPs convey the above expanded
scope of PPPs, as well as the growing emphasis on the third 'P': the 'partnership' approach that is essential for developing an efficient project team and a
successful PPP. For example, the Hong Kong SAR introductory guide to PPPs
(Efficiency Unit, 2003) conveys that:

- PPPs are based on a partnership approach, where the responsibility for the delivery of services is shared between the public and private sectors, both of which bring their complementary skills to the enterprise.
- PPPs bring together public and private sectors in a long-term relationship, with the private sector moving on to become a 'long-term service provider' rather than a 'simple upfront asset builder'.

Singapore too stresses the 'partnering relationship' and 'services' elements in introducing a PPP (Ministry of Finance, 2007) as 'a long term partnering relationship between the public and private sectors to deliver services'. It proceeds to describe a PPP as part of their 'best sourcing framework' where the public sector engages the private sector to deliver those services which the latter can be more effective and efficient in providing the public at the best VFM.

8.1.1   Accentuated team-building needs

Such higher-level expectations, and indeed demands, by the general community and the public sector, require a rethinking of approaches to PPPs. The increasing popularity of PPP-type procurement of infrastructure signals the need for both public and private sector organisations and personnel to develop new knowledge bases, skill sets and mind sets, as required for structuring and implementing successful PPPs. For example, those trained in traditional modes of infrastructure procurement would be well-versed with say, short-term and focused design–bid–build or design–build projects to deliver a building complex, road or bridge to their client, who would then take over the operation and maintenance (O&M). Even governmental public works departments, who may plan, design or supervise the design, and oversee the construction by virtue of their infrastructure procurement expertise, would often hand over the O&M to the end-user department/organisation, e.g. Highways Department to Transport Department, or Buildings/Architectural Services Department to Education Department. However, in a PPP, the consortium engaged to plan, design and construct the infrastructure asset would 'transition' from construction project management to asset management.

Asset management calls for additional knowledge pools and skill sets, as well as a much longer time horizon that in turn demands appropriate mind sets as required for building and sustaining a multi-disciplinary team for the extended project duration. For example, the O&M experts to be entrusted with the asset management would be mobilised early, to contribute valuable inputs into the design, that would focus more upfront attention on durability, maintainability and whole lifecycle performance and costs.

Figure 8.1 provides a basic illustration of how asset management may be integrated with the construction project management functions of infrastructure procurement and delivery. The extent/parts of the 'total project management' transferred to the 'project consortium' (PC – the private sector group, sometimes called the special purpose vehicle (SPV), sponsor or concessionaire) by the 'client' (the public sector granting authority, sometimes called the promoter), could vary with the type of PPP adopted. For example, even within PPPs that involve private financing, the Hong Kong SAR Efficiency Unit (2003) identified a spectrum of possible arrangements. However, given

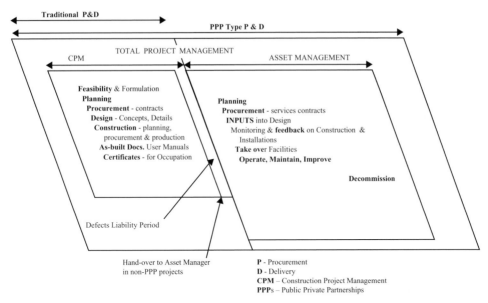

Figure 8.1   Integrating construction project management with asset management in PPPs.

the need for the PC to deal with at least some financial and commercial issues, as well as with social, legal, economic, environmental, political and techno-logical (e.g. 'SLEEPT' as described by Eaton *et al.*, 2006), it is expected that the PC would share in the risks and rewards of some parts of the 'total project management' that is indicated, although not detailed, in Figure 8.1.

Although the PC could commission a turnkey contractor for handling part of the construction project management, and a facilities/asset management company for the asset management, it would undertake some of the functions such as in the upfront technical and financial feasibility analysis, planning and procurement, as well as in the interface management, e.g. in resolving conflict-ing priorities and perceptions of design and operations functions/personnel.

This chapter identifies the need for new approaches to integrating PPP project teams, both of the PC and the client within themselves, as well as together in the total PPP project team. It then conveys how such needs are being approached, by summarising relevant insights gleaned from previous studies, as well as from a series of interviews with key players in recent initiatives in Hong Kong and Singapore. Finally, a conceptual framework is developed as a basic aid towards structuring the selection and eventual development of the evidently more integrated and sustainable teams needed for successful PPPs.

8.2   Integrating and Sustaining PPP Teams

8.2.1   Relational integration and longer time horizons

While Figure 8.1 projects a broad brush visualisation of some of the functions and responsibilities of the team, more detailed descriptions may be obtained if needed, depending on the function or discipline/profession being focused upon. For example, the Hong Kong Institute of Surveyors has compiled a list of potential professional services to clients of PPP projects (HKIS, 2004). However, it appears that the uniqueness of the risk allocation and responsibility profile in each PPP calls for more interactive multi-disciplinary team-working, flexible approaches and broader mind sets than would be expected in even the most complex non-PPP project.

The much wider demands on PPP teams, coupled to the imperatives to sustain their relationships, effectiveness and efficiencies through much longer timeframes, pointed the authors to the need to investigate how 'winning' teams are being assembled, and indeed may be assembled better, in the new wave of PPPs.

The critical needs for more integrated teams for achieving the desired step-change improvements in the procurement and delivery of infrastructure projects in general, have become the battle cry of many industry reports (e.g. Latham, 1994; Egan, 1998; Construction 21, 1999; CIRC, 2001; Constructing Excellence, 2004); as well as the mantra of many academic 'advisories' (e.g. Belbin, 2004; Kumaraswamy and Rahman, 2006). Not surprisingly, an appreciation of such imperatives and industry readiness to move towards 'relationally integrated' (rather than merely 'structurally integrated') teams has been noted, for example in a survey across Australia, Hong Kong, Netherlands, Singapore and the UK (Kumaraswamy et al., 2005; Rahman et al., 2005). The need for 'relational integration' should be much greater in PPPs, given the broader and longer interactions and commitments as described above.

8.2.2   Relevant findings from PFIs/PPPs in Australia, UK and in general

In the context of the above sub-section, it is noted that certain aspects of the importance of team relationships in PPPs seem to have been appreciated, studied and documented to some extent, in the past. Some examples are given below:

- In Australia, a set of unpublished speaker's notes on a case study of the PPP on the 'Victoria Country Court Project' as presented by the project director to a visiting study team of Hong Kong officials in June 2004, conveyed that 'the quality of the project team is also important, and should include project management (process) skills, as well as specialist knowledge (content) skills; and content knowledge of the client area is critical'; 'a self-starting project co-ordinator' for the client area must 'provide necessary understanding and networks to facilitate project delivery'.
- In the UK, a recent report (Partnerships UK, 2006) conveyed the results of a survey of operational PFI projects (thereby including non-construction projects) that covered many aspects. Interestingly, 66% of public sector respondents rated the performance of these service providers as either 'very good' or 'good'; while 72% of public sector contract managers rated their relationships with the service providers as either 'very good' or 'good', and 25% as 'satisfactory'; 79% of users were satisfied 'always' or 'almost

always' with services received. More interestingly, a 'clear correlation' was noted between 'good' or 'very good' relationships and high levels of performance. Although not so strong, a positive link was also discerned between good relationships and user satisfaction (Partnerships UK, 2006). Key factors that influence relationships between public sector and private sector teams were said to include 'communication' (by over 30% of respondents), trust (20%) and shared objectives (17%). Clarity and understanding of roles and motivations of others, flexibility and shared information were also seen to be important. On the other hand, high levels of staff turnover on either side were one of the factors straining relations. This evidence thus reinforces the need for more relationally integrated as well as sustainable teams in PPPs.

- In a critical review of published research in PFI/ PPPs in construction, Pantouvakis and Vandoros (2006) found that of 78 PPP journal papers in four selected leading journals over the period 1996–2006, although 42% were on stakeholder relationships at the contractual level, there has been a shift of interest towards financial management related issues after the late 1990s. It is also noted that the emphasis seems to have been on contractual relationships, which could be more on the structural and legal arrangements, rather than the relational integration which arises from 'relational contracting' type approaches (Kumaraswamy *et al.*, 2005). Much therefore remains to be studied in this respect.

8.3   Hong Kong Perspectives of PPP Teams

8.3.1   Historical background

Hong Kong can take credit for formulating a successful BOT-type PPP for the Cross Harbour Tunnel in the late 1960s, even before the word BOT was reportedly coined in the early 1980s. This tunnel has been transferred back to the government following the BOT cycle, while four other tunnels have since been procured on a similar BOT basis. All these have contributed to a knowledge base of strengths and some weaknesses (e.g. inappropriate toll adjustment mechanisms) of this type of PPP in Hong Kong (Kumaraswamy and Morris, 2002). Approaches to the concessionaire (PC) selection in these projects have been detailed by Zhang *et al.* (2002). For example weightings of 60:20:20 have been applied in combining three designated packages of (1) financial and general, (2) engineering and (3) operation and transport planning. Sub-package criteria include consortium structure, strengths and experience in package 1, consortium ability and environmental proposals under package 2. However, the sub-criteria weightings did not indicate high priorities for such sub-criteria at that point of time.

Nevertheless, the importance of teamworking in these PPP ventures was conveyed in interviews conducted with (1) the project manager of the PCs construction project, for one of the more recent tunnels, who had also worked on one of the previous PPP tunnels; and (2) a representative from the client's team who had worked on the recent Hong Kong tunnel projects as well, and was involved in the take-over (transfer) of the first of these tunnels (the Cross Harbour tunnel):

- The PC construction project manager conveyed, for example:
  - □ Given the many parties constituting the joint venture PC, how he became 'almost a professional project manager with no strict allegiance to any one of the partners'.
  - □ How this reputation for independence helped him to focus on the best interests of the project, as well as to build a coherent team from the diverse participants.
  - □ How he always tried and managed to succeed in developing a very strong team spirit where everybody identified with the project. Their interest was to get the project built and put their best into it.
  - □ How he achieved this, for example by assigning tasks to people, giving them enough 'range' to do them, but not letting them 'fall over'.
- The second interviewee on this project was from the client's team. When asked to cite the strengths, he said 'the whole process goes smoothly because of the team spirit and the motivation of the private sector', and added that the operational responsibility with the franchisee 'ensures that the quality is of acceptable standard' for the long term, thus bringing in the sustainability angle as well.

8.3.2   Recent initiatives and experiences

The Hong Kong government handbook on PPPs highlights relevant specifics such as: (1) 'partnership attitude', 'right skills mix at the right times', and 'desire to make partnership work in practice'; and (2) 'select a private partner that you will be able to work well with throughout the life of the project'.

On the other hand, it is noted that there seems to be a reluctance in some line departments to initiate PPPs in spite of general encouragement, and indeed despite some exhortations for PPP procurement from policy makers in the last few years. Although many conferences and study visits have disseminated the potential advantages of PPPs to many public sector players, very few projects have graduated to the planning stage, while at least one was shelved after initial planning due to fears of public sector job losses. Of those that were initiated, the West Kowloon Cultural District project (WKCD) has been suspended, while the AsiaWorld-Expo (AWE) seems to be quite successful.

The WKCD was conceived a few years ago as a PPP mega project to establish a world-class integrated arts, cultural and entertainment complex, involving a 50-year land grant to a single consortium. This would grant residential and commercial development rights, in exchange for the financing, planning, design, construction and operation of the cultural hub for 30 years in the 40 hectare West Kowloon Reclamation site (Lo, 2006). However, the project is now 'on hold' after proceeding through initial stages of a concept design competition, pre-qualification of consortia, and design concept proposals submission by the pre-qualified consortia. Debate and controversies arose over the single developer model, as well as over the funding model and potential conflicts of interest in entrusting cultural development and dissemination

Part One

to traditional developers (Lo, 2006). This project will no doubt be revived in some form in due course.

Meanwhile, Lo (2006) conducted 18 structured interviews with a range of well-positioned stakeholders, to identify critical success factors (CSFs) for such projects. A cluster of CSFs was derived from the international literature and the local interviews, and grouped into five elements in a CSF framework. The broad 'elements' are (1) macro environment, (2) government–project relationship, (3) construction–project relationship, (4) government–consortium relationship, and (5) project; while the number of CSFs identified within each element are 8, 8, 5, 7 and 3 respectively, in that study.

Some of the above CSFs involved team building not just within the PC team, but also with stakeholders, e.g. 'communication with stakeholders' as well as specifics like 'culture of partnership/trust' and 'right skills mix'. Specific shortcomings in the CSFs perceived as having 'held up' the above project were found to be:

- Problems with sector policy and its implementation
- Inadequate communication with stakeholders
- Unclear and inadequate output specifications
- Lack of justification of VFM
- Inappropriate sharing of skill sets and risks
- Lack of protection of public interest
- Political pressure
- Social opposition

From this it is seen that team-building for PPPs should be approached in a wider context. It should commence upfront with the client, and involve key stakeholders at the overall project level, before moving into the selection and assimilation of a 'winning' PC team. In fact the latter could be considered as a key sub-team within the overall PPP project team, which in turn needs to be relationally integrated and sustainable, as discussed in previous sections.

On the other hand, the project for the new international exhibition and conference centre AWE (AsiaWorld-Expo) has been said by the HKSAR Chief Executive to be 'an excellent example of a successful public private partnership', as reported in a newsletter of one of the key project players. In this project, the client invested substantial equity, but also mobilised private sector finance, as well as design–build and operational expertise. AWE has reportedly been doing better than expected since it opened in end 2005.

Three key project players were interviewed to tap into the experiential knowledge gained in team building for this project. The organisational structure and arrangements were interesting, although not unusual for such PPPs. For example, the international construction group based in France that led the PC mobilised a Hong Kong investment company as an equity partner, and invested together with the government and the Airports Authority (AA) to form a joint venture company (JVCO). The JVCO contracted the design–build package to the Hong Kong arm of the international construction group; and contracted the operation to 'AWE Management' which also involved the same Hong Kong arm, as well as a European electrical and mechanical (E&M) subsidiary and the same local private equity partner. The Project

Services Division of the AA were the 'employer's representative' who were tasked with the project management, including approvals.

There was no formal partnering in AWE, unlike in some other AA contracts, but the team did 'pretty well' and developed good relationships despite pressures. An example cited of the benefits of such relationships, was the provision for stockpiling excavated materials that was provided by AA. In any case it was said that even formal partnering would have eventually depended on personalities and could easily suffer when problems arose.

It was noted that the general experience in PPPs gathered by this international construction group and its key staff in other countries, was a valuable resource that could be leveraged on projects such as this. While they had in-house financial and legal expertise, they brought in specialists where needed, in this case on convention centre planning and operation from the UK. The 'whole lifecycle focus' was said to be refreshingly different from non-PPP projects, although it introduced interesting tensions between the design–build and operational sub-teams, since the latter often pushed for better quality and lower maintenance elements. Their demands on this account were often more stringent and strident than the client's, given their experience and knowledge of alternatives. In this project, such situations created interesting multi-functional and inter-disciplinary teamworking scenarios, where it was seen that all the professionals must strike a good balance between compromising on their functional standards and the expected economies, while remaining 'open minded' to the problems and issues faced by others. A focus on the project and open honest 'round table' discussions (without any dominant pressures) were found to achieve good results.

This project also provided an interesting example of cross-team 'migration' of a key team player from (1) the front-end of 'business development' in Asia in general, who led the tendering for this consortium, to (2) the CEO of AWE Management who now operate AWE. This is perhaps indicative of how well the 'business case' of PPP projects can be identified by key team members while sustaining successful operations in the long term. At the operational end as well, it is believed their strength lies mainly in their people ('99% of our assets'), especially those who are ready to take up the various challenges and to strive for the best interests of their functions, while developing integrated solutions. In terms of what more may be done on future PPP projects of this nature, it was suggested that more key operational staff may be brought on stream earlier, since inputs into the design could be more demanding than planned. Overall, the experiences exemplified an observation at one of the interviews, that working on even one good PPP project can significantly change a person's mind set and approaches, so that they would be much better teamworkers, even on future non-PPP projects.

In the context of the overall PPP team, the flexibility of the client was also seen to be valuable, for example in not adopting a 'take it or leave it' approach at the initial stages, but instead taking on board useful suggestions from the pre-qualified consortia, for example for phased construction and operation, for greater equity injection, and in allowing adequate flexibility in the operations, e.g. in the control of the events calendar in the case of a convention centre. Although the scale and composition of the AWE is totally different from the proposed WKCD, it is possible that some lessons may be drawn as regards what may and may not be viable in similar project elements.

From a financing angle, it has been noted that the government's share of equity has been larger than in some other types of PPPs. This could be a feature of the entertainment PPP type models. Even in the case of the Hong Kong Disneyland, the government invested a great deal in the infrastructure and land provisions, and possibly more than some critics thought was necessary at that time. No doubt, indirect and spin-off benefits were also considered important in the total 'equation'.

8.4   Singapore Perspectives of PPP Teams

8.4.1   Background and build-up

Neither Singapore nor Hong Kong government has been unduly constrained by a lack of funds for infrastructure development. Therefore they have some similar objectives in their approach to PPPs, although focus areas, thrusts and programme roll-outs are naturally region-specific. The Singapore Ministry of Finance (2007) views a PPP as 'part of the Best Sourcing framework' and as 'a long-term partnering relationship', as described in more detail in the introduction to this chapter. Version 1 of its *Public-Private Partnership Handbook* issued in October 2004, summarises 'key factors in a successful relationship' as (a) mutual respect and understanding, (b) open communication and (c) recognition of mutual aims. It also summarises 'key factors in the management structure' in a way that continues to convey the importance of relationships, for example 'the relationship at the senior management level sets the tone of the PPP relationship', 'the governance arrangements should be equitable and relationships are peer-to-peer', 'a large number of formal management levels should be avoided', 'formal committee structures should not be seen as overly rigid' and 'clear roles and responsibilities should be set and the staff empowered for the different structures in place to manage the relationship at different levels'; it does not neglect practicalities, e.g. 'clear escalation procedures should be instituted and followed'.

From a financing angle, a law firm, Lovells Lee and Lee, in a note to clients, indicates that 'PPP is the name given to an extensive and disparate collection of constructive relationships between the public and private sectors'; and that PPPs provide a structure which takes on board some of the lessons learnt from privatisation, for example that 'privatisation retained too many elements of monopoly; PPP provides greater potential for competition'; and also that 'PPP should create a fair distribution of wealth, avoiding the financial windfalls often criticised in privatisation'.

8.4.2   Recent initiatives and experiences

Having carefully planned their ventures into PPPs, the Singapore government rolled out the initiatives, while preparing the PPP Handbook Version 1 issued in October 2004. Four PPP projects have already been awarded (Ministry of Finance, 2007):

- A BOO type desalination plant awarded in January 2007
- A DBOO type water treatment plant awarded in December 2004
- A DBOO type incinerator plant awarded in November 2005
- A one-stop integrated logistics information port called TradeXchange awarded by the Singapore Customs in December 2005, which includes development of software, maintenance and operation of the required IT system for 10 years from 2007

A few other PPP projects are being formulated, for example the 'Sports Hub', an institute of technical education, a university residential complex and also some defence sector PPPs.

However, some industry participants felt that the number of potential PPP projects, or PPP 'deal flow' was insufficient to attract sustained serious attention from big international PPP players. In the same context, it was felt that the SG$50m (US$1 ≈ SG$1.6) threshold above which public projects should be considered for PPPs, was too low to attract serious players. The fact that there was no great need for Singapore to rush into PPPs was also acknowledged by both private and public sectors, given the availability of public funds, developed infrastructure and what was said to be a reasonably efficient public sector. Some opinions surfaced that the first couple of designated PPP projects were not 'pure PPPs' e.g. that the waste water treatment plant, although tendered on a PPP basis, was not financed with a typical PPP model.

Eight in-depth interviews were conducted with well positioned government officials, legal and financial experts, a PPP client/promoter and three PC partners/potential partners, with two on the 'design and construction' side and the third on the 'operations' end. In addition, supplementary general views on the uptake and potential for PPPs were derived from a series of subsequent interviews on the development of the Singapore construction industry, and how it sustained itself over periods of recession, given that it has recently emerged from a relatively long recession, while Hong Kong is also looking for lessons to be learned during such downturns.

The following paragraphs convey relevant perspectives from the eight specific semi-structured interviews on PPPs in Singapore that incorporated questions on how PPP teams were selected and/or trained/developed in both public and private sectors, and on what more could be expected from both sectors to make designated PPPs work 'better' and yield more VFM.

To start with, the public sector approach to choosing projects that can be better procured by PPPs, is essentially based on the questions of: what can the private sector do better than us?; and what do they need from us to help them to do it better? Furthermore, market testing and stakeholder 'buy-in' is considered essential, before 'rolling (any project) out to the market'. Being a relatively new procurement approach, the public agency/government department client must ask and answer the above two questions during the market testing as well, while drawing on a vast store of knowledge from the Ministry of Finance. It is recognised that new skill sets and changes in mind sets are needed. More open and flexible mind sets targeting win–win scenarios need to be developed. This is the same in other countries with a history of tightly regulated public infrastructure procurement that hitherto assumed the need for stringent controls of providers of design, construction and other services, because their interests were assumed to diverge from those of the client. Of course an appropriate balance is needed, so as not to 'give away' too much to the private sector and to safeguard public interests.

Awareness sessions for relevant public officers and specific training for those involved in PPPs are provided. It is expected that after more PPP projects come on stream, a body of local experiential knowledge would enable the development of more standardised PPP models and contract clauses. This should in turn assist in the training and development of the human resources needed for these endeavours.

Two interviewees felt the need for accelerating the learning curves and PPP awareness of lower level public officers, so that they may be more confident to provide the quick feedback needed by private sector partners/potential partners under this 'new' procurement regime. They felt that even some banks and developers in Singapore were still wary of the PPP model. In fact, initial opinions on PPPs in Singapore, as asked for at the supplementary (non-PPP focused) interviews, projected a general view that it was still 'early days' in this respect, and some big consultants and contractors were waiting to see how things worked out in this form of procurement.

Some saw the usefulness of a high profile 'champion' who would lead the drive for better PPPs, where they are proven useful, whereas an alternative view was that good systems should preclude the need for a 'champion' who could be misconstrued as being autocratic. However, there can be little dispute on the need for driving personalities in both public (client) and private (PC) sector groups of a PPP project team. They could inspire and shape the vast changes in mind sets and working arrangements that were unanimously seen to be at the core of any PPP success, given the innovative approaches required.

For example, the contractors and facilities managers interviewed saw the value in PPPs providing more innovative, much better and more efficient financial plans, designs and facilities management. But to achieve these, they recognised the criticality of teamworking across disciplines and functions, for example for empowering facilities management 'feed-forward' (apart from feedback) into designs, i.e. from monitoring the inputs in 'real time', rather than only waiting for the outputs to provide feedback (Kumaraswamy, 1996). It was said that approaches must change from reactive to proactive, e.g. to preventive rather than 'fix-it' in operations management. Interesting examples of benefits from knowledge transfers across functions, disciplines and, indeed, sectors, were provided. For example, a facilities management team were gratified to have picked up some specific management techniques on planning and control systems and on inventory optimisation, from the CEO of a client, and they were now applying these techniques on other projects as well.

Lessons learnt included the need to include insurance advisors in the up-front planning of the client, i.e. along with the legal and financial advisors, rather than attempt to 'bolt-on' complex insurance inputs after finalising the legal and financial frameworks. The additional work and rework involved provided an example of the importance of bringing together at an early stage, the many functions and disciplines needed on a PPP project.

8.5   Integrating PPP Project Teams

Recent experiences and expectations in Hong Kong and Singapore underscore the need for greater integration in PPP team structures, relationships and working arrangements. The needs and potential for relational integration and construction project teams in general have been mapped by many researchers, e.g. as outlined by Kumaraswamy and Rahman (2006). The more pronounced imperatives in PPPs are emphasised in this chapter, given the larger number of interactions, and the fact that these need to be sustained for much longer than in non-PPP construction projects.

In terms of improving relationships, increasing productivities and enhancing results, 'relational' approaches have been postulated as necessary alternatives to the adversarial approaches in traditional construction contracts. For example, the benefits of superseding rigid dispute-generating traditional contracts with relational contracting (Macneil, 1974), have been shown to empower joint risk management between partners (Rahman and Kumaraswamy, 2002). These have also been demonstrated in practice, by successful partnering and alliancing arrangements (e.g. CII-HK, 2004). Relational contracting decreases recourse to contract documents, reduces transactional friction and ensuing disputes (Rahman and Kumaraswamy, 2002). It enables a focus on common objectives, including VFM, and fuels the often elusive in-depth cooperation between project participants (Phua and Rowlinson, 2004). Relational contracting approaches could thus counteract the 'push apart' force fields of the classical contracting approaches by 'pulling together' each pair of team members. They could work closer together and cooperate better, if the relational forces overcome the traditional contracting forces that generate distrust and adversarial attitudes, thereby tending to push parties apart. This idea could be extended for drawing together the many organisations involved in a PPP project into a truly integrated project team as in Figure 8.2, which has been developed for this chapter on the basis of the foregoing discussions.

8.6   Selecting Project Consortium Teams

The potential for eventual integration and the probabilities of sustaining team performance may be boosted by incorporating critical success factors for such relationships and sustainability into team selection criteria. PC team selection systems have previously been fine-tuned for evaluating 'technical' potential, e.g. within packages of financial and general criteria, engineering criteria and operation and transport planning criteria (Zhang et al., 2002) for Hong Kong BOT tunnel projects, as detailed in sub-section 8.3.1. What remains is to introduce realistic 'relational' and 'sustainability' performance criteria, and weight them appropriately for a more comprehensive assessment of the

Part One

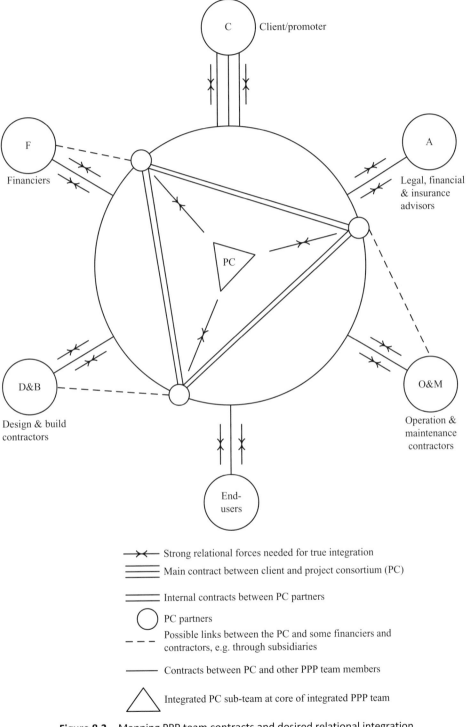

C Client/promoter

F

Financiers

A

Legal, financial
& insurance
advisors

PC

D&B

Design & build
contractors

O&M

Operation &
maintenance
contractors

End-
users

→✕← Strong relational forces needed for true integration

Main contract between client and project consortium (PC)

Internal contracts between PC partners

◯ PC partners

Possible links between the PC and some financiers and
contractors, e.g. through subsidiaries

Contracts between PC and other PPP team members

△ Integrated PC sub-team at core of integrated PPP team

Figure 8.2 Mapping PPP team contracts and desired relational integration.

potential success of a candidate PC team, including its expected contributions to an upcoming PPP project.

Such an exercise has been launched, incorporating the needs and drivers identified in preceding sections, together with criteria and success factors, in the first instance being those collected from previous research exercises in team relationship building and sustainability enhancements (Kumaraswamy *et al.*, 2005; Ugwu *et al.*, 2006). The development of a framework for integrating relational factors and sub-factors, and the computation of a relational index for each PC, is described at length by Kumaraswamy and Anvuur (2008). While this chapter provides a more holistic picture covering the whole PPP team, space limitations and the undesirability of repetition, prompt the authors to point to the information in the above journal paper, on the proposed framework, relational index and its parallel sustainability and technical indices. However, the latter focused on prequalifying candidate consortia, and focused on 'past performance scores', whereas this chapter's expanded proposal is to move this forward into the next stage of evaluating PPP tender packages, i.e. for selecting an optimal PC sub-team, which would be at the core of the PPP team, as pictured in Figure 8.2.

Therefore, while methodologies for developing and applying similar scoring and weighting sub-systems in general have been described before (Zhang *et al.*, 2002; Kumaraswamy and Anvuur, 2008), Figure 8.3 projects the broader proposal in this chapter for building up a comprehensive tender score, 'S'. As shown, S is built up from four packages: (1) team criteria; (2) financial criteria; (3) infrastructure criteria, including both technical and sustainability parameters of the infrastructure itself, as against the technical and sustainability potentials of the team which would be incorporated in package 1, as for example described by Kumaraswamy and Anvuur (2008); and (4) operational and usage criteria.

The numbers of criteria in the above four packages are taken as u, v, x and y. The combined consortium criterion score for any given tendering consortium would be computed as t_1 for team criterion 1, and so on. This would be based on the combined strengths of the consortium partners in respect of each criterion. It should be noted that team criteria in package 1 would include technical criteria (e.g. 'successful' experiences in providing the required type of infrastructure'), as well as relational criteria (e.g. 'proven capacity for successful partnering or alliancing etc. on previous projects') and sustainability potential criteria (e.g. 'demonstrable sensitivity to sustainability issues in previous designs of this type of infrastructure'). Examples of relevant factors and sub-factors are presented in Kumaraswamy and Anvuur (2008).

Financial criteria in package 2 would include initial and lifecycle costings, toll rates and adjustment formulae where applicable, refinancing and risk-sharing mechanisms etc. Infrastructure criteria in package 3 would include all important engineering, durability and other such parameters. 'Operational and usage' criteria in package 4 would include operational and contingency plans, comprehensive maintenance strategies and protocols etc.

Each tender package score, for example T for 'team' score for each consortium, is built up by weighting each constituent criterion score by a client chosen weighting factor, w_a before adding the weighted scores together as in

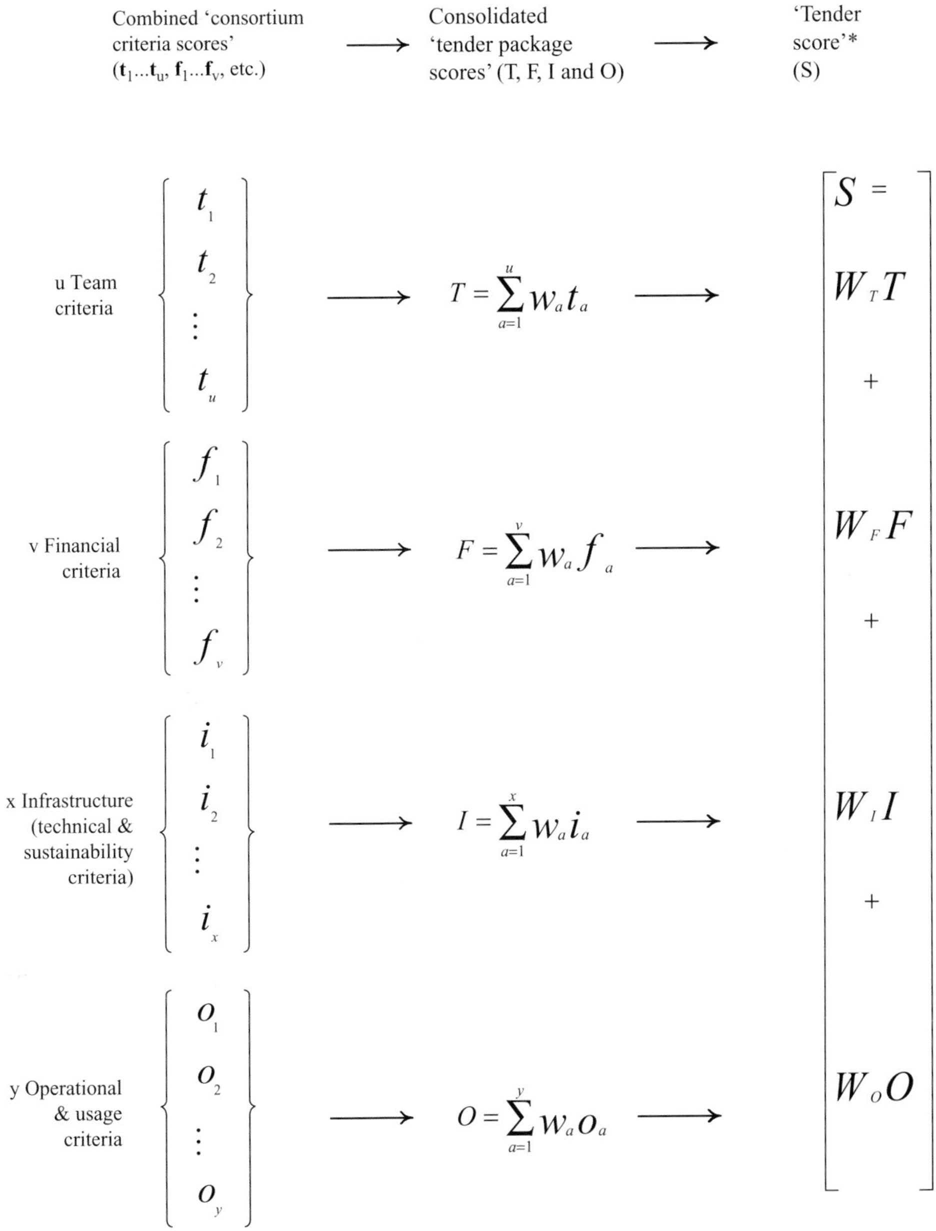

Combined 'consortium criteria scores' $(t_1...t_u, f_1...f_v, \text{etc.})$	$\longrightarrow$	Consolidated 'tender package scores' (T, F, I and O)	$\longrightarrow$	'Tender score'* (S)

$$S =$$

u Team criteria $\left\{\begin{array}{c} t_1 \\ t_2 \\ \vdots \\ t_u \end{array}\right.$ $\longrightarrow$ $T = \sum_{a=1}^{u} w_a t_a$ $\longrightarrow$ $W_T T$

$+$

v Financial criteria $\left\{\begin{array}{c} f_1 \\ f_2 \\ \vdots \\ f_v \end{array}\right.$ $\longrightarrow$ $F = \sum_{a=1}^{v} w_a f_a$ $\longrightarrow$ $W_F F$

$+$

x Infrastructure (technical & sustainability criteria) $\left\{\begin{array}{c} i_1 \\ i_2 \\ \vdots \\ i_x \end{array}\right.$ $\longrightarrow$ $I = \sum_{a=1}^{x} w_a i_a$ $\longrightarrow$ $W_I I$

$+$

y Operational & usage criteria $\left\{\begin{array}{c} o_1 \\ o_2 \\ \vdots \\ o_y \end{array}\right.$ $\longrightarrow$ $O = \sum_{a=1}^{y} w_a o_a$ $\longrightarrow$ $W_o O$

* computed separately for each tendering consortium

NOTES: (1) w_a is the variable weighting factor chosen for the ath criterion in each package, as shown
(2) W_T, W_F, W_I and W_O are the weightings chosen for the T, F, I and O packages, respectively

Figure 8.3 Structuring the compilation of a comprehensive tender score.

Part One

Figure 8.3. The four package scores would be weighted in turn by W_T, W_F, W_I and W_O before combining them into the tender score of each tendering consortium. The foregoing descriptions and illustrations are intended as a preliminary demonstration of a framework for a more comprehensive selection of the crucial core PC sub-team that would be the base building-block of the overall PPP team. More work is of course needed to flesh out the framework, and also to incorporate appropriate criteria, factors, sub-factors and indicators.

8.7   Concluding Observations

The underlying needs for integrated and sustainable PPP teams that thread through this chapter underscore the presently perceived gaps, discontinuities and conflicts between functions (such as between the design and operation functions), as well as between sub-teams (such as between the financing and construction). Any targeted alignment of objectives of sub-teams must be preceded by mutual understanding of mind sets, motivations and capacities. For this, the stereotyping of professions and disciplines such as architects, bankers, lawyers, engineers and quantity surveyors erects barriers or 'walls', that need to be replaced by 'bridges'. These bridges should ideally be built both bottom-up during education and training, as well as top-down through more open-ended and flexible top management interactions.

Endeavours to inculcate undergraduate level cross-disciplinary appreciations are emerging, for example at the Department of Civil Engineering of The University of Hong Kong, in both virtual and real modes, in Year 2 and Year 3 courses in 'Engineering Design & Communication' and 'Inter-disciplinary Design Project' respectively. While the former involves multi-functional and multi-disciplinary role playing by civil engineering students alone, the latter brings together architectural, building services, civil engineering and quantity surveying students in group projects. They soon appreciate the needs for teamworking, and adjust their initial unitary approaches, in order to reach holistically optimal outcomes. This also generates mutual respect for the contributions and functional needs of different disciplines, and for moving over to multi-disciplinary and multi-functional approaches to complex issues. Such valuable interface knowledge development should of course continue with real-life experiences and supplementary CPD (continuing professional development) training in industry.

Given the clear needs and accelerated trends towards integrated teams for infrastructure projects in general and PPPs in particular, it seems useful to develop such hitherto *ad hoc* teaching–learning and training initiatives, under the umbrella of a more comprehensive and integrated strategy, and with more coordinated programmes across academia and industry. The results of such programmes should in turn yield benefits, for example in improving the probabilities of selection of organisations (and key individuals) that (and who) have participated in such real or simulated exercises; as well as in their enhanced performance in PPP teams, if so selected. Enhanced performance would, for example, be expected to follow naturally from the efficiencies generated by the closer and longer-term relational integration and collaborative working arrangements that have been described in this chapter.

Acknowledgements

Grant HKU 7138/05E from the Hong Kong Research Grants Council, as well as the Universitas 21 Fellowship that facilitated the lead author's study visit to Singapore, are gratefully acknowledged for assisting in much of the reported research.

References

Akbiyikli, R. and Eaton, D. (2006) A value for money (VFM) framework proposal for PFI road projects. *CIB W92 Conference Proceedings*, Salford, November, 18–35.

Akintoye, A., Beck, M. and Hardcastle, C. (2003) *Public Private Partnerships – Managing Risks and Opporunities*. Blackwell Science, Oxford.

Anvuur, A. and Kumaraswamy, M.M. (2006) Making PPPs work in developing countries: major issues and challenges. *CIB W107 Construction in Developing Economies International Symposium*, 18–20 January, Santiago, Chile, CD ROM.

Belbin, R.M. (2004) *Management Teams – Why They Succeed Or Fail*, 2nd edn. Elsevier Butterworth-Heinemann, Amsterdam, Netherlands.

Branco, F., Ferreira, J. and Branco, M. (2006) Specifications to achieve quality in B.O.T. projects. *CIB W107 Construction in Developing Economies International. Symposium*, 18–20 January, Santiago, Chile, CD ROM.

CII-HK (2004) *Partnering – A Comparative Study of Project Partnering Practices in Hong Kong*. CII-HK Report 1, September, Construction Industry Institute, Hong Kong.

CIRC (2001) *Construction Industry Review Committee Report*. HKSAR Govt., HK.

Clifton, C. and Duffield, C.F. (2006) Improved PFI/PPP service outcomes through the integration of alliance principles. *International Journal of Project Management*, 24, 573–586.

Constructing Excellence (2004) *Strategic Forum sets new targets for change*, http://www.constructingexcellence.org.uk. (Accessed 20 June 2006)

Construction 21 (1999) *Reinventing Construction*. Ministry of Manpower and Ministry of National Development, Singapore.

Eaton, D., Casensky, M., Sara, P. *et al.* (2006) UK PFI model of procurement: improvements based upon current practice in UK schools and hospitals. *CIB W92 Conference Proceedings*, Salford, November, 146–157.

Efficiency Unit (2003) *Serving the Community – By Using the Private Sector*. Government of the HKSAR.

Egan, J. (1998) *Rethinking Construction*. HMSO, London.

HKIS (2004) *The Professional Services For Public Private Partnerships*. Pamphlet/booklet of the Hong Kong Institute of Surveyors (HKIS).

Kumaraswamy, M.M. (1996) Contractor evaluation and selection: a Hong Kong perspective. *Building and Environment*, 31(3), 273–282.

Kumaraswamy, M.M. and Anvuur, A.M. (2008) Selecting sustainable teams for PPP projects. *Building and Environment*, 43(6), 999–1009.

Kumaraswamy, M.M. and Morris, D. (2002) BOT-type procurement in Asian megaprojects. *Journal of Construction Engineering and Management*, 128(2), 93–102.

Kumaraswamy, M.M. and Rahman, M.M. (2006) Applying teamworking models to projects. In: Pryke S., and Smyth H. (eds.) *The Management of Complex Projects – a Relationship Approach*. Blackwell Publishing, Oxford, pp. 164–186.

Kumaraswamy, M.M., Ling, F.Y.Y., Rahman, M.M. and Phng, S.T. (2005) Constructing relationally integrated teams. *Journal of Construction Engineering and Management*, 131(10), 1076–1086.

Latham (1994) *Constructing the Team*. HMSO, London.

Lo, C.M. (2006) *Critical success factors for public private partnerships at West Kowloon Cultural District Project*. Unpublished Final Year Project Report, Dept. of Civil Engineering, The University of Hong Kong.

Macneil, I.R. (1974) The many futures of contracts. *Southern California Law Review*, 47(3), 691–816.

Ministry of Finance (2007) Public Private Partnership. http://www.mof.gov.sg/policies/ppp.html. (Accessed on 19 January 2007)

Pantouvakis, J.P. and Vandoros, N. (2006) A critical review of published research on PFI/PPPs in construction. *CIB W92 Conference Proceedings*, Salford, November, 410–419.

Partnerships UK (2006) *Operational PFI Projects*. Report, March 2006, 127.

Phua, F.T.T. and Rowlinson, S. (2004) How important is co-operation to construction project success? A grounded empirical quantification. *Engineering, Construction and Architectural Management*, 11(1), 45–54.

Rahman, M.M., and Kumaraswamy, M.M. (2002) Joint risk management through transactionally efficient relational contracting. *Construction Management and Economics*, 20(1), 45–54.

Rahman, M.M., Kumaraswamy, M.M., Karim, K. *et al.* (2005) Cross-country perspectives on integrating Construction Project Teams. 6th *Construction Specialty Conference of the CSCE*, 2-4 June, Toronto, Canada CD ROM.

Ugwu, O.O., Kumaraswamy M.M., Wong A. and Ng, S.T. (2006) Sustainability appraisal in infrastructure projects (SUSAIP): part 1. Development of indicators and computational methods. *Automation in Construction*, 15(2), 239–251.

Zhang X.Q., Kumaraswamy, M.M., Palaneeswaran, E. and Zheng, W. (2002) Concessionaire selection of BOT tunnel projects in Hong Kong. *ASCE Journal of Construction Engineering and Management*, 128(2), 155–163.

Part Two
PPP Finance

9

PPP Infrastructure Investments: Critical Aspects and Prospects

Demos C. Angelides and Yiannis Xenidis

9.1 Introduction

In the 1980s and 1990s, the PPP infrastructure development scheme had been widely used for upgrading the level of existing infrastructure or to develop new infrastructure in both developing and developed countries. The motivation to adopt PPP instead of other traditional techniques had mainly two objectives:

- To confront the shortage of public funds required for ambitious, yet necessary, public infrastructure programmes.
- To introduce higher standards for the infrastructure and for effective management approaches.

The substantial contribution of the private sector was vital to pursue both objectives; therefore PPP was developed and used as the project delivery scheme that could serve the interests of both sectors, public and private. The result was the emergence of a new market (i.e. PPP construction projects) that could be described as the interface between the construction industry and the finance and insurance markets. PPP came into fashion and this resulted in the rapid development of infrastructure. In 1990–1997 the increase of the annual infrastructure investments at the international level was up to 30%, while in almost the same period (1990–2001) the infrastructure investments of the private sector concerned about 2500 projects and US$755bn distributed in 132 countries around the world as shown in Figure 9.1 (Harris, 2003).

Despite these large numbers, a downward trend of private sector investments in infrastructure became visible in the late 1990s. Figure 9.2 presents this trend, which, according to the World Bank's Private Participation in Infrastructure Project Database (2008), was not uniform, neither with regard to the various sectors (see Figure 9.3), nor with regard to geographical regions (Harris, 2003).

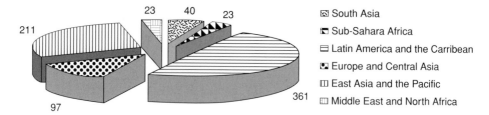

Figure 9.1   Investment (in 2001 US$ billion) in infrastructure projects with private participation, 1990–2001 (adjusted from Harris, 2003).

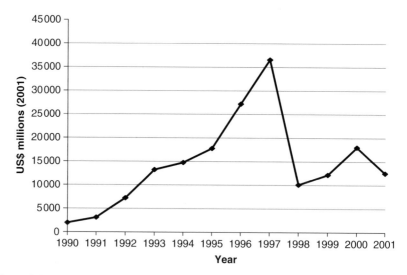

Figure 9.2   Investment in infrastructure projects with private participation in low and middle income countries, 1990–2001.

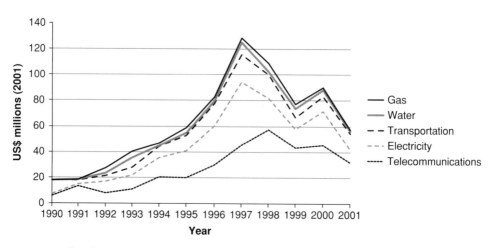

Figure 9.3   Private investment in infrastructure by sector, 1990–2001.

The onset of the decline of the private sector's investments in infrastructure was, undoubtedly, the Asian currency crisis in 1997, which had a multiple impact on the global economy (Asian Development Bank, 2000):

- Economic growth forecasts were dramatically revised, thus leading to an analogous revision of infrastructure investment programmes.
- Partially completed privately financed public infrastructure projects suffered from deterioration and stagnation.
- The optimism for profits, which was the driving force for the vast majority of PPP, was replaced by the realisation of the significant exposure of government and private sector investors to great financial risks.

The crisis resulted, reasonably, in a halting of large investments and, what was more important, to a re-evaluation of the PPP project delivery scheme. Constructive criticism took place next to the excessive optimism that was prevailing in the 1990s concerning the benefits of private investments in public infrastructure. This resulted in a fair judgement of PPP as a tool for infrastructure development and in the identification of critical success factors for PPP. The scope of this chapter is to summarise the main issues of concern with regard to the financial issues and risks in PPP and explore the prospects for the future of this particular project delivery scheme.

9.2   Critical Issues in Financing PPP Projects

Initially, the private sector had been involved in the financing of infrastructure projects in two ways: as a project sponsor and as an institutional bond investor. The fact that there were considerable risks for both types of investment led to funding partnerships between the private and the public sector of several types: bonds and bank loan transactions between commercial and government-owned institutions; bonds issued directly by governments; government-owned enterprises and private companies contracted by government authorities to provide a public service etc. (Asian Development Bank, 2000). Such partnerships require a stable fiscal and political environment, a well-structured sector (e.g. transportation, energy, telecommunications etc.) in terms of the related market's operation, and a strong legal and juridical system. Lamech and Saeed (2003) presented a ranking of the priorities of investors (requirements) for engagement with PPP in the power sector. This ranking is presented in Figure 9.4 and it is generally representative for PPP in all sectors.

Figure 9.4 includes the critical points of a so-called 'healthy investment environment', which obviously was not met in many cases of PPP in the recent past. The consequence of not meeting these requirements was either not to attract investors at all or establish PPP where the public sector assumed a great part of the risks that, normally, should be allocated to the private sector. Even in these cases, a number of serious problems occurred that seriously endangered or cancelled several PPPs throughout the world. Schur *et al.* (2006) refer to re-negotiation of an estimated 40% of the contracts for infrastructure projects and around 160 cancelled or distressed projects over

Part Two

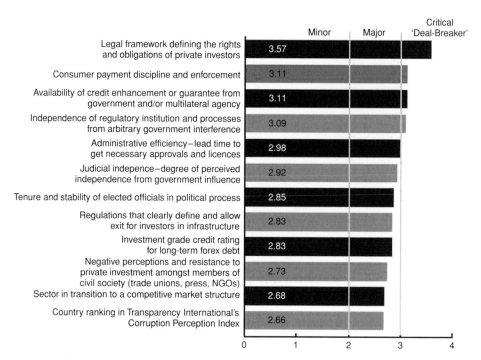

Figure 9.4   Investors' priority issues for engagement with PPP (adjusted from Lamech and Saeed, 2003).

the period 1990–2004. A great number of these problems were related to financing. In the following sections (9.2.1–9.2.4) the most important issues of this type of problem – as determined from several real cases – and lessons learned from them are highlighted.

9.2.1   Lack of strong domestic capital markets

PPPs have a long lifecycle that on average exceeds a period of 25 years; therefore capital markets are essential for a sustainable supply of funds in the phases of operation and maintenance of the infrastructure facilities. Domestic capital markets have the advantage of a better influence on the development of a PPP project in terms of the project achieving targets, ensuring profits and being supported and operated properly. Fewer risks (e.g. currency convertibility, multi-lateral guarantees) are involved in raising funds from domestic capital markets than foreign ones. Therefore, domestic capital should be the first option for infrastructure funding; however, foreign capital should not be excluded, as it is also of major importance, especially because of the transactions between foreign partners that often exceed the financing capability of the local debt market.

9.2.2   Limited raising of institutional funds

Institutional funds generated by pension funds and insurance companies constitute a large pool of funds for both developing and developed countries. This type of funds, which is currently available and will also be in the future, can ensure viability of infrastructure projects, since they can guarantee correspondence with the projects' long-term debt requirements. The main hindrances to use of this type of funds are: the low level of maturity of institutional debt markets even in many developed countries and the regulatory restrictions and risk-averse policies engaged in investments of institutional funds. The social role and primary scope of such funds result in a conservative exploitation of them and use in investments of low-risk projects with a previous record of successful rate of returns. In this way, the promising contribution of institutional funds to infrastructure development remains unexplored and unfairly limited.

9.2.3   Non-dependable project revenue streams

A stable and dependable project revenue stream is essential for: payment of the debt service; operation and maintenance costs; and generation of profits. Although such a stream is considered as being guaranteed, in many cases this is not true. This is because the revenue from infrastructure projects consists mainly of user fees where the user is usually the public or a state or local organisation, institution or administrative unit (municipality, prefecture, ministry etc.). In all cases, payments of these fees are not regular for several reasons (Streeter *et al.*, 2004):

- Poor organisation of local governments or enterprises in collecting revenue from end-users
- Unstable transfers from the central government to the local authorities (a main income of local governments in many cases)
- Weak public acceptance for user fees

9.2.4   Improper assessment of the value of government guarantees

The participants from the private sector seek the greatest possible insurance against losses during the project's operation. Therefore, they do not depend solely on the project's revenue stream but also on guarantees provided, almost always by the host government. The main types of these guarantees are as follows (Fishbein and Babbar, 1996):

- Equity guarantees, which allow the investor to be bought out by the government with a guaranteed minimum return on equity.
- Debt guarantees, which ensure loans repayment either in all cases or just in cases of cash-flow deficiencies. As with an equity guarantee, a debt guarantee entails no public cost as long as the project generates sufficient cash flow to service debt.
- Exchange rate guarantees, which provide the investor with compensation for increases in the local cost of debt service due to exchange rate movements.
- Minimum revenue guarantees, which provide the investor with compensation in cash if revenue falls below a specified minimum level (e.g. 10–30%).

Part Two

Typically, the minimum revenue threshold is set below the expected level in order to reduce government exposure, while providing sufficient coverage to support the debt component of the capital structure.

- Grants and subordinated loans, which are furnished by the government to strengthen project economics while reducing exposure to risks from the other types of guarantees. Grants and subordinated loans are not dependent on the project's performance. The subordinated loans can be repaid after debt service on senior loans and before returns to equity.
- Concession extensions and revenue enhancements, which do not provide capital but time flexibility to the project to allow the investors to recover revenues that have not been collected for several reasons.

The appropriate type of guarantees for a certain project is a case-based decision both for the government and the investor. For example, the government should not provide exchange rate guarantees to avoid the related risk in cases where the investor can raise funds from a domestic capital market. Equity and debt guarantees are extremely helpful for investors since they ensure minimum return on equity and repayment of loans, but at the same time they may significantly reduce the investor's performance incentives (Fishbein and Babbar, 1996). Figure 9.5 illustrates the range of guarantees usually offered from governments to the investors. The order of the several types of guarantees in this range is determined by the significance for the investor (i.e. the impact on the investor's ability to raise financing) and the significance for the government (i.e. the government's exposure to financial risks).

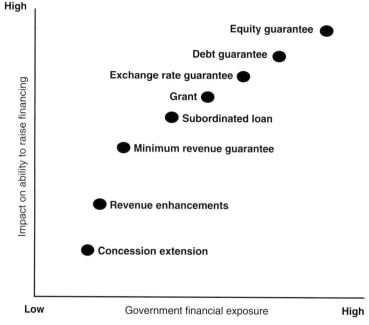

Figure 9.5   Range of guarantees in PPP with regard to significance for investors and governments (adjusted from Fishbein and Babbar, 1996).

Investors should evaluate the guarantees offered with regard to:

- The time and process of their use. Guarantees are brought into force either: proactively or retroactively; in an automatic way or subject to appropriation as part of the guarantor's budgetary process; or following a time-framed or open-ended process of review and evaluation of the guarantee utilisation request (Streeter *et al.*, 2004). A proactive and automatic utilisation after a time-framed review process is the best alternative, because it prevents default from the payment from occurring and ensures the project's full and continuous operation. A retroactive utilisation, subject to appropriation after an open-ended review process, is the worst alternative because the payment default occurs and the deposit in the debt-account on behalf of the guarantor follows a time-consuming process and according to the provisions of planning which is not related to the project (guarantor's budgetary process). This alternative may result in significant delays and jeopardise the full project's operation.
- The guarantee's priority of payment with respect to other government obligations.

Based on the results of such an evaluation, the value of the guarantees can be determined and this, consequently, leads to agreement or re-negotiations before the guarantees are included in the financing plan.

9.3   Prospects for PPP Infrastructure Development

There is strong evidence that the decline of private participation in infrastructure investment cannot be permanent. The basic elements that triggered the boom of PPP in the 1990s still exist:

- Provision of infrastructure remains the main consideration for governments both in developing and developed countries, since it is acknowledged as the most significant element of human and economic development. Especially for developing countries, Harris (2003) refers to estimates of population lacking basic infrastructure needs:
  - □ Around 2 billion lack an electricity connection
  - □ Around 1.1 billion lack a safe water supply
  - □ Around 2.4 billion do not have access to improved sanitation
- Drawn from the Global Development Finance (The World Bank, 2004), Table 9.1 presents a comparison of multi-sectoral infrastructure stocks and service access between developing and developed countries, which explicitly shows the high demands for infrastructure of the developing countries.
- World economy is coming out of a long recession coupled with significant availability and liquidity of private funds. According to Harris (2003) the average annual investment in developing country private infrastructure projects since 1990 has been around US$60bn per year. Annual global savings were around $7 trillion in 1999. A simple comparison proves that resources for investment are available.

Part Two

Table 9.1 Comparison of infrastructure stock between developing and developed countries (adjusted from The World Bank (2004))

	Installed capacity per 1.000 (kW)	Electricity consumption per capita (kWh)	Average telephone mainlines per 1000 persons	Road density (km/sq. km of land)	Access to improved water source (% of population)
Year	2001	2001	2001	2000	2000
Developing countries	272	1054	95	0.15	78
Developed countries	2044	8876	501	0.58	99

- The governments of the developing countries still require substantial funds from external financial resources to meet perceived infrastructure needs. Gomez-Ibanez *et al.* (2004) refer to the World Bank's estimates that the developing countries need to double their rate of current investment in water and sanitation infrastructure. Noel and Brzeski (2005) present an estimate of the required investments for the new country members of the European Union to comply with the Integrated Pollution Prevention Control (IPPC) Directive (see Table 9.2). Considering the state of the economy of the listed countries and the need for implementation of the Directive, it is certain that private sector investors' involvement is essential to meet the goals set by the Directive.
- Even if international funding institutions and organisations expand their lending for government infrastructure projects, there is still substantial private funding required to meet infrastructure needs (Gomez-Ibanez *et al.*, 2004).

Judging from the above, the issue is not if but when PPP will be brought to the limelight again. In fact, as Figures 9.6 and 9.7 depict, a shift towards increased private funds in infrastructure may have already began. According to the data from the World Bank's Private Participation in Infrastructure Projects Database (2008), globally the total investment in greenfield projects has continuously risen since 2003, with significant increase of invested funds

Table 9.2 Investment needed to comply with the IPPC Directive for selected EU countries (adjusted from Noel and Brzeski, 2005)

Country	Investments needed in € millions
Bulgaria	3261
Czech Republic	3725
Estonia	489
Hungary	1761
Lithuania	44
Poland	6927
Romania	806
Slovak Republic	1596
Total	18 609

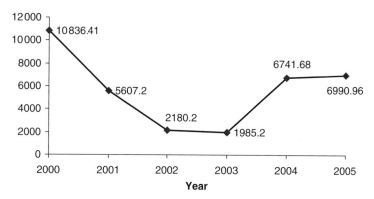

Figure 9.6 Worldwide total investments in greenfield projects, 2000–2005.

from 2003–2005 (i.e. from US$1985.2m to US$6990.96m respectively). A similar trend with increased amounts of invested funds is also observed for concessions from 2004 onwards.

The emphasis should be given to the identification of the specific issues that require particular consideration to properly implement PPPs and avoid the numerous problems and risks that led to their decline in the past. In the next sections, based on lessons learned, the key considerations with regard to financial features of future PPPs are presented.

9.3.1   Integration of alternative funding sources

A clear conclusion drawn from Section 9.2 is the necessity for a shift towards new models of financing that will exploit the potentials of domestic capital markets (DCMs). This can be achieved first by strengthening of DCMs and second by integration with other funding resources. A very beneficial direct consequence of strong DCMs on infrastructure development would be, for example, the issuing of bonds with longer maturities than the existing ones. Currently, even robust DCMs like those of Mexico or South Korea issue

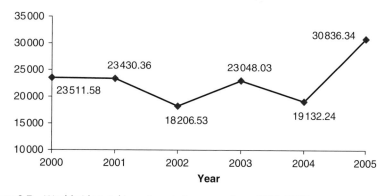

Figure 9.7 Worldwide total investments in concessions, 2000–2005.

(a) Talca-Chillan, Route 5   (b) Tramo Santiago –   (c) Santiago – Valparaiso –
                                         Los Vilos, Route 5   Vina del mar, Route 68

Figure 9.8   Overview of toll roads developed as PPP in Chile (Adjusted from: Ministry of Public Works, Chile, www.concesiones.cl)

bonds of short- or medium-term maturities that lead to the requirement of a refund after 3–7 years; this could address the risk that revenue growth and financial margins may not be able to accommodate interest rate volatility (Harris, 2003; Streeter *et al.*, 2004). Overcoming these risks requires more decisive steps to allow sustainability of infrastructure funding.

Funding from DCMs can be assisted by currency loans and hedging products offered by international financial institutions (e.g. the International Finance Corporation). The International Bank for Reconstruction and Development and the Inter-American Development Bank offer currency loans to governments for investments through PPPs. Funding from international financial institutions coupled with alternative sources such as institutional investors (e.g. insurance companies, investment funds etc.) and private pension funds provides a new pool of funding resources that is characterised by: high liquidity levels; lower levels of risks; and high levels of sustainability of infrastructure funding. A good example of funding infrastructure through integration of new resources of funds is seen in Chile, which was the first Latin American country that allowed investments of pension funds in infrastructure projects in the transportation sector (Fay and Morrison, 2005). Figure 9.8 provides an overview of some toll roads developed with pension funds. To allow these investments, the infrastructure bond, a new financial instrument that offers long-term maturities (20 years) was introduced for the first time in 1998 for the Talca-Chillan, Route 5 PPP project. Until 2003 infrastructure bonds issued in Chile averaged about US$1bn a year, an amount that was more than half the country's total issuance (Fay and Morrison, 2005).

New infrastructure funding instruments have also been designed and used in developed countries as well. In the US, for example, the state revolving funds (SRFs) model, which was used to create many wastewater and water supply projects allowed for capitalisation grants to be set aside in a debt-service reserve fund and invested in collateralised guaranteed investment contracts (GICs) with highly rated financial institutions (Streeter *et al.*, 2004). SRFs can also be used to make direct loans, the repayment of which can be pledged against future leveraging or issue bonds (Streeter *et al.*, 2004). Another instrument to increase the funding resources is pooling credit risk. As described in Streeter *et al.* (2004), a promising ultimate recovery value of a loan portfolio can lead to the capitalisation of the fund by the country that processes multilateral bank grants, with reserves against the expected cash flow deficiencies within the loan portfolio. Then, interest income from the collateral can be used to reduce the borrowing costs of the entities within the infrastructure pool. A single debt emission by the bank on behalf of the pool participants will also create liquidity within the domestic debt market on the theory that the market has more appetite for the larger debt issuance of the bank than for the smaller individual project loans of the bank's participants. Liquidity in the capital markets also lowers borrowing costs for the participants. This cheaper access to pooled capital greatly increases the resources available to meet local infrastructure needs.

The above examples prove the demand for integration of funding resources and further development of new investment tools for promoting PPP infrastructure projects.

9.3.2   Involvement of local investors

As mentioned in Section 9.2, the decline of PPP was accompanied by the withdrawal of international firms from almost 160 cancelled or distressed projects from 1990–2004. A trend for re-involvement of these firms to new PPPs has not yet been established and it is highly unlikely that, even if re-involvement happened, it would have the characteristics of the period before 1997. The new era in infrastructure development should be based on greater participation of local investors in PPP. Experience of local firms earned from their participation in PPP in the past, often as minority partners with developed country investors, has increased their expertise in infrastructure investments; in many cases, this expertise rendered these local firms capable of undertaking new projects as primary contractors. Schur (2005) presents several results from studies focused on specific sectors (e.g. water and electricity) that prove a trend for larger participation of local investors in medium- and large-scale infrastructure projects. Ettinger *et al.* (2005) have compared the periods of 1998–2000 and 2001–2003 and found that the share of main project sponsors from developed countries fell from 57% to 50% from the first to the second period. The slack was taken up both by local investors (whose share rose from 33% to 36%) and foreign investors from developing countries (from 7% to 12%).

The advantages that local firms present against foreign ones in the field of the infrastructure market are related, but not confined, to the critical issues identified in Section 9.2. These advantages include:

- The impact of the economic environment on local investors. Investors in developing countries are largely dependent on the socio-economic structures and mechanisms of the host country. Furthermore, they are mostly benefited by the development of modern infrastructure, which can effectively serve their business activities. Both of these issues provide some sort of guarantees with regard to the level of motivation, commitment and compliance to quality standards for the development of infrastructure projects.

- The interdependency between the investors and the domestic capital markets. Both, investors and markets in emerging economies are, in fact, the

Part Two

two sides of the same coin. Broadening and deepening of capital markets in developing countries results in greater capacity of local investors to raise funds, which in turn are invested again to feed the stream of economic development. Therefore, local investors involved in infrastructure development significantly support strong capital markets, which are requirements of successful PPPs.

- Better understanding and, therefore, dealing with the political economy issues in their own country (Ettinger *et al.*, 2005).

Judging from the above, the potential role of local investors can be critical in future infrastructure PPPs. There are of course implications, which are not so clear for the time being. For example, although assumed, it is not yet proven that local investors are well equipped to deal with the political economy issues raised by private participation in infrastructure (Schur *et al.*, 2006). Another issue is that if local capital markets are undeveloped, the advantages of better access to local currency equity capital and debt are relatively modest (Schur *et al.*, 2006). In the long run, however, mobilisation of local investors can make the positive difference in future PPPs.

9.3.3   Use of new instruments for mitigating risks

The mobilisation of domestic funds and local investors as well as the improvement of the effectiveness of well-known tools for financing risk mitigation (e.g. government guarantees) will create a new framework for mitigating risks in future PPPs. In this framework, new instruments will be required and financial institutions, banks and governments will have the responsibility of developing them. There is already a new generation of risk mitigation instruments that requires further development and testing of appropriateness. Some of these instruments are:

- The Liquidity Facility developed by Overseas Private Investment Corporation in 2001 to protect bondholders against foreign exchange risks (Fay and Morrison, 2005). This one-of-a-kind instrument is based on a revolving credit line that becomes available to the project company, if a depreciation of the local currency makes the issuer unable to meet its debt service obligations. The instrument was very well received by rating agencies and institutional investors, despite the inability to deal with extreme foreign exchange shocks. However, the provision of liquidity to the project company permits repayment of loans, even in cases where tariffs are not timely adjusted to a new exchange rate for reasons such as avoidance of social agitation.
- Partial risk guarantees (PRGs) provided from multi-lateral institutions such as the International Finance Corporation and the Inter-American Development Bank. This instrument has been used to support issuance of long-term local currency bonds in Latin America (Fay and Morrison, 2005). PRGs can protect lenders or bond holders against perceived risks, providing the credit enhancement that project companies require to raise adequate financing. The credit enhancement significantly reduces the cost

of debt issued to finance infrastructure projects; also it provides higher credit ratings that can open up local capital markets and a broader range of financial investors to infrastructure project companies (Fay and Morrison, 2005). PRGs can be provided by facilities especially established for this purpose by the government, maximising the attractiveness of local infrastructure market to private investors by protecting private project debt from political and regulatory risk (Fay and Morrison, 2005).

- Debt covenants, which in accordance to the governing legal and institutional framework for the PPP provide contractual protection and protection of the creditors' interests (The World Bank, 2004). Covenant provisions typically take the form of (The World Bank, 2004): restrictions on dividends, mergers and acquisition transactions; asset disposals; limitations on indebtedness; requirements of third-party guarantees; maintenance of good regulatory standing etc. Covenants can protect the safety and seniority of debt holders' claims, ensure repayment of principal, and provide legal remedies in the event of default.
- Self-hedging of infrastructure projects to overcome difficulties of underdeveloped DCM and low tariffs. Self-hedging, according to Harris (2003), can be achieved by focusing on smaller projects that stand a better chance of being financed through local capital markets or by using less debt, which would provide more of a cushion in difficult times.

9.3.4   Effective design of tolls and tariffs

The unique real income from PPP infrastructure projects is the payment of tolls and tariffs by the entities that use the infrastructure, i.e. the public or public services. Therefore, the most critical issue of a financial plan is the effective design of the mechanism for adjusting tolls and tariffs during the lifecycle of the infrastructure project.

Lessons learned from the PPPs launched in the 1990s highlight several issues to be considered for appropriate designing of a project's tolls and tariffs. These are:

- Tariffs for infrastructure services that have been provided by the government are always significantly lower than those required to achieve sustainability of financing to the project (Gomez-Ibanez *et al.*, 2004). While this is done for obvious social reasons, in terms of financing, it results in loss-making public-owned infrastructure. Therefore, when the private sector is involved with one of the variations of the PPP scheme (i.e. from management to divestiture), there are inevitably substantial tariffs to ensure, at least, an income for cost recovery of the provided services. To avoid public dissatisfaction and mistrust, which often results in loss of government support due to the threat of political cost, the tolls and tariffs need to be very well justified and relatively close to those existent prior to the PPP. Another helpful instrument can be well-designed social tariffs for the poor consumers like those used in Latin America (Fay and Morrison, 2005).

Part Two

- Acceptability of tariffs depends on the affordability for consumers of the public services offered through PPP infrastructure projects. The fact that in many cases the infrastructure serves basic living needs (e.g. water supply) is not a reason for excessively high tariffs. Even if the public is temporarily forced to pay for using the infrastructure, in the long run the dissatisfaction and non-affordability may cause serious flaws in the development of the investment (e.g. high cost of essential services and tariff hikes effectively reduce real income, drawing more people closer to or under the poverty line, thus exacerbating existing poverty). Well-designed tolls and tariffs are those which are consistent with the real income of the users and the state of the host county's economy.

- Another important issue is the provision for revisions of tariffs on a regular basis. According to Guasch (2004), 'tariff revisions should normally occur at 5-year intervals and must follow a formula that applies to the average tariff that is billed by the concessionaire, but the circumstances under which an extraordinary tariff revision is permitted should be narrowly defined'. Tariff revision does not necessarily mean an increase of the tariffs each time these are revised; the provision, in fact, allows for a steady and sound process of tariff adjustment that the investors may rely on in case the risk levels to which they are exposed change unfavourably to them. Although, the whole concept is not quite exceptional, according to Estache and Serebrisky (2004), only a very few countries in the world (Australia, UK, Mexico) have provisions to support this kind of process. Therefore, a larger implementation of this tool should be introduced into the PPP model to foster investors to commit for the long run.

The central element of the design of tolls and tariffs is ensuring full recovery of the cost for supplied services. This is a key issue for achieving financial viability and sustainability of the utilities; the respective cost for using the infrastructure must, primarily, comply with this need, leaving behind excessive profit chase.

9.4   Conclusion

Demand for infrastructure remains at the top of the list for both developing and developed countries. Global demographics, public health and safety needs, as well as economic development goals, translate into infrastructure requirements far in excess of currently available financing resources. Therefore, the private sector retains a vital role in financing the required infrastructure along with a substantial contribution from the public sector. Recent experience has shown progress as well as deficiencies in the development of infrastructure through PPPs. These deficiencies coupled with some major economic events resulted in a temporary decline of PPP as a project delivery scheme. The considerable experience and the lessons learned, however, can help to identify critical issues for successful PPPs and the required steps towards an enhancement of this system for infrastructure development.

Several issues summarised below were identified in this chapter as critical with regard to financing:

- Lack of strong domestic capital markets
- Limited raising of institutional funds
- Non-dependable project revenue streams
- Improper assessment of the value of government guarantees

The deficiencies associated with these issues intensified financial risks and failed to provide the required safety to private investors. These deficiencies can be addressed with new mechanisms identifiable in association with aspects of infrastructure financing and in anticipation of future PPPs for infrastructure development. Such mechanisms are: integration of alternative funding sources; involvement of local investors; use of new instruments for mitigating risks; and effective design of tolls and tariffs. These mechanisms and the associated instruments, which are highlighted in this chapter, are very promising and able to support a new generation of PPPs for infrastructure development.

References

Asian Development Bank (2000) *Developing Best Practices for Promoting Private Sector Investment in Infrastructure – Power*. Asian Development Bank, Manila, Philippines.

Estache, A. and Serebrisky, T. (2004) *Where Do We Stand on Transport Infrastructure Deregulation and Public-Private Partnership?* World Bank Policy Research Series, Working Paper 3356.

Ettinger S., Schur, M., von Klaudy, S. *et al.* (2005) *Developing Country Investors and Operators in Infrastructure*, Trends and Policy Options Series, No. 3. Public–Private Infrastructure Advisory Facility (PPIAF), Waschington DC.

Fay, M. and Morrison, M. (2005) *Infrastructure in Latin America and the Caribbean: Recent Developments and Key Challenges*, Vol. 1 (Main Report), Report No. 32640-LCR. Finance, Private Sector and Infrastructure Unit, Latin America and the Caribbean Region, The World Bank, Washington DC.

Fishbein, G. and Babbar, S. (1996) *Private Financing of Toll Roads*, RMC Discussion Paper Series No. 117, Report No. 16437. World Bank, Washington DC.

Gomez-Ibanez, J.A., Lorrain, D. and Osius, M. (2004) *The Future of Private Infrastructure*, Working Paper. Taubman Center for State and Local Government, Kennedy School of Government, Harvard University, Cambridge, MA.

Guasch, J.L. (2004) *Granting and Renegotiating Infrastructure Concessions: Doing it Right*. The International Bank for Reconstruction and Development/The World Bank, World Bank, Washington DC.

Harris, C. (2003) *Private Participation in Infrastructure in Developing Countries. Trends, Impacts and Policy Lessons*, Working Paper No. 5. The International Bank for Reconstruction and Development/The World Bank, World Bank, Washington DC.

Lamech, R., and Saeed, K. (2003) *What International Investors Look for When Investing in Developing Countries: Results from a Survey of International Investors in the Power Sector*, Energy and Mining Sector Board Discussion Paper No. 6. World Bank, Washington DC.

Part Two

Ministry of Public Works (Ministerio de Obras Publicas) (undated) *Concession Projects*, http://www.concesiones.cl/index.php?option = com_content&task = blogcategory&id = 0&Itemid = 105

Noel, M. and Brzeski, W.J. (2005) *Mobilizing Private Finance for Local Infrastructure in Europe and Central Asia – An Alternative Public Private Partnership Framework*, Working Paper No. 46. The International Bank for Reconstruction and Development/The World Bank, World Bank, Washington DC.

Private Participation in Infrastructure Projects Database (2008). The World Bank Group, Public-Private Infrastructure Advisory Facility (PPIAF): http://ppi.worldbank.org/ index.aspx. (Accessed July 2008)

Schur, M. (2005) *Developing Country Investors and Operators in Infrastructure Projects: Prevalence, Emerging Trends and Possible Policy Implications for the African Region*. Public-Private Infrastructure Advisory Facility (PPIAF), Washington DC.

Schur, M., von Klaudy, S. and Dellacha, G. (2006) *The Role of Developing Country Firms in Infrastructure*, Gridlines No. 3. Public-Private Infrastructure Advisory Facility (PPIAF), Waschington DC.

Streeter, W., Zurita, G.R., Dell, J.C. et al. (2004) *Public-Private Partnerships: The Next Generation of Infrastructure Finance*, Special Report. Fitch Ratings, Fitch, Inc. and Fitch Ratings, Ltd., New York.

The World Bank (2004) *Global Development Finance – Harnessing Cyclical Gains for Development – I: Analysis and Summary Tables*. The International Bank for Reconstruction and Development/The World Bank, World Bank, Washington DC.

Wood, W.G. (2002) Private Public Partnerships in Toll Roads in PRC. In: *Proceedings of ADB/PPIAF Conference on Infrastructure Development – Private Solutions for The Poor: The Asian Perspective*, 28–30 October, Manila, Philippines.

Part Two

Patterns of Financing PPP Projects

Sudong Ye

In the development of infrastructure, the term public-private partnership (PPP) describes a kind of contractual arrangement under which the public and private sectors join together to utilise the best skills and capabilities of each in order to serve the public better. Projects developed under that kind of contractual arrangement are referred to as PPP projects. In fact, the term PPP itself evolved from various practices of infrastructure development. These include various contractual arrangements such as build, transfer (BT), build, operate, transfer (BOT), build, own, operate, transfer (BOOT), build, own, operate (BOO), etc. for the development of greenfield projects, and transfer, operate, transfer (TOT), sold, modernisation, transfer, etc. for existing projects.

Usually PPP projects are financed by the private sector on a non- or limited recourse basis. According to Nevitt and Fabozzi (2000), the key to a successful project financing is to structure the financing of a project with as little recourse as possible to the sponsor, while at the same time lenders are satisfied with sufficient credit support. This chapter investigates patterns of financing for PPP projects. It begins by analysing the requirements involved in the development of PPP projects, in order to identify the key components of project financing, and then investigates possible approaches to deal with each of these components. Different approaches to deal with these components result in different patterns for financing PPP projects.

10.1.1   A framework for developing PPP projects

The development of PPP projects usually requires the private sector to be involved in almost all the phases of a project lifecycle. Take a project developed by BOT. The private partner finances, designs and builds a facility, and operates the completed facility for a specified period of time under a concession agreement to realise a reasonable return on its investment through user

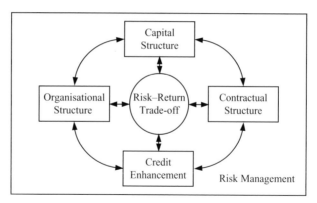

Figure 10.1   A framework for financing PPP projects.

charges. At the end of the concession period, the private partner transfers the facility back to the public sector free of charge. Thus, the private partner is usually responsible for the five key tasks required for the development of the project, namely management, financing, design, construction and operation.

Each of the five key tasks has its own requirements. The task of project management requires the sponsor to design an organisational structure to carry out the project, i.e. what form of economic entities should be established to carry out the project. The task of financing requires the sponsors to decide the capital structure of the project: how much should be provided by the sponsors in the form of equity financing and how much should be borrowed from lenders in the form of debt financing, especially on a non- or limited recourse basis. The tasks of design, construction and operation require the sponsor to design a contractual structure to complete the project on time, within budget and in conformity with sound technical performance standards, and operate the completed facility efficiently. When a majority of funds is provided by lenders on a non- or limited recourse basis, the lenders' concern with debt security requires the sponsors to provide credit enhancement measures in order to secure debts in favour of the lenders.

Decisions on these tasks are based on risk–return trade-offs. Different decisions on each task will lead to different risk allocation. Assume that participants are risk averse. The benefit received by a participant should be sufficient to compensate risk assumption by the participant. Therefore, all the decisions on dealing with the tasks should meet participants' risk–return trade-off requirements. As the five tasks are interrelated, financing a PPP project mainly involves four interdependent aspects, namely the optimisation of capital structure, the design of organisational structure, the design of contractual structure and the enhancement of creditworthiness, based on risk–return trade-offs, as shown in Figure 10.1.

There are various options for each component. Their combinations may lead to different financing structures for the development of PPP projects. The following sections investigate possible approaches to deal with each component.

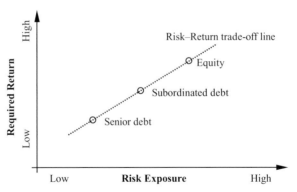

Figure 10.2   Risk–return trade-offs of financial instruments.

10.1.2   The optimisation of a capital structure

There are three general categories of funds used in financing a project: equity; subordinated debt (sometimes called mezzanine financing or quasi-equity); and senior debt. Different kinds of funds are exposed to different level of risks and therefore their providers require different returns. Equity providers assume a higher degree of risk, and therefore require a higher level of return, while lenders assume a relatively lower degree of risk, and require a lower level of return. Subordinated debt is somewhere between equity financing and debt financing. Figure 10.2 qualitatively illustrates the risk–return trade-offs of different funds.

The optimisation of capital structure mainly deals with financial sources, ratios of different types of funds, and timeframe of fund usage. There are various sources of funds for PPP projects. Usually, equity financing, including subordinated debts, is provided by sponsors, and sometimes by institutional investors, the host government and the public. Debt financing may be provided by a syndicate of commercial banks, financial institutions, export credit agencies, international agencies (e.g. the World Bank, Asian Development Bank etc.), and so on. Different sources have different requirements, especially different risk–return trade-offs. The optimal capital structure should meet the fund providers' risk and return trade-offs through appropriate financial instruments such as commercial bank loans, export credit financing, various notes or bonds.

The capital structure falls into the range from total equity financing to total debt financing. A project may be financed with equity financing or debt financing, but more often than not this will be done with a combination of both. Since investment in PPP projects is not their core business, it is not practical for sponsors to use 100% equity financing. If a project is financed by debt financing on the basis of non- or limited recourse, it is difficult to obtain 100% debt financing for the reasons of financial prudence. Increase in equity financing can enhance the financiability of the project because equity financing ties the sponsors to the project so that the sponsors would not shy away when the project is in difficulty, and provides a debt cushion so that lenders' losses would be reduced if the project company went bankrupt. The

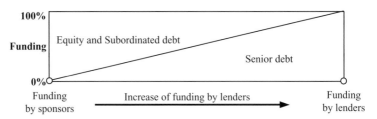

Figure 10.3   Funding for financing projects.

capital cost may be increased, however, due to higher returns required by equity providers. In general, PPP projects are usually financed using a combination of equity financing and debt financing with varying ratios of equity to debt. Usually, debt financing exceeds 70%. Sometimes debt financing reaches nearly 100%, for example, the third Dartford River Crossing of the Thames River was funded by debt financing with only £1000 (US$1700) paid in equity capital. Figure 10.3 qualitatively depicts funding sources for the development of projects.

The capital structure also includes the factor of time. Usually, a major portion of funds for a PPP project is financed with long-term project financing. Sometimes a project is first financed by construction loans with higher interest rates and then refinanced by long-term loans with lower interest rates after completion to reduce the total capital cost.

10.1.3   The design of an organisational structure

Since a PPP project is usually large in size and is exposed to various risks, a single sponsor may not be able to fund the project or is unwilling to assume all the risks alone. As a result, a PPP project is often sponsored jointly by a consortium or joint venture of interested parties. The interested parties may include construction contractors, equipment vendors, facility operators, fuel suppliers and so on. Each of them has their own core competency and is interested in works related to their core business. For example, construction contractors are interested in the construction work of the project; equipment vendors are interested in selling equipment; operators are interested in the operation of facilities. But none of them may wish to own the resulting facility because investment in PPP projects is not the core business of sponsors. Therefore, an entity is required to act as the client of the project.

There are various forms of entity which are theoretically contenders for the role of project client. In practice, the following four basic forms of entity are encountered in the development of PPP projects: (1) incorporated companies; (2) contractual joint ventures (unincorporated joint venture); (3) partnerships (general and limited partnerships); and (4) trusts. Each form of entity has its advantages and disadvantages. The incorporated company is an independent legal entity with limited liability, which can provide a high degree of insulation for a sponsor from the risks and liabilities of a project, but the sponsor cannot directly control project cashflows. The unincorporated joint venture can provide a high degree of flexibility for internal management through

Part Two

freely writing rules in joint operating agreements. However, it does not in itself provide any form of limited liability. Partnerships, both general and limited, cannot provide insulation for a sponsor from the risks and liabilities of a project, but can provide some tax benefits. A trust can be used to hold title to a project and raise funds for the project, but it is rarely used to manage a project.

The design of organisational structure is concerned with the question as to which form of entity should be used and how many entities should be used. Among the four forms of entity, the incorporated company with limited liability is the most popular form for the development of PPP projects. The unincorporated joint venture may occasionally be used to take the advantage of management flexibility. In this case, sponsors usually participate in unincorporated joint ventures through companies with limited liability, which are established especially for this purpose. For the same reason, the sponsors usually join a partnership through a specially formed limited liability company. A trustee can be an independent, nominally capitalised corporation, or a financial institution. The most popular organisational structure is a single economic entity as the project owner. Sometimes a more complicated organisational structure may be required to optimise the project development. For example, two or more economic entities may be developed: one for the management of the project, one for the financing of the project, and possibly one more for other purposes such as equipment leasing.

10.1.4   The design of a contractual structure

A project can be viewed as a bundle of rights, obligations and risks to be priced and allocated among the project participants. The design of contractual structure is about pricing and allocating these rights, obligations and risks among the participants. If it does not have the ability to design, construct and operate the project, the entity for the development of project has to contract with external participants for the construction and operation of project. There are various contractual arrangements for the construction of projects, for example traditional construction contracts, design and build or turnkey contracts etc. The most popular contractual arrangement for the construction of a PPP project is a fixed-date, lump sum turnkey contract.

The entity may have two options for the operation of project: operating the project by itself or contracting the operation out to specialised operators. The majority of transportation projects are operated by the entity itself. In contrast, the operation of most independent power projects and water treatment projects are usually contracted out to a specialised operator. Moreover, the entity may enter into other contracts such as off-take contracts, fuel supply contracts and so on.

10.1.5   The enhancement of project creditworthiness

In project financing, the residual risks with the economic entity determine the risk profile that will be assumed by lenders when the economic entity

Part Two

is a specially formed limited liability company. Thus, lenders of project financing require further measures to increase debt security, the realisation of anticipated cash flows and the control of project revenues.

Debt security is the centre of a lender's concerns. The common view of security is that lenders take security over an asset in order to sell it if their loan is in default and to apply the proceeds against amounts outstanding under the loan. In project financing, the asset of a project is used as collateral for the loans. However, the asset is not in place when the credit agreement is signed. Therefore, lenders usually require borrowers to provide a completion bond, which guarantees the completion of project facilities.

Under the arrangement of project financing, when a lender provides a loan to a borrower, it will initially look at project revenues for the repayment of debt and their reliability rather than its creditworthiness. A project cannot generate revenues unless it is completed. Therefore, the project should be completed on time, within budget and in conformity with the design specifications. When the construction of a project is contracted out to a construction contractor, project creditworthiness is dependent on the performance of contract parties. It is recognised by Strzelecki (1990) that a contractual guarantee does not provide strong support if the party to the contract does not have the resources to backstop its guarantees. Therefore, the project financing usually requires guarantees of contractor's parent companies and/or bank bonds for the performance of contract parties.

Furthermore, the completed facility should be operated to generate expected revenues efficiently. If project revenues come from offtakers under offtake contracts or from lessees under lease agreements, project creditworthiness is mainly dependent on the credit of offtakers and lessees. Any measures increasing the credit of offtakers and lessees will increase the creditworthiness of project. Therefore, lenders usually require guarantees and/or performance bonds for the performance of offtakers/lessees in order to reduce risk exposure. In addition, lenders may want to increase the security of project cashflows by using escrow account or appointing indenture trustees (or security trustees).

10.2   General Patterns of Financing PPP Projects

In the development of PPP projects, the majority of funds required for a project is obtained through project financing. The term project financing, in its narrow sense, means to finance 'a particular economic unit in which a lender is satisfied to look initially to the cash flows and earnings of that economic unit as the source of funds from which a loan will be repaid and to the assets of the economic unit as collateral for the loan' (Nevitt and Fabozzi, 2000). The definition highlights three key points. Firstly, the financing is carried out through an economic unit, that is, project financing involves an organisational decision on the creation of an economic entity to own the asset. Secondly, debt service depends upon the anticipated cash flows of that economic unit. This implies that all the relevant contracts are important to ensure the realisation of anticipated cash flows. Finally, liquidation is limited

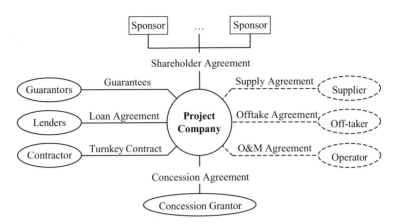

Figure 10.4   A mono-entity structure for project development.

to the assets of the economic unit, that is, debt financing is based on a non- or limited recourse basis. All the three points are related to the economic unit. Therefore, the organisational decision is the centre of the financing structure.

There are three types of organisational structure to develop PPP projects. They are: (1) a mono-entity organisational structure, (2) a dual entity organisational structure, and (3) a multi-entity organisational structure. The three organisational structures will lead to three patterns of financing PPP projects: financing structure based on a single entity, financing structure based on two entities and financing structure based on multi-entities.

10.2.1   A financing structure based on a single entity

In a mono-entity structure, a single economic entity is established for the sole purpose of developing a project, which acts as the project client for both financing and managing the project. Theoretically an economic entity can be any form of entity mentioned above (i.e. incorporated companies, contractual joint ventures, partnerships and trusts). Usually, the economic entity is an incorporated company, a legally independent entity. Sometimes, other forms of entity may be used. For example, in Shajiao B power project in China, a joint venture was established between Shenzhen Energy Corporation and Hopewell Power (China) Ltd for the development of the power project.

In this organisational structure, project sponsors establish an economic entity by which various contracts may be entered into with different participants for the financing, design, construction and operation of the project. For example, it may enter into a loan agreement with lenders; an engineering contract with a designer and a construction contract with a construction contractor, or an engineering, procurement and construction (EPC) contract with a construction consortium; an operation and maintenance agreement with an operator; a supply agreement with suppliers; and possibly an offtake agreement with offtakers or a usage/lease agreement with users. The typical mono-entity structure for project development is shown in Figure 10.4.

If the production process is simple, the economic unit can be capable of operating the facility by itself. The majority of transportation projects (e.g. toll roads, tunnels, bridges etc.) are developed by owner–operator companies, specially established by project sponsors. For example, the sponsors of Malaysia South–North Highway project established a project company for the development of the highway and the operation of the completed highway. Another example is the Channel Tunnel project. Eurotunnel (the project company) was established as owner–operator company, which entered into a series of contracts with various contractors, suppliers etc., but operated the tunnel by itself.

If the production process is complex, the economic unit may not be able to operate the facility by itself, and it has to employ a specialised operator. In this case, the project company plays a role of owner, an owner company. The operation of most independent power projects and water treatment projects are usually contracted out to a specialised operator. Take Laibin B power project in China for example. The project company entered into an operation and maintenance agreement with a joint venture of Electricite de France International, Guangxi Power Industry Bureau and Guangxi Development and Investment Co. Ltd.

Case Study 10.1   The Financing Structure of Paiton Power Project in Indonesia

The Paiton power project, the first and the largest independent power project in Indonesia, is a coal-fired power plant with an installed capacity of 2×615 MW. Its total capital investment is over US$2.5bn, which was funded by senior debt funds (72.8%) and equity and subordinated loans (27.2%). The project was developed on a BOO basis over 30 years.

Paiton power project was initiated by the Indonesian government in 1991, but sponsored by a consortium led by Edison Mission Energy (EME), a US public utility holding company owning many power-generating facilities throughout the world. The consortium consists of EME, Mitsui & Co Ltd (Mitsui), P.T. Batu Hitam Perkasa (BHP) and General Electric Capital Corporation (GECC). The project sponsors provided about 27% of the base project costs in the form of equity and subordinated debt, and agreed to provide $300m (12% of the base project costs) to cover the funding for contingencies and cost overruns.

After winning the concession, the sponsors established Paiton Energy Co. (the project company) on 11 February 1994. Then, the project company entered into a power purchase agreement (PPA) with Perusahaan Listrik Negara (PLN). Since the Indonesian government did not enter into implementation or concession agreements to contractually determine the basic rights and obligations of the project company, the PPA is the key agreement that provided that PLN would 'take-or-pay' for a certain quantity of electricity at an agreed tariff. So, the Indonesian government was requested to issue the support letter in which the government agreed to cause PLN to discharge its payment obligations under the PPA which are due and payable and unsatisfied by PLN.

After the PPA was in place, the project company entered into a fixed-price, turnkey construction contract with the consortium consisting of Mitsui & Co Ltd (Mitsui), Duke and Fluor Daniel International Services and Toyo Engineering Corporation on 30 March 1995; an operations and maintenance (O&M) agreement with PT Mission Operations and Maintenance Indonesia (MOMI) on 31 March 1995; and a fuel supply agreement with BHP on 14 April 1995. The project company also entered into the Common Agreement, the Trust Agreement, the Inter-creditor Agreement, the Collateral Agency Agreement, the Equity Support Agreements, the JEXIM (the Export-Import Bank of Japan) Facility Credit Agreement, the USEXIM (the Export-Import Bank of the United

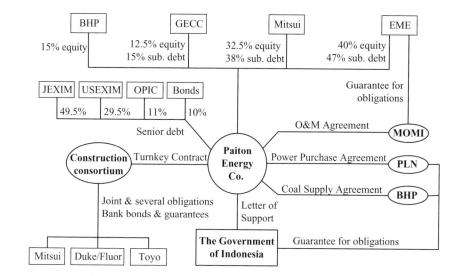

Figure 10.5   The financing structure of the Paiton Power Project.

States) Agreements, and the OPIC (the Overseas Private Investment Corporation) Facility Credit Arrangements to finance the project on the basis of limited recourse. Moreover, the parent companies of the consortium were asked to provide bank bonds and guarantees to assume joint and several obligations for the completion of project; the USEXIM was asked to provide guarantee for credit facilities; and EME provided guarantee for the obligations of the operator. Figure 10.5 shows the financing structure of the Paiton power project.

The case study shows that the mono-entity structure has many advantages. First, there is only one communication hub which provides clear relationships between participants. Second, it is easier for lenders to evaluate the creditworthiness of the project because the residual risks with the economic entity reflect the risk profile that will be assumed by lenders.

10.2.2   A financing structure based on two entities

In a dual-entity structure, two economic units are established to carry out different tasks or different parts of a project. There are various situations which require a dual-entity structure. Among them, one is that the project involves a lot of lenders/investors with different requirements, and another is that the project can be split into two parts for one or another reason.

Dual-entity structure: separating funding from construction

When a project involves a lot of lenders/investors with different requirements, the project sponsor may want to separate the task of funding from the other task. In this case, two economic units are established: one for financing (e.g. a trust borrowing vehicle) and the other for managing the project (e.g. a project company). An economic entity is specially established as a borrowing vehicle to raise funds for the project so that the project company can avoid dealing with a lot of lenders/investors directly. Then the project company enters into

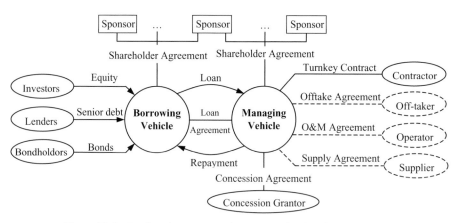

Figure 10.6   Dual-entity structure (separating funding from construction).

a loan agreement with the borrowing vehicle. Similar to the mono-entity structure, the project company in a dual-entity structure may act as an owner–operator company or an owner company. The types of the two economic units depend on project characteristics. Usually a limited company is established for management, and a trust for funding. A typical dual-entity structure to separate funding from construction is shown in Figure 10.6.

Case Study 10.2   The Financing Structure of Hills M2 Motorway in Australia

Hills M2 Motorway is a 21 km, four-lane motorway that links the lower north shore and the northwest regions of Sydney, Australia. This $644m toll road opened to traffic in May 1997 and is now a key part of the Sydney motorway network.

In Hills M2 Motorway, the sponsors and institutional financiers have provided equity through a combination of shares in the Australian Stock Exchange, infrastructure bonds and a 15-year syndicated bank loan. Two economic entities were established for the development of project: Hills Motorway Limited (HML) and Hills Motorway Trust (HMT). HML is a listed company, which was granted a concession (the project deed) and was responsible for the implementation of the project. HMT is a listed unit trust, which was the sole borrower for the construction and project loan facilities provided by the lenders, and the issuer of the CPI bonds. HMT issued CPI indexed bonds in two tranches of A$100m each in December 1994 and June 1996, with terms of 27 and 25.5 years respectively, and also borrowed a traditional bank debt facility of around A$120m. Then the proceeds of bonds and debt facility were lent to HML for the construction of Hills M2 Motorway. Upon completion of the construction phase the project sponsors will jointly invest A$30m in equity. HML entered into a turnkey contract with a contractor for the construction of the motorway and an operation contract with an operator for the operation of the motorway. This dual corporate structure was developed to meet the different needs of the debt and equity providers. Figure 10.7 shows the financing framework of Hills M2 Motorway.

In this dual-corporate structure, there is a trap for unwary lenders. If the trust allows its funds to be linked to the project company without any security, debt security will suffer. In Hills M2 Motorway, the trust-and-company structure hid the inevitable losses by allowing the trust funds to be linked to the company without any security. While the company and trust are distinct legal entities, those entities must effectively be controlled by people at the board level.

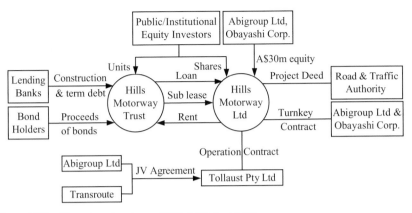

Figure 10.7 Financing structure of Hills M2 Motorway.

Dual-entity structure: dividing a project into two parts

When a project can be divided into two interrelated parts, the project sponsor may want to establish two separate economic entities for the development of the two parts in order to maximise project profit. In this case, one part is developed by an economic entity (referred to as leasing company) and the other part is developed by another economic entity (referred to as project company). The two parts are connected through a lease agreement. This arrangement allows the project company to take advantage of tax deductions for lease payments. In this dual-entity structure, the leasing company may raise funds based on the lease agreement with the project company, that is, a leveraged lease, while the project company raises debt financing based on the anticipated cash flows, generated from user charges or offtake contracts. A typical dual-entity structure of this type is shown in Figure 10.8.

This dual structure is usually employed in projects with a significant part of equipment. Project sponsors want to separate equipment from construction to take the advantage of tax deductions for lease payments. The leasing company purchases equipment from an equipment vendor and leases them to the

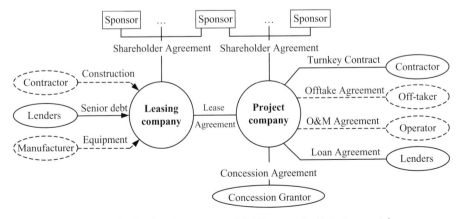

Figure 10.8 Dual-entity structure (dividing a project into two parts).

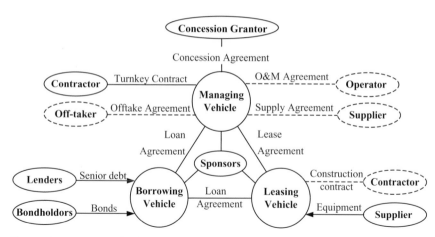

Figure 10.9   An example of three-economic units structure for project development.

project company, and the project company is responsible for the management of the whole project.

10.2.3   A financing structure based on multi-entities

In a multi-entity structure, more than two economic entities are established: one for financing and the others for managing different parts of a project respectively, or each entity for developing one part of the project. When a project is complex or very large in size, the project may be broken down into more than two parts, which can be developed by different entities. For example, a PPP project is developed using three separated economic units: one for financing, one for leasing and the other for managing the project. An example of a multi-entity structure is shown in Figure 10.9.

Case Study 10.3   The Financing Structure of TermoEmcali in Colombia

TermoEmcali power project in Colombia is a 233.8 MW natural gas fired power plant. It was financed in the Rule 144A private placement market with a back-up commercial loan commitment. The project was sponsored by Emcali (the local utility) and Bechtel Enterprises Engergy BV. In order to develop the project, three economic units were established, namely TermoEmcali, TermoEmcali Leasing Ltd (Leaseco) and TermoEmcali Funding Corporation. TermoEmcali was a specially established project company to own the project. Leaseco was established to purchase equipment from Bechtel International Corporation and then lease it to TermoEmcali pursuant to the financial lease agreement. TermoEmcali Funding Corporation was established by Leaseco under the laws of the State of Delaware in the US for the sole purpose of issuing the notes. A portion of the proceeds was lent to Leaseco for the acquisition of equipment, and the remainder was lent to TermoEmcali for the design, construction, start-up, testing, initial operation and related activities of the project pursuant to the participation agreement between the funding company and Deutsche Bank Trust Company Americas. TermoEmcali entered into a PPA with Emcali for dispatch of project output, a fixed-price, turnkey contract with Bechtel Overseas Corporation for the construction of power plant, an O&M contract with Stewart & Stevenson for the operation of power plant, a 16-year firm gas supply agreement and a gas transportation agreement with Ecopetrol for supply of fuels, and a

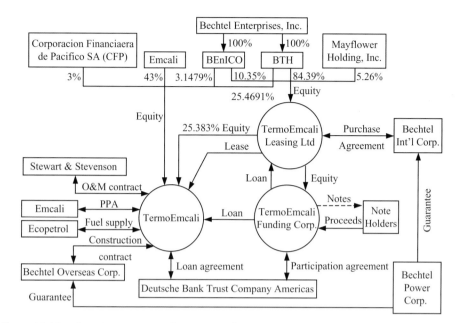

Figure 10.10   Financing structure of TermoEmcali Power.

loan agreement with Deutsche Bank Trust Company Americas for payment of notes. Figure 10.10 shows the financing structure of TermoEmcali Power Project.

In this project, project sponsors developed a three-economic-unit structure: a project company, a leasing company and a funding company. The project company is responsible for the management of the project. The leasing company purchases equipment from an equipment provider and leases them to the project company. The funding company issues notes to note-holders and then lends the proceeds of the notes to the project company and the leasing company. This arrangement allows the project company to take advantage of tax deductions for lease payments. Moreover, a bank debt-service letter-of-credit facility replaces the government guarantee.

10.2.4   A mixed financing structure

Both the dual-entity structure and the multi-entity structure can be applied by concession grantors. When a project can be divided into two or more interrelated sub-projects (sub-project 1, sub-project 2, and so on), the concession grantor may want to use different procurement strategies to develop these sub-projects. The sub-projects may be connected through lease agreements or other agreements, depending on the relationships between the sub-projects. A typical dual entity structure for a two-part project is shown in Figure 10.11.

This multi-entity structure is applicable to the following two types of projects: a project requiring a huge capital investment, and a project with negative profit. For a large-sized project requiring a huge capital investment, it is difficult for a private firm or consortium alone to bear this responsibility, in addition to the difficulty in obtaining competitive tenders. In this case, the

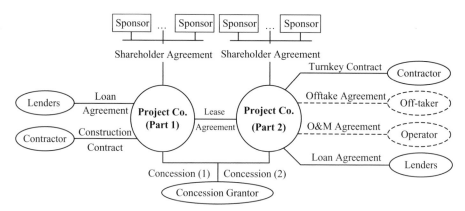

Figure 10.11   A mixed financing structure.

project may have to be divided into two (or more) interdependent or independent sub-projects. For a less profitable project, the project may be broken down into a profitable sub-project and a less profitable one. The profitable sub-project can be developed using BOT procurement strategy; while the less profitable one can be developed using other strategies. Beijing Metro Line 4 project illustrates the application of a dual entity structure.

Case Study 10.4   The Financing Structure of Beijing Metro Line 4 in China

Beijing Metro Line 4 is 28.65 km in length, running from the south to the north, with 24 stations along its way, in the west region of Beijing. Its total investment is about RMB15.3bn (equivalent to S$1.9bn). Its development required a creative financing strategy.

The current subway transportation systems in Beijing suffer losses and their operation is subsidised by Beijing municipal government. Due to competition from bus transportation systems, a subway transportation system cannot be profitable. It is therefore impractical for a private firm to develop the project without government support. Line 4 has a huge socio-economic benefit because it will be the main north–south traffic artery of Beijing. So, Beijing municipal government was willing to fund up to 70% of the capital investment. Moreover, the government wanted to utilise the management efficiency of private developers. Therefore, a mixed financing pattern was applied to meet these requirements.

The project was divided into two parts: A and B. Part A (about 70% of the total investment) consists of civil works and relevant facilities such as tunnel, tracks, stations and service lifts, including land-requisition and relocation of residents; and part B (about 30% of the total investment) consists of rolling stock, traffic control systems (communications systems and signals) and power supply facilities. The municipal government of Beijing, via Beijing Metro Line 4 Investment Co. Ltd (BML4C), a state-owned company, was responsible for the construction of part A. Only part B was granted to private developers under a 30-year concession contract through negotiated bidding.

A concession-winning consortium comprising Hong Kong Mass Transit Railway Corporation (HKMTRC), Beijing Capital Group (BCG) and Beijing Infrastructure Investment Co. Ltd (BIIC) won the contract. On 16 January 2006, Beijing Jinggang Subway Co. Ltd (or Beijing Mass Transit Railway Corporation Limited), with a registered capital of approximately RMB1.5bn (equivalent to S$188m), was established for the implementation of the contract. According to the concession contract, the project company will lease the facilities of part A from BML4C under a 30-year lease agreement, install traffic control systems and then operate the whole facility for 30 years. During the concession period, the project company will charge riders (tickets) to recover its capital and earn a reasonable

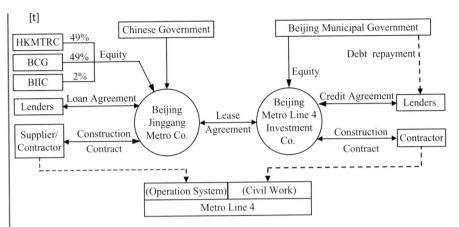

Figure 10.12 Financing structure of Beijing Metro Line 4.

return. Beijing municipal government guarantees the minimum traffic flow. If the annual average revenue is less than the projected level, the government will make up the difference. If it is higher than the projected level, the exceeded part will be transferred to the government according to the pre-specified sharing formula or to the developer of part A as increased lease payments, subject to the approval by the government. In this way, project risks are spread among the three main participants. Figure 10.12 illustrates the financing structure of Beijing Metro Line 4 project.

This two sequential parts arrangement provides many advantages. First, the concession period for each part can be different. For example, part A of Beijing Metro Line 4 can have a longer period than that of part B because the facility of part A has a long lifespan. Second, the government can provide substantial support to a project without complicating the relationships between the public sector and the private sector. In Beijing Metro Line 4 project, BML4C is a government-supported company. That is why there is no clearly defined concession period for part A and surplus revenues (if any) may be paid to the part A developer as an increased lease payment. Third, the developer of the second part can enjoy tax benefits because lease payments are tax deductible to the project company as a part of its normal operating costs. In addition, this arrangement can spread the financial burden and risk among the participants so that any participant's financial burden and risk will not be onerous, but combination of efforts/commitments from the various parties will drive the success of the project.

10.3   Choice of Financing Patterns

The choice of financing patterns depends on various factors. Among them, the complexity of construction and the characteristics of fund providers are two key determinants. Projects can be roughly divided into two categories of simple projects and complex projects according to their construction complexity, though there is actually a continuum from extremely simple to extremely complex. If the construction of a project is simple, the project is viewed as a simple project; otherwise, it is a complex project. A project may be financed by a few fund providers with similar requirements or by a lot of fund providers with different requirements. The former is regarded as a simple financing source, and the latter, a complex financing source. These two perspectives result in four combinations to form four scenarios.

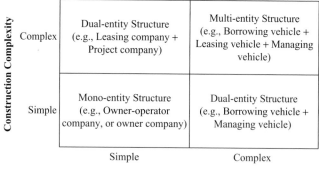

Figure 10.13   Choice of financing patterns for PPP projects.

In general, most projects can be developed using a mono-entity financing structure, and only some complex projects may employ a dual-entity structure or multi-entity structure to take the advantage of these financing patterns. If both the project construction and its financing source are simple, the mono-entity structure may be good enough to develop the project. If the project construction is complex, but its financing source is simple, the dual-entity structure consisting of a leasing company and a project company may be employed to take the advantage of tax deduction for leasing payment. If the project construction is simple, but its financing source is complex, the dual-entity structure consisting of a trust and a project company may be employed to take the advantage of trust for raising funds from the public. If both the project construction and its financing source are complex, the multi-entity structure (including mixed financing patterns) may be needed to develop the project. Financing patterns are proposed for the four scenarios, as shown in Figure 10.13.

10.4   Summary

Financing PPP projects involves the optimisation of capital structure, the design of organisational structure, the design of contractual structure, and the enhancement of creditworthiness, based on risk–return trade-offs. Among them, the organisational structure plays an important role in the financing structure. There are three general financing patterns: mono-entity structure, dual-entity structure and multi-entity structure. The most popular pattern is the mono-entity structure. The other patterns are occasionally used for some special situations.

The concept of a multi-entity structure may be applied at the level of awarding concessions. A project may be divided into two or more sub-projects, each of which may be developed using different procurement strategies. As a result, the development of a project may involve two or more procurement strategies. This results in a mixed financing pattern.

The choice of financing patterns depends on various factors. Among them, the complexity of construction and the characteristics of fund providers are two key determinants. In general, most projects can be developed using a mono-entity financing structure, and only some complex projects may employ a dual-entity structure or multi-entity structure to take the advantage of these financing patterns.

References

Nevitt, P.K. and Fabozzi, F. (2000) *Project Financing*, 7th edn. Euromoney Books, London.

Strzelecki, R. (1990) *The Investor's/Supplier's Viewpoint*. Private Sector Participation in Power through BOOT Schemes, Industry and Energy Department Working Paper, Energy Series paper No. 33. The World Bank Industry and Energy Department, PRE.

Part Two

11

PPP Financing in the USA

Arthur L. Smith

11.1   Introduction

The use of PPPs to implement infrastructure projects is not a new development in the USA. Early examples of PPPs include the Lancaster Turnpike, the first long-distance stone and gravel road in the country, which was built by the Philadelphia and Lancaster Turnpike Company from 1792–1795; the Erie Canal, which opened in 1823; and the Transcontinental Railroad, completed in 1869.

PPPs were not commonplace, however, and remained a controversial means of achieving development goals. As Jonas Platt, a New York State Senator and leading proponent for building the Erie Canal was to recall in an 1828 letter after the Canal's successful debut, 'Powerful and appalling obstacles . . . were presented, in the honest doubts and fears of many sensible and reasonable men . . . in rival and hostile political interests . . . and in the political hostility . . . The leading advocates of the canal were objects of ridicule throughout the United States; hallucination was the mildest epithet applied to them' (Hosack, 1999). Concerns centred on the engineering feasibility of the project and the assessment of the potential economic benefits, as well as the project financing. Assemblyman Peter Sharpe summed up the views of many opponents when he warned that the 'most respectable and opulent of (the city's) merchants are daily becoming bankrupts . . . (The state) will sink under (the debt whose) magnitude is beyond what has ever been accomplished by any nation' (Bernstein, 2005). In fact, the completed canal reduced travel time from the east coast of New York to the Great Lakes by half, reduced shipping costs by half, and made New York City the busiest port in the US. It is still in use today, albeit with several enlargements over the years to accommodate changes in traffic volume and vessel size.

The use of PPPs accelerated in the US in the latter half of the twentieth century, spurred by examples such as Union Station in Washington, DC. In 1981, the US Congress enacted the Union Station Redevelopment Act, which authorised the US Department of Transportation to enter into a PPP to renovate the then-shuttered historic train station and return it to financial self-sufficiency.

The station reopened in 1988 as a redeveloped inter-modal transportation facility and retail centre. Today, it is the most visited destination in Washington, with over 25 million visitors a year. Throughout the 1980s, 1990s and into the 2000s, PPPs became more frequently employed to provide transportation, water/wastewater, power, and academic and recreational facilities.

A PPP may be defined as 'any arrangement between a government and the private sector in which partially or traditionally public activities are performed by the private sector' (Savas, 2000). More expansively:

> 'A Public-Private Partnership is a contractual agreement between a public agency (federal, state or local) and a private sector entity. Through this agreement, the skills and assets of each sector (public and private) are shared in delivering a service or facility for the use of the general public. In addition to the sharing of resources, each party shares in the risks and rewards potential in the delivery of the service and/or facility.' (National Council for Public-Private Partnerships, 2007)

Contracting for public goods and services is complex under the best of circumstances, with many competing interests: government agencies (often multiple agencies at different levels of government), private sector providers, taxpayers, consumers and special interest groups. It is also a very large market. In the US, public sector expenditures for goods and services, at all levels of government, comprise 9% of GDP. For this reason, there are comprehensive sets of procurement regulations, designed to ensure open, equitable acquisitions and accountability on the part of the civil servants who procure goods and services and from the private firms which provide them.

PPPs for infrastructure, however, are more complex than traditional construction contracts. The differences include:

- Risk allocation: traditional contracts have a fairly straightforward risk allocation model. PPPs require analysis and allocation of a broader spectrum of risks, which may include, but are not limited to, design and construction risk, operational risk, demand risk, technological risk, regulatory risk, political risk, *force majeure* and others. The risk allocation is more complex than in traditional construction contracts, where demand risk, for example, would typically not be borne by the developer. Identification, disclosure and appropriate allocation of risk are therefore critical to the PPP environment.
- Duration: PPP contracts frequently extend for 30 years or longer; the longest example in this chapter is a 99-year contract. This greatly complicates the difficulty of projecting service demand, and quantifying other risks such as technological and regulatory change and currency fluctuation.
- Financing: traditional government contracts are government-funded. PPPs typically entail financing predominantly or in whole from the private sector. This requires establishing a tendering and contract administration environment in which investment opportunities can be quantitatively assessed, and prudent investment opportunities can reasonably expect to be awarded with an appropriate return on investment.

Discussion of PPP financing in the US is further complicated by the fragmented nature of the US PPP market. Unlike many countries (e.g., the UK,

Part Two

Ireland, the Czech Republic etc.), the US has no single federal agency with oversight of PPP policy and issues. Authority to undertake PPPs is typically granted to agencies by the Congress on an agency-specific basis, or even function-specific or project-specific basis. There is no standard approach to federal PPPs for infrastructure analogous to, for example, the UK's Private Finance Initiative (PFI). Thus, entering into a PPP to provide and operate a wastewater treatment plant on a US defence installation would fall under different statutes than to provide a similar facility at a national park or forest. On a state level, a recent analysis showed that only 27 of the 50 states had implemented laws authorising transportation PPPs. Differences in the state laws exist as well, and in local government approaches.

On the positive side, this fragmented environment makes the US a virtual 'PPP laboratory' in which a number of varied approaches to PPP structure and finance have been attempted, with varying degrees of success, providing valuable lessons learned. Less positively, this fragmentation makes the US a more complex investment environment. In addition, although hundreds of PPPs are created in the US each year, the lack of a central agency to track and report these transactions makes it more difficult to demonstrate the full extent of PPP activity.

11.2   PPP Financing Models in the US

Given the breadth of PPP models in the US today, this chapter will focus on one particular area: transportation partnerships. Transportation is the largest area of PPP activity in the US in terms of dollars (the number of water/wastewater projects is larger, but the average transaction size is smaller). Transportation PPPs are supported as a tool for infrastructure development by the Department of Transportation (DOT) and several of its constituent agencies, in particular the Federal Highway Administration (FHWA), which have attempted to stimulate the use of PPPs. At the state and local levels there has been a high level of activity, with the Commonwealth of Virginia's Public-Private Transportation Act of 1995 frequently cited as a model for other jurisdictions.

The increasing reliance on PPPs may be attributed to several factors, but the most compelling has been a rapid growth in the demand for service, which has exceeded the public sector's financial capacity. For example, in the US today there are over 4 million miles of public roads and highways, which experience roughly 2.9 trillion vehicle miles per year of use. In 1956, there were 65 153 810 motor vehicles registered in the US, with the average vehicle driven 9623 miles per year (US Department of Commerce, 1958). In 2005, there were 241 193 944 motor vehicles registered, an increase of 370%, while the average vehicle was driven 12 190 miles, an increase of 27%, for a total increase in road utilisation of 470% (US Department of Transportation, 2006). Over the same period, public expenditures on roads and highways increased by only 20% in real terms, to $150bn in annual expenditures. In the absence of public sector capacity to fund the need for infrastructure, private sector financing and expertise can greatly assist federal, state and local

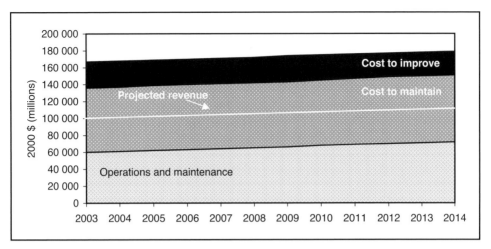

Figure 11.1 Comparison of projected highway revenue with investment requirements. (From *Highway Finance and Public–Private Partnerships – New Approaches to Delivering Transportation Services*, January 2005. Federal Highway Administration, Publication Number FHWA-HOA-05-003, p. 5.)

governments in meeting the public's needs. Today, public road ownership in the US, in miles, is 78% local government, 19% state government and 3% federal, although the federal government's role in construction of new roads and highways is more extensive.

One unique aspect of financing highway construction in the US is that under the Federal-Aid Highway Act of 1956, the use of tolls on new highways has been prohibited (earlier-constructed toll roads are exempt from this provision). Of the 53 000 miles of interstate highways, 49 500 miles, or 93.3%, is untolled (Perez and March, 2006). Instead, the Interstate Highway System is funded by a national fuel tax, currently set at 18.4 cents per gallon, and a vehicle excise tax. These revenues contribute to the National Highway Trust Fund, which pays the majority of highway construction and improvement costs, with the remaining funds provided by the states. Since 1956, the real per-gallon value of the fuel tax has declined by 25%; Congress last authorised an increase in the tax in 1993 (Basso, 2005). This has limited the availability of federal funds to meet the need for new highway construction; the majority of the trust funds have been focused on maintaining and improving the existing highways. State funds for new construction are limited as well. The FHWA estimates that the financial resources required to make capital investments in highway construction and expansion, while continuing to operate and maintain existing highways, exceed current federal, state and local highway funding by 40% (Figure 11.1). The inadequacy of available funding at both the federal and state level, has led to an increased focus on PPPs to leverage the limited government funding.

The Federal Department of Transportation has long sought additional authorities from Congress to encourage and enhance the use of PPPs for transportation infrastructure. On 10 August 2005, the President signed into law the Safe, Accountable, Flexible, Efficient Transportation Equity Act: A Legacy for Users (SAFETEA–LU), Public Law Number 109-59. SAFETEA–LU, the latest in a series of federal transportation bills, provides new and expanded mechanisms for the private sector to participate in highway infrastructure projects. These improvements include:

- Enhanced use of private activity bonds
- Enhanced ability to toll interstate system
- Increased flexibility in design–build contracting
- Streamlined environmental processes
- Enhancements to innovative finance programme

The following paragraphs will discuss some of the key funding programmes which support transportation PPPs in the US, and specifically address where their capacity has been impacted by SAFETEA–LU.

11.2.1   63-20 Public Benefit Corporation

An important facilitator of PPPs in the US has been the Internal Revenue Service (IRS) Ruling 63-20. This Ruling establishes that state and local governments can issue tax-exempt toll revenue bonds through either established conduit users or creation of not-for-profit corporations. This type of debt keeps interest costs low and generates attractive opportunities for non-public investors. Recently, a number of highway and transit projects have been funded by debt issued by non-profit corporations, which, pursuant to IRS Rule 63-20 and Revenue Proclamation 82-26, are able to issue tax-exempt debt on behalf of private project developers. Examples of PPP projects which have utilised 63-20 public benefit corporations include the Pocahontas Parkway in Virginia (completed in 2002), the Las Vegas Monorail (2004) and Massachusetts Route 3 North (2006).

There are two primary models for using 63-20 tax-exempt debt to finance PPPs for infrastructure. In the first model, for revenue generating projects, the 63-20 corporation issues debt by leveraging future user fee-based revenues, with the public benefit corporation entering into an agreement with a private contractor to design, build, operate and maintain the project for a predetermined period. In this case, the private partner usually assumes responsibility for arranging financing, but does not actually issue the debt; the financing package is submitted to the board of the 63-20 corporation for approval and then issued on its behalf by a brokerage agency.

In the second model, a lease-back arrangement is used as a revenue source to back the 63-20 corporation's debt. In this case, a department of transportation or a transit agency agrees to lease the transportation asset to be developed by the 63-20 corporation for a designated period of time. The 63-20 corporation then leverages the future lease payments to issue its debt. Typically, the private partner plays an important role in assembling the financing package for this type of lease-back Design-Build-Operate-Maintain (DBOM) transaction.

Under US law, a non-profit corporation is a private, non-stock corporation formed under the Non-profit Corporation Act of a state. The formation of a non-profit corporation does not require special legislation, nor does it require a referendum in the local or sponsoring jurisdiction. A non-profit corporation may be formed for any lawful purpose other than for pecuniary profit, including, without limitation, any charitable, benevolent, educational, civic or scientific purpose. Non-profit corporations are regulated by the State Attorney General for compliance with the Non-profit Corporation Act, by state tax authorities for compliance with the requirements relating to their state income tax exemption, and by the IRS for compliance with the use of a non-profit project sponsor, which may also enable a project to receive public funds since the revenues generated by the project will not benefit any private party. It may also be possible for the non-profit sponsor to issue public or privately-placed debt if it can enter into long-term contracts for the use of the facility, or if the facility generates revenues from direct user fees.

In order for a non-profit corporation to issue tax-exempt debt, it must satisfy the following criteria established by the IRS:

- The corporation must engage in activities which are essentially public in nature.
- It must not be organised for profit, except to the extent of retiring indebtedness.
- The corporate income must not inure to any private person.
- The state or a political subdivision thereof must have a beneficial interest in the corporation while the indebtedness remains outstanding.
- The corporation must be approved by the state or a political subdivision thereof, which must also approve the specific obligations issued by the corporation.
- Unencumbered legal title in the financed assets must vest in the governmental unit after the bonds are paid.

The rules for determining whether the governmental unit has the requisite 'beneficial interest' in the non-profit corporation are also defined by the tax code:

- The governmental unit must have exclusive beneficial possession and use of at least 95% of the fair market value of the assets; or
- If the non-profit corporation has exclusive beneficial use and possession of 95% of the fair market value of the facilities, the governmental unit appoints 80% of the members of the board of the corporation and has the power to remove and replace members of the board; or
- The governmental unit has the right at any time to acquire unencumbered title and exclusive possession of the financed facility by defeasing, i.e., by paying off or providing for payment of the bonds.

11.2.2 Transportation Infrastructure Finance and Innovation Act of 1998 (TIFIA)

The TIFIA Credit Program provides federal credit assistance to large-scale projects of regional or national significance that might otherwise be delayed or not constructed at all because of risk, complexity or cost. There are three forms of credit assistance available (secured (direct) loans, loan guarantees and standby lines of credit) for surface transportation projects of national or regional significance. These credit instruments may offer more flexible re-payment terms and more favourable interest rates than would be available from other lenders. The fundamental goal of the TIFIA Credit Program is to leverage federal funds by attracting substantial private and other non-federal co-investment in critical improvements to the nation's surface transportation system. In general, public or private entities seeking to finance, design, con-struct, own, or operate eligible surface transportation systems can seek TIFIA assistance. Under SAFETEA–LU, TIFIA loans may now be used to refinance long-term project obligations or federal credit instruments if such refinanc-ing provides additional funding capacity for the completion, enhancement or expansion of the project.

From 1999–2005, TIFIA provided more than $3.6bn in credit assistance to projects representing more than $16bn in infrastructure investment, at a cost to the government of less than $200m. Utilising federal funds to provide credit assistance leverages the limited available resources. For private investors, TIFIA benefits include:

- The Department of Transportation lends TIFIA funds at the US Treasury's borrowing rate, with no premium for risk. This results in cost savings compared to the likely rates on alternative financing instruments.
- TIFIA credit can have a final maturity date as much as 35 years after the date of substantial project completion.
- TIFIA offers extensive flexible payment features. TIFIA allows projects backed by user charges to structure debt service based on project cash flows, e.g., to defer interest payments during both construction and ramp-up of operations. The TIFIA Program also allows borrowers to prepay at any time without penalty.
- TIFIA credit can be subordinated to that of senior lenders, thus enhancing the creditworthiness of the remaining senior-lien capital market financing.

11.2.3   GARVEE bonds

A Grant Anticipation Revenue Vehicle, or GARVEE, is a debt-financing in-strument where debt service and related financing costs can be reimbursed by federal-aid highway funds. PPP projects are eligible for GARVEE bonds. GARVEEs can be issued by a state, a political subdivision of a state or a pub-lic authority. States can receive federal-aid reimbursements for a wide array of debt-related costs incurred in connection with an eligible debt-financing instrument, such as a bond, note, certificate, mortgage or lease. Reimbursable debt-related costs include interest payments, retirement of principal and any other cost incidental to the sale of an eligible debt instrument.

Candidates for GARVEE financing are typically projects, or a programme of projects, that are large enough to merit borrowing rather than pay-as-you-go grant funding, with the costs of delay outweighing the cost of financing. GARVEE candidates do not have access to another revenue stream, such as local taxes or tolls, and other forms of repayment are not feasible. The sponsors are required to reserve a portion of future federal-aid highway funds to satisfy debt service requirements.

11.2.4   Private activity bonds

The term 'private activity bonds' under the IRS Code §103(2) refers to any bond issued as part of an issue for which: more than 10% of the proceeds are to be used for any private use and the payments of the principal of, or the interest on, more than 10% of the proceeds of such issue is secured by an interest in property used to or to be used for a private business purpose or payments in respect to such property. Alternatively, bonds of an issue are private activity bonds if more than the lesser of 5% or $5m of the proceeds of the issue is to be used (directly or indirectly) to make or finance loans to a non-governmental entity. Interest on private activity bonds is taxable, unless exempted by law ('qualified'). Public benefit corporations, however, may issue tax-exempt bonds on behalf of a state or local government for the promotion of trade, industry or economic development. This provision has allowed 63-20 corporations to raise financing for PPPs at favourable rates.

SAFETEA–LU facilitates PPPs by authorising the issuance of tax-exempt (qualified) private activity bonds up to a national cap of $15bn. Eligible projects are already receiving federal assistance for surface transportation projects, surface freight transfer facilities, highway facilities, international bridge or tunnel projects and facilities for the transfer of freight from truck to rail or rail to truck.

11.2.5   Section 129(a) loans

US tax law, codified as Section 129(a)(7) of Title 23, US Code, allows states to loan some of their federal-aid funds to pay for projects, including PPPs, with dedicated revenue streams. A state may directly lend apportioned federal-aid funding to projects generating a toll or that have some other dedicated revenue such as excise taxes, sales taxes, property taxes, motor vehicle taxes and other beneficiary fees. The state must receive a pledge from the project sponsor to use those revenues to repay the loans.

These loans, commonly referred to as Section 129(a) loans, can be subordinate to the other debt, so that other investors can have a first or senior lien on project revenues. In this manner, the state shares the demand risk of the project, i.e. that project use and therefore revenues, will fall below projections. By reducing the amount of senior debt, the state increases the likelihood of it attaining an investment grade rating, and thus lower interest rates.

11.2.6   Special Experimental Project 15

In October 2004, the FHWA implemented Special Experimental Project 15 (SEP–15), based on the successes of prior, recently completed programmes, such as SEP–14 and Test and Evaluation Project No. 045, which provided enhanced used of design–build techniques. SEP–15 allows FHWA to experiment in four major areas of project delivery: contracting, right-of-way acquisition, project finance and compliance with the National Environmental Protection Act and other environmental requirements. SEP–15 is specifically designed to promote the use of PPPs, and encourages states to identify current laws, FHWA regulations and practices which inhibit the use of PPPs and private investment in transportation improvements. State departments of transportation may submit proposals to FHWA identifying the project, the experimental techniques proposed, and the reason(s) why they are necessary or beneficial to the project. If use of a special process or waiver is approved, the public-private sponsors must submit, upon completion of major milestones, an independently prepared report evaluating the experiment undertaken and documenting lessons learned. A number of projects have already been approved for use of SEP–15 experimental procedures, including the Trans-Texas Corridor, the Texas Open Road Tolling System and the Oregon Innovative Partnerships Program.

11.2.7   Other PPP-related programmes

The financing mechanisms and programmes discussed above are among the most significant for transportation PPPs in the US today. However, there are many other PPP-related initiatives in place. These include an allowance for states to transfer a portion of their Federal Highway Trust Fund allocations to state infrastructure banks (SIBs). SIB loans can be used to finance state transportation projects, but under this new programme, they will now retain their character as federal funds. After repayment, SIB loans can be reloaned to support other projects. There is also new flexibility to use tolls on federally supported projects. The US government is aggressively exploring these and other avenues to maximise PPPs to leverage government funds.

11.3   Case Studies

The following paragraphs describe four implemented PPPs in the US. Two of the four utilise financing tools made available by the federal government: 63-20 public benefit corporations and TIFIA. The remaining two PPPs were implemented at the state and municipal levels, using private financing.

Case Study 11.1   Massachusetts Route 3

Massachusetts Route 3 North is an existing 21-mile limited access highway between the Boston metropolitan area and the New Hampshire border to the north. The project was designed to alleviate a number of significant transportation problems on this heavily utilised highway. The project involved:

- The addition of a third travel lane in each direction
- Creation of median shoulder and a 30-foot clear zone
- Improvements to 13 interchanges
- Replacement of 40 bridges

In August, 1999 the Massachusetts Legislature approved legislation authorising creation of a PPP to finance, design, build, operate and maintain the project. The legislation enabled creation of a 63-20 public benefit corporation.

A request for tender was issued, proposals evaluated, and a firm selected for the project development. Following contract award, the 63-20 corporation (the Route 3 North Transportation Improvements Association) was formed by MassHighway and the developer, and project financing of $394.5m was secured through its issuance of tax-exempt lease revenue bonds. The bonds were secured by lease payments pledged by MassHighway over the 34-year term of the lease (4 years for construction, and 30 years of operation and maintenance). At the end of the lease, responsibility for road operation and maintenance will transfer to MassHighway. Lease payments are derived from an annual appropriation of the Legislature; there are no tolls on this highway. Construction began in October 2000, and the full length of the improved highway opened to traffic in October 2004. The final construction, to include landscaping, drainage and installation of bridge joints, was completed in late 2006.

The contract helped contain cost by granting the developer rights to pursue surface, subsurface and air rights development as sources of non-project revenues. The project includes installation of fibreoptic cable, with the developer sharing in resulting revenues, and planned development and sublease of a service plaza, also with shared revenues; 40% of ancillary development revenues go to the developer.

Several other innovative approaches helped to reduce cost. Annual lease payment due dates were set well into the Commonwealth's fiscal year, which reduced risk due to budgetary delays; this, in turn, reduced the need for a liquidity debt service reserve. In addition, project risk insurance was purchased, with the developer acting as co-insurer. A requirement of the policy was that the developer establish a contingency fund equal to 10% of the design–build price.

The Route 3 North Project demonstrates a successful non-toll approach for a major infrastructure requirement. By spreading the construction cost over the 34-year lease term, the Commonwealth was able to attain the benefits of this project years before it would otherwise have been affordable.

Case Study 11.2   Chicago Skyway

The Chicago Skyway is a 7.8-mile elevated toll road connecting I-94 (The Dan Ryan Expressway) in Chicago to I-90 (The Indiana Toll Road) at the Indiana border. The facility includes a 3.5-mile elevated mainline structure crossing the Calumet River. Built in 1958, the Skyway was operated and maintained by the City of Chicago Department of Streets and Sanitation. By 2004, the facility carried approximately 50 000 vehicles per day, and generated $44m in annual toll revenues. The Skyway was a significant infrastructure asset, and had been operating at a profit for roughly a decade.

The City of Chicago announced on 1 March 2004 that it would seek to create a PPP under which a private entity would pay the City for the right to levy tolls on the Skyway for at least 50 years, and would maintain, develop and operate it over this period. As the City stated, 'While the Skyway is a significant source of stable cash flows to the City, the City considers that a private entity may be able to derive substantially more economic value from the asset while providing excellent service for Skyway users' (Samuel, 2004a).

The City received ten responses to its request for tender, and in May 2004 selected five firms as qualified bidders based on the City's weighted evaluation of their financial capacity and experience in operating a tollway. Proposals were submitted in October 2004, and on 27 October the City signed a contract with the selected bidder.

Under this agreement, the Skyway Concession Company, LLC (SCC), a joint venture formed specifically for this project, paid the City $1.83bn for the right to operate the Skyway and retain all of its toll and concession revenues for a 99-year lease period. The joint venture partners used $882m in equity and approximately $950m in bank loans to finance the transaction. The SCC assumed operations on 24 January 2005. Subsequently, the SCC refinanced its capital structure to reduce the investors' equity holdings, primarily through capital accretion bonds with a 21-year maturity and 12-year floating rate notes.

Part Two

The size of the $1.83bn lease at 41 times the annual toll revenue exceeded both City and industry expectations (Samuel, 2004b). The toll schedule is fixed by the lease agreement (with inflation and congestion pricing provisions), so the realism of SCC's traffic projections may determine the long-term success of this PPP from the private partner's perspective.

For the City, the PPP appears highly successful. Of the $1.8bn, $463m was used to pay off the Skyway's debt, and $392m was used to pay off the City's debt. The lease payment from SCC also funded a $500m long-term and a $375m medium-term reserve for the City of Chicago, and a $100m neighbourhood, human and business infrastructure fund to be drawn down over 5 years (Replogle, 2006). It is notable that the City now receives more in annual interest from its investment of the lease proceeds than its annual revenue from operating the Skyway itself.

Case Study 11.3   Indiana Toll Road

The Indiana Toll Road (ITR), in operation since 1956, stretches 157 miles across the northernmost part of Indiana from its border with Ohio to the Illinois state line, where is provides the primary connection to the Chicago Skyway and downtown Chicago. The ITR links the largest cities on the Great Lakes with the Eastern Seaboard. The facility varies from four to six lanes and in 2005, carried approximately 46 000 vehicles per day on its western end and 25 000 vehicles per day in the east. For the past 25 years, the ITR has been operated by the Indiana Department of Transportation (INDOT).

In early 2005, the Governor of Indiana, Mitch Daniels, directed the Indiana Finance Authority (IFA) to analyse the feasibility of leasing the facility to a private entity. This interest was spurred in part by the adjacent example of the Chicago Skyway, as well as the Governor's past experience. Prior to his election as Governor, Daniels had served as Director of the US Government's Office of Management and Budget, which oversees some of the federal partnership programmes.

Based on the IFA's assessment, the state issued a request for tender in September 2005. Four proposals were received in October 2005, and evaluated by the state. On 23 January 2006 the state announced its intent to award a 75-year lease to Statewide Mobility Partners (SMC), a joint venture whose principals were the same two firms which constituted the SCC, the joint venture which was awarded the Chicago Skyway lease. Under the terms of the ITR lease, SMC would operate, maintain and develop the ITR for 75 years in exchange for all toll revenues over the lease term. In exchange for this right, SMC would pay the state $3.85bn. The $3.85bn offer was funded by 10% in equity from each partner, and 80% ($3.03bn) in bank loans, from several European banks.

The Governor had proposed a major initiative to develop the road system in Indiana over a 10-year period (approximately 200 transportation projects in total), and the $3.85bn offer from SMC was critical to the funding of this proposal. The Governor also proposed that one-third of the funding ($1.3bn) be spent on projects in the seven counties (out of 92 in the state) near the ITR (i.e. those counties whose residents were most likely to be frequent ITR users).

Despite these incentives, the proposed contract proved controversial. There was significant local opposition to the idea of foreign firms managing a state road, although SMC comprised the same two firms that were already managing the Chicago Skyway, and non-US management had not been a significant issue in Chicago. In addition, legislators representing the part of the state where the ITR was located contended that a greater share of revenues should be devoted to projects in their area (Goldstein, 2006). After extensive lobbying by the Governor, and a broad public education campaign, the legislature narrowly approved the contract in March 2006. The opponents contested this award through the state judicial system; however, on 20 June 2006, the Indiana Supreme Court affirmed the legality of the contract. The contract was awarded to the IRT Concession Company LLC (formed by the SMC partners) on 12 April 2006. The firm assumed operational responsibility for the ITR on 29 June 2006.

Case Study 11.4   Pocahontas Parkway

At the state level, the Commonwealth of Virginia has one of many established programmes. The Commonwealth of Virginia's Public-Private Transportation Act (PPTA) of 1995 is a legislative framework enabling the Virginia Department of Transportation (VDOT) to enter into agreements authorising private sector entities to develop and/or operate transportation facilities. Private sector entities may identify a need, such as a new connector highway or light-rail system, and submit an unsolicited proposal to VDOT. Alternatively, VDOT may identify a requirement which may be appropriate for a PPP solution, and issue a request for tender. Proposal/project evaluation is then a six-phase process:

- Quality control. Does the proposal address needs identified in the appropriate local, regional, or state transportation plan? Will it provide a more timely, efficient, or less costly solution than the public sector? Is there appropriate risk sharing?
- Independent review panel (IRP). The proposal is reviewed by an IRP with members from the State Transportation Board (STB), VDOT, transportation professionals, academics and representatives of the affected jurisdictions. The review is based either on the basic criteria established by the law (which is available on the Internet) or on a modified version of these criteria, as provided in the state's published request for tender. Public meetings and input are part of the process.
- STB recommendations. The STB reviews the proposals and recommendations of the IRP and recommends to VDOT whether to proceed with the project. A decision to proceed means that VDOT will advance to a public request for detailed proposals. Such requests are advertised on the state's procurement website and are open to participation by any responsible party. The request for tender will identify the state's evaluation criteria.
- Submission and selection of detailed proposals. VDOT forms a proposal review committee and requests detailed proposals. Based upon its review, VDOT may select none, one, or more proposals for further negotiation.
- Negotiations. If the quality of proposals merits, VDOT will negotiate for the interim and/or the comprehensive agreement which will, among other things, outline the rights and obligations of the parties, set a maximum rate of return to the private entity, determine liability and establish dates for termination of the private entity's authority and dedication of the facility to the state.
- Agreement. The negotiated agreement undergoes final legal review, and is then submitted for signature and implementation. State law also provides for debriefings of unsuccessful bidders and an appeals process.

This process has been successful in generating effective partnerships, and led the Commonwealth in 2002 to pass the Public-Private Partnerships Education Act (PPEA), which enabled the use of PPPs for educational and other facilities.

The first project to be completed as a result of the PPTA was the Pocahontas Parkway, in 2002. This is a 14.1 km, four-lane road, including a high-level bridge over the James River, which connects two major commuting routes in the Richmond, Virginia area. The business model was based on the premise that commuters would be willing to pay a modest toll to reduce their commuting time.

The project was initiated through creation of a 63-20 public benefit corporation, which designed, built, financed and operated the project. Financing included $354m in tax-exempt toll revenue bonds sold by the 63-20 corporation (the Pocahontas Parkway Association), and $27m in government funds. The bonds were to be paid through toll receipt, over the 30-year term of the PPP, after which the Parkway would revert to state operation and maintenance.

From the outset, revenues were below the expectations generated through motorist services and county growth projections; in the first year of operation, toll revenues were 42% below the projection. Revenue shortfall appeared to be due to slower than predicted economic growth in the Richmond, Virginia area and, in particular, at the Richmond International Airport (Regimbal, 2004; Samuel, 2005). In 2004, the Commonwealth approved an increase in the toll, 18 months before it was scheduled, in an attempt to increase revenues. The bonds were downgraded, and VDOT made a loan to maintain the project's viability. In consequence, the Commonwealth began to consider the possibility of refinancing the project.

This led the Commonwealth to explore a Chicago Skyway type of transaction. Although not as large a revenue generator as the Skyway, the Pocahontas Parkway, with $11m in toll revenue in 2005, was a candidate for a long-term lease partnership. A request for tender was issued and the Commonwealth began to consider developer proposals.

On 29 June 2006 the Commonwealth signed a 99-year lease agreement with an Australian toll road operator, under which the firm paid $611m to operate, maintain and develop the Pocahontas Parkway in exchange for the rights to toll revenues, with revenue sharing, based on a series of calculations tied to real net cash flow and internal rate of return. The developer's proposal included a plan to link the Parkway to the Richmond International Airport through a 2.53 km, four-lane extension, contingent upon award of $150m in TIFIA credit; this credit was subsequently approved by US DOT. The TIFIA funds are being used to refinance the long-term senior bank debt and upgrade the electronic tolling systems, as well as contribute to the airport connector financing. The developer also defeased all of the Pocahontas Parkway Association's underlying debt. The financing plan included $195m in equity and subordinated debt provided by the developer, and $416m in bank loans.

11.4   Conclusions

The demand for infrastructure operation, maintenance and improvement transcends the availability of government funds at the federal, state and local level in the US. As a consequence, government agencies are increasingly turning to PPPs to accelerate infrastructure acquisition and maintenance. The transportation sector has been particularly active, with the federal government implementing a succession of new initiatives to facilitate private participation in transportation projects. State and local governments have also been active in this regard, and there has been a number of highly successful PPPs, encouraging broader use of this approach. Success is not universal, however, as demonstrated by the need to refinance the Pocahontas Parkway project. Government entities at all levels will continue their efforts to identify and implement improved PPP-enabling mechanisms.

References

Basso, J. (2005) *Prospects for Funding and Reauthorization of TEA-21 and its Impacts on the States*. Presentation to the National Conference of State Legislatures.

Bernstein, P. (2005) *Wedding Of The Waters: The Erie Canal And The Making Of A Great Nation*. W.W. Horton & Company, New York.

Goldstein, A. (2006) Privatization Backlash In Indiana. *The Washington Post*, Washington DC, June 18.

Hosack, D. (1999) University of Rochester (2007) Memoir of Dewitt Clinton, http://www.history.rochester.edu/canal/bib/hosack/APPDX.html. Rochester, New York.

National Council for Public-Private Partnerships (2007) *Public-Private Partnerships Defined*. http://www.ncppp.org/howpart/index.shtml#define

Perez, B., and March, J. (2006) *Public Private Partnerships and the Development of Transport Infrastructure: Trends on Both Sides of the Atlantic*. Banff, Alberta.

Part Two

Regimbal, J., Jr. (2004) *An Analysis of the Evolution of the Public-Private Transportation Act of 1995*. Prepared for the Southern Environmental Law Center.

Replogle, M.A. (2006) *High Performance Corridors: Emerging Transportation Management Framework?* Presented at First International Conference on Funding Transportation Infrastructure, Banff, Alberta, Canada.

Samuel, P. (2004a) Chicago Skyway – buyers wanted. *Toll Roads News*, Frederick, MD, March 3.

Samuel, P. (2004b) Cintra-Macquarie to take over Chicago Skyway for $1.8b. *Toll Roads News*, Frederick, MD, October 15.

Samuel, P. (2005) Transurban moves to buy troubled Pocahontas Parkway VA. *Toll Roads News* Frederick, MD, June 16.

Savas, E. (2000) *Privatization And Public-Private Partnerships*. Seven Bridges Press, LLC, New York.

US Department of Commerce, Bureau of Public Roads (1958) *Highway Statistics 1956*, U.S. Government Printing Office, Washington DC.

US Department of Transportation, Federal Highway Administration Office of Highway Policy Information (2006) *Highway Statistics 2005*, http://www.fhwa.dot.gov/policy/ohim/hs05/index.htm. Springfield, Virginia.

Part Two

12

Financial Modelling of PPP Projects

Ammar Kaka and Faisal Alsharif

12.1   Introduction

Private Finance Initiative (PFI) is the name given to the policies announced by the Chancellor of the Exchequer in his autumn statement of 1992 (RICS, 1995). It is a type of PPP where project financing rests mainly with the private sector (Akintoye *et al.*, 2001). It has become a major procurement method in the UK and worldwide. Since its launch in 1992, the UK government has supported PFI and encouraged local authorities to use PFI where it is applicable and can provide VFM. Since then, many projects have been provided through PFI.

According to NAO (2001), in 1999 there were 163 236 firms in the construction sector, of which 95% had 1–13 employees, 4% had 14–79 employees and only 1% had over 80 employees. This means that 99% of the construction firms in the UK are small and medium-size contractors based on employee number classification and are therefore excluded from PFI which is limited to large-size contractors. Bing *et al.* (2005) found that only 15% of construction cost and 13.2% of the operation net present value (NPV) cost of the 53 PFI projects they surveyed were less than £10m. Consequently, the scale and complexity of PFI projects put them beyond the capabilities of small and medium contractors. Similarly, they do not have the financial and managerial capabilities to complete a competitive PFI bid. The complexity of relationships, negotiation, arrangements, agreements and long-term engagement are also barriers for small construction organisations.

The bidding cost is considered to be high in PFI projects. Both the public and private sectors are required to hire technical, legal and financial consultancies to ensure the project's affordability, quality and VFM for the public sector. Ahadzi and Bowles (2004) stated that the bidding and advisory costs to both the private and public sectors were equally high, ranging from £0.1–2.0m depending on project type. The reason for bidding costs often being highlighted in the context of PFI is that the costs associated with bid preparation are an inherent cost of doing business and if the client awards the project contract to a competitor or does not award it at all, the contractor will not be compensated for these costs (Rintala, 2004).

Public sector authorities are required to ensure procurement should be based on VFM – defined as the optimum combination of whole-life costs and quality to meet the customer's requirements – rather than initial purchase price. A VFM assessment requires a public sector comparator (PSC) in which the authority should compare other options against the PFI procurement method. As authorities have completed many projects using traditional procurement methods, comparative data is accessible. In the case of PPP, consultants must provide estimated data and results and will have to spend a reasonable amount of money before they know if they are going to go forward with the PFI option or not, thereby raising the project cost, whatever the selected option may be.

An integrated cost and cash flow model for PPP projects is required to overcome some of these problems. A tool is needed to model the project cost, cash flow, and to assess the affordability and viability of the project investment. If the tool provides the flexibility to change inputs and check outputs, this may facilitate the assessment of alternatives, which is important for the feasibility studies and VFM in PFI projects.

12.2   Research in PPP Financial Modelling

Whilst PFI is considered to be a new procurement system its share of the total construction industry outcome is growing. By August 2004, only seven PhD research studies into PFI/PPP had been undertaken in the UK. Al-Sharif and Kaka (2004) reported that only 2.61% of the total papers (1314 papers) published in four of the top journals of the construction sector related to PFI/PPP. Although the study was limited to only four journals over a 6-year period (1998–2003), it suggested a need to involve academic researchers in this field to find solutions and overcome problems so as to attract construction firms to bid for PFI/PPP projects and to further ground PFI/PPP research whilst enhancing its quality.

Research into PPP financial modelling in particular has been limited. Handley (2002) classified PFI financial tools as follows:

- Pre financial close: an evaluation tool. Is the deal worth entering into? Is it known how much the procuring authority can afford to pay? Will this amount cover the anticipated costs and provide a return to investors?
- Post financial close: a monitoring and control tool used during the build and operational phases, for compensation on termination, changes and refinancing.

Daniel (2002) listed the features of best practice in PFI financial modelling as follows:

- Design the output first: consider the purpose and audience at each stage of the bid.
- Keep things simple: do not model irrelevant things.

- Calculate things in one place: the best layout for calculation is not best for printing, so:
  - Ensure logic easy to check
  - Speed recalculation
- Maintain a 'house' structure and methodology: repeatable, improvable and transferable, because modelling is central to the bid.

MacMillan (2002) stated that PFI models currently conform to funders rather than inform the client. According to Fox (2002), funders require models to provide the following:

- A demonstration that the project is compatible with basic commercial terms in terms of maturity, cover ratio, margins, equity internal rate of return (IRR).
- A demonstration that the project is robust enough to cope with economic and performance sensitivities (e.g. inflation, operating and lifecycle cost, payment deduction).
- A robust forecasting and monitoring tool for the next 30 or more years.
- A product that consumers other than the modellers can understand.

Whilst there has been limited academic research into PPP financial modelling, the situation is different in the case of traditional construction projects. There has been considerable research in cost modelling, whole lifecycle costing and cash flow forecasting. PPP as a procurement system is concerned with the integration of the different phases of construction projects (design, construction and facilities management). In the same way, financial modelling in PPP projects would entail the integration of different aspects of financial management research. The following, outlines some of these areas of research that will form the basis for a PPP project level financial model.

12.3   Cost Models

Unlike the economists' world of mass production, where repetitive production costs and price levels are assumed to be accurately known, the construction–contracting situation demands that estimates be made of the price level and future production costs for a product, and in an environment which may only loosely resemble previous products and environments. Consequently the situation must be viewed as non-deterministic and events will need to be considered in terms of their probability of occurrence (Skitmore, 1989).

Cost estimation could be described as the technical process or function undertaken to assess and predict the total cost of executing an item of work at a given time using all available project information and resources (Kwakey, 1996). The purpose of cost estimating activities in the inception and feasibility phases is to determine the cost range together with some indications of quality or advice on the owner's cost limits. Cost planning in this phase refers to the process in which it is decided whether or not construction of the project is suitable under the prevailing physical and legal conditions (Tas and Yaman,

2005). In order to control the cost within an acceptable level, it is necessary to have an appropriate and accurate measurement of various project-related determinants and the understanding of the magnitude of their effects (Chan and Park, 2005).

A construction project is unique, wide in scope and high in cost; creating a prototype is not only uneconomical but also impractical. Therefore, it is crucially important to produce a forecast of the probable total building cost (Tas and Yaman, 2005). Cost is one of the measures of function and performance of a building and should therefore be capable of being 'modelled' so a design can be evaluated (Ferry *et al.*, 1999). Construction costs and cost modelling are among the subjects most often dealt with in construction management research studies. Some studies have aimed to survey how cost models are applied in reality. These studies were summarised by Fortune (1999) who listed 32 different types of cost models available.

Ashworth (2004) classified cost models based on their purpose and stated that, while there is overlap between them, their chrematistics will be different. The main cost model purposes he classified were design optimisation, tender prediction and resources based. Ferry *et al.* (1999) classified cost models based on the design stage they were undertaken at, ranging from briefing stage to working drawing.

Akintoye (2000) identified seven main factors that influence the contractor's cost estimate: project complexity, technological requirements, information, team requirements, contract requirements, duration and market requirements.

In cost estimation, construction companies generally produce cost data after the completion of contractors' work and settlement of final accounts. Historical cost data derived from similar projects provides feedback to assist with future designs. Consequently, larger construction companies tend to have larger cost databases (Tas and Yaman, 2005).

PPP projects are tendered for based on output specifications. Ideally, this is expressed in terms of functions of the product to be provided rather than its design specification. Unfortunately, no research has been carried out to develop a cost model based on building functions. Furthermore, no data is available to correlate functions with cost. In these circumstances the authorities, and later the project team, must translate functions to design specification and use a cost model based on traditional projects given the limited data available on historical PPP projects, to produce a cost estimate.

12.4   Occupancy Cost

Whole lifecycle cost (WLCC) is defined as the total cost of an asset over its operating life, including initial acquisition and subsequent running costs (Flanagan and Norman, 1983). Seeley (1996) defined lifecycle cost (LCC) as a technique of cost prediction by which the initial constructional and associated costs and the annual running and maintenance costs of a building, or part of a building, can be reduced to a common measure. This is a single sum representing the annual equivalent cost or the present value of all costs over the life of the building. In other words, the LCC of an asset is the present value of the total cost of that asset over its operating life, including initial capital cost, occupation costs, operating costs and the cost or benefit of the eventual disposal of the asset at the end of its life (RICS, 1999).

Operating costs are the cost associated with operating the building itself (RICS, 1986) and include estimates of rent, rates, energy costs, cleaning costs, building-related staffing costs and other costs (RICS, 2006). Maintenance costs are the cost of keeping the building in good repair and working condition, including painting, decorating, repairs and renewals (Al-Hajj and Horner, 1998).

Occupancy costs are the cost of performing the function for which the building is intended. They are distinguished from operation costs as they relate to the costs attributable to a specific process undertaken by the client, which may change within the life of the building (RICS, 2006).

According to Flanagan and Norman's (1983) definition of LCC, running costs are any cost included in the LCC apart from initial acquisition cost. Many terminologies are used in literature to describe this cost including operating cost, maintenance cost, running cost, occupancy cost, maintenance and operation cost and facilities management cost.

Several mathematical modelling techniques are available for calculating the LCC cost in building investments. Boussabaine and Kirkham (2004) and Kishk *et al.* (2003) listed many mathematical equations used to calculate the whole-life cost by applying the different investment decision-making methods, such as present worth, NPV, discounted payback period, internal rate of return, and others. These models employ mostly the NPV (Boussabaine and Kirkham, 2004). The WLCC decision-making exercise can be done at any stage of the project, however, it is most beneficial during the early design stage (Kishk, 2005).

The LCC mathematical models are simple to apply in the main, although Kishk *et al.* (2003) stated that their accuracy lies outside the expected range. This could be because the mathematical equation relies on variables that require judgement that could be the source of inaccuracy.

There are many computer-based models listed by Kishk *et al.* (2003) for the calculation of LCC and comparisons of alternative options. One example is the Internet-based software provided by Ampsol Ltd. to estimate the costs over the life of the equipment, taking into consideration the purchase price, running costs and overheads. The model requires the user to enter the capital cost of the item of plant in the set-up cost, yearly consumable cost, yearly maintenance cost, refurbishment cost and its interval, and occasional cost and its interval. Inflation rate per annum should also be considered for each cost element. The analysis period (project duration) and discount rate should then form the basis for the NPV of the LCC cost. The result appears in a graphical format in a small window in the screen of the trial version of the software.

Another model was developed by the National Institute of Standard Technology (NIST) to assess the WLCC of bridges. The model allows the user to compare the LCC between a number of alternatives. The 'new project wizard' allows the user to define the project, alternatives and dates and allows for the setting of inflation and discount rates. The user then defines physical elements in the structure, the classifying and quantifying dimensions of each structure alternative. The final step is to determine the input optional cost and service life data. The cost data entry includes construction cost, operation, maintenance and repair costs, years between repairs and the disposal cost. The model then calculates the PV of LCC cost of each alternative.

The occupancy stage is critical in the PFI contract due to its length and potential for different types of risk. The prediction of WLCC helps in investment decisions and comparison with PSCs. This information forms the basis for VFM assessment.

12.5   Cash Flow Models

It is often said that cash is king. In construction contracting cash is the contractors', and sub-contractors', number one concern (Kaka, 2001). Cash flow forecasts are of great importance to construction contractors and the client to help prevent the unsavoury consequences of liquidation and bankruptcy. However, the accurate forecast of construction cash flow has been a difficult issue owing to the risks and uncertainties inherent in construction projects (Odeyinka *et al.*, 2002). Research into cash flow has, in the main, concentrated on two main processes: how to forecast cash flow (Kaka, 1994; Kenley and Wilson, 1986; Navon, 1995) and how to manage cash flow (Kenley, 1999; Kaka and Lewis, 2003; Cheetham *et al.*, 1997) with the former receiving significantly more attention than the latter.

Cash flow forecasting at the tendering stage needs to be fast and simple considering the short time available and the associated cost (Kaka and Cheetham, 1997). Accurate cash flow forecasting is an essential activity during the bidding stage for all successful construction contracting organisations. It provides contractors with information such as the amount of capital required to perform a contract, the amount of interest to be paid to support any overdraft required and evaluation of different tendering strategies (Kaka and Fortune, 2002).

These risks and uncertainties are unquestionably greater in PFI projects. Cash flow forecasting in PPP projects is essential at the bidding stage, not only to enable the project team to ensure adequate funding for the project but, more importantly, the cost of finance has to be estimated and added to the tender value. Cost of finance in PPP projects is significant and often forms a large proportion of the unitary value tendered for. Traditional cash flow forecasting models are focused on the construction phase and their use in PPP projects is plausible; however, the model here must span beyond the construction phase to include pre-construction and the very long period of time during occupancy and operation of the PPP facility.

12.6   PFI Financial Modelling in Practice

The following highlights the outcome of a series of interviews undertaken with members of the industry, examining the practice of financial modelling in the early stages of a PFI project. It was found that contractors and client organisations are unable to prepare the financial models in-house. Consequently they use independent consultants and financial modellers to analyse the financial activities of PFI projects. They estimate the project costing and rely on the financial modeller to do the rest.

12.6.1 Financial models

Practitioners acknowledged the importance of computer-based financial models in encouraging private sector participation in public projects. There are concerns associated with the modelling process as many PFI bidders and their sub-contractors are able to design the project, define the facilities management (FM) processes and calculate the cost input. However, the difficulty is in gathering together the financial information in a model package. This can be an expensive exercise and only a few companies can provide this type of service. Indeed, the effective participation of the private sector in public project investment took off when computer-based financial models became available. This is critical to the analysis of the impact of various scenarios and outputs on cash flow. Computer-based financial models are sophisticated in terms of their ability to analyse uncertainty and 'what if...' questions.

The long duration of a PFI contract makes the financial model the cornerstone of decision making and negotiation between all parties. It also enables the contractors to make the ultimate decision – whether to bid for the project or not. None of the interviewed practitioners were aware of any commercially available software to assist with this process. In the main the software used to model the financial aspects of PFI projects by either the public or private sector is mainly in the form of spreadsheets. In most cases the financial advisor produces the model and each financial advisor has his own model that may differ significantly from what others use but targets the same output. The models are designed to support the decision during the early stages of the project for both the public and private sector. In practice these models are designed to put everything in place for the 'financial close' and are much more complicated than the models used to control the financial process of PFI projects during the operational stages. The latter are mostly traditional accounting and cash-flow software widely applied on other non-PFI projects.

Bidding costs entail the SPV paying the bank advisors, lawyers, financial advisors and contractors their costs for their time and efforts in putting the bid together. As stated, the cost of bidding is often seen as one of the main barriers to PFI projects. Contractors face a great deal of expense before they are awarded the project which is only valuable if they win. Otherwise it is a considerable waste of money. It is claimed that high bidding cost is the reason for the number of competitors for PFI projects declining from six to eight companies to two to three on average. Even with the increase in PFI projects, there was only an average of 1.7 competitors for each project in Scotland during the last 2 years. The Highland school project entailing the building of 11 new schools on ten sites under a 31-year PFI concession is a prime example of where only one bidder tendered for the project.

Financial models are usually checked by a legal advisor to ensure that the inputs and assumptions match all of the contract documents and the risk allocation fully agreed in the contract. Lawyers are not concerned with the mechanics or the output of the model; these are the concerns of the clients and their financial advisors. Nevertheless, the model should be simple enough to be read and understood by all parties in order for them to give the right advice or contribution.

12.6.2   Project costing

Contractors depend on the superficial size of the project, in square metres, to determine their initial project cost. This information comes from client requirements detailed in the output specification. They start their arrangements for the project using these preliminary numbers to predict the project value and requirements. Simultaneously, they start the design, either by using in-house engineers or often by employing a design consultant. The client issues an invitation to negotiation (ITN) and normally allows 12–14 weeks for proposal submission. This is a short period for the completion of design documentation normally used as the basis for an accurate cost estimate. The FM contractor provides the FM cost at this stage on a square metre basis for each element of the services using historical data and data from indices such as the Chartered Institution of Building Service Engineers (CIBSE) Guide for energy efficiency in buildings. The average FM rate is then generated from the total FM cost in relation to the building area. The lifecycle cost is provided by the cost consultant, again as a cost per square metre, encompassing the cost of changing building elements or conducting building refurbishment work during the life of the building.

In estimating the project cost, the main items included are the cost of bidding, construction and facilities management. The debt arrangement roll-up and diligence cost are then added, together with the SPV's management and insurance costs. The total of these costs form the basis for the unitary charge for the project. In terms of costing, estimating the cost of the construction project and simply adding further factors or items to the running and maintenance cost for a PFI project is not particularly difficult if one can develop a format that is compatible with the classification of categories founded in practical cost indices. The difficulty arises when attempting to develop a standard format that suits all. This could be done based on historic data with some recognition of the risks and assumptions associated.

Many factors affect the total cost of a PFI project including the type of project, interest rate, taxation, inflation and risk assessment. These factors should all be assessed and incorporated into the model. The cost of financial models and other costs of consultancy services during bidding stage are high which is why the abortive cost is very high. For these reasons few contractors and major players have pulled out of bids altogether.

To produce a cost estimate for a project before a complete design has been prepared project teams rely on the use of historic cost models. In the case of a school project, the size of the facility will usually drive cost figures based

Part Two

on the knowledge of past school projects and what they cost per square metre. This is used to calculate a lump sum figure per school using the given information on the number of pupils and the internal floor area per person.

Another way to cost the project and decide whether or not to bid is to acquire general information from the client before even purchasing the project documents. In a school project, the contractor can search for information about the project such as the number of schools to be built and the number of pupils involved and use this to calculate the size of the facility.

Shortly after the Labour Party came to power in 1997, it was made explicit that assets would remain owned by the public sector throughout the project lifecycle. Until recently the practice was to lease out assets to the private sector and then lease them back to the public sector leading to the perception that the private sector owned the assets. Now the approach used is 'contract debtor accounting' which provides the right to occupy and provide services. When it comes to projects such as electric generation, it would be much less important for the public sector to own those assets because the interest is purely in the output, i.e. electricity. Policy seems to be moving towards assets being placed within the public sector but it is suspected that the dynamics are still not clear. It makes little sense for local authorities to own huge waste disposal units in 25 years' time when the PFI contract comes to an end.

12.6.3   How do the financial models work?

A financial advisor will be appointed by the SPV and will be responsible for the provision and running of the financial model forming the basis for submitting the tender. As shown in Figure 12.1, the financial advisor relies on other parties to provide the financial data needed for the model. The project company SPV will provide the initial cost of the project and its management cost. The construction contractor will provide the construction cost and the LCC on a monthly basis. The FM company will provide operation and maintenance costs (FM cost) on a 6-monthly basis and the banks will provide financial information related to project financing. The financial advisor will collect all cost estimates for the project and feed them into the model together with adjustments to the number of occupants and variable rates to suit the service provider target.

These models are commercially very sensitive. They can make a significant difference to the final bidding offer and the chances of winning the project. By optimising the model the contractor can make their bid more competitive by reducing the capital required for the project giving them a large commercial advantage. For a school project, it is claimed that using the financial model can make a large saving (5–10%) to the client organisation, and/or a substantial profit margin to the SPV.

The client will provide data on the budget available to the public sector for the project including the first year's unitary charge and the overall NPV of the project. The financial advisor will then be able to use the indexation provided in the contract to schedule how much money will be available each year. The financial advisor compares these with the minimum unitary charge

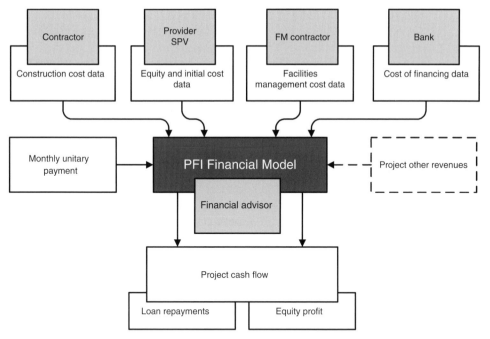

Figure 12.1 Current use of financial models in PFI projects.

that could be accepted by the equity provider, and these inputs are adjusted in coordination and negotiation with those sub-contractors and other parties who provide the figures.

Lifecycle cost

There are three parts to LCC: the lifecycle of the building and its fabric; the lifecycle of the furniture (tables, chairs etc.); and the maintenance of the building fabric. The distinction between the lifecycle of the building fabric and the maintenance of the building fabric is that the latter relates to things like window cleaning, cleaning gutters, cleaning carpets, or servicing the boiler plant. A window may last 20 years and should be replaced at that time. In the intervening years if a pane of glass is broken or it has to be re-painted, this will be undertaken by the FM contractor working for the builder. The same pertains for other elements of each school spread out over the 20–25 year operating life of the project.

There are no specific details included in the pricing. There are many assumptions and the pricing is based on these assumptions. Contract conditions include a process by which the detailed design is developed to confirm the assumptions, much of which depends on established working relationships based on past experiences. In the majority of cases, the people involved in the negotiation of the contract, such as the facilities manager and other contractors, belong to the same group of companies facilitating this process. The

Part Two

agreed pricing will indicate the extent to which the LCC has been taken effectively into account and the level of profit margins expected. The interviewees suggested that whether PFI projects will result in adequate profit margins will only be known in a few years time when there is a significant number of projects with 10–15 years' maturity. Only at that time will it be possible to evaluate the reliability of the lifecycle estimates. Typically technical advisors produce an estimated schedule of when each building item will need to be replaced and this is used to calculate the LCC.

Risk assessment

The models currently being used have no specific provision for costing risks within a bid. Construction costs normally include a margin to cover for risks and uncertainty and the same applies to FM costs where input of the risk is hidden in the operation and maintenance costs. The only time the financial advisor puts a particular risk premium within the construction cost would be when an item of a section of the building has been identified in the contingency section of the contract. In cases where uncertainty is high, the impact of the risk on the project cost is separately assumed and taken up by the client. For example, in historical cities, archaeological risks are often taken up by the council. In addition to the risk associated with cost of uncertainty, the schedule of works will also need to be examined and assumed so as to ensure completion of the construction work by the agreed time scale provided in the contract.

Risk assessment in PFI projects is rather less scientific than might be imagined. The client identifies risk in three parts: the risk associated with construction cost, resources cost and delivery risks. In terms of the cost of construction, the risk is the same as any other construction project. To build a school, hospital or office block, one looks at the design, the schedule programme that is required to build the school, resources availability, ground and site conditions, surrounds and the overall timescale in which the project is going to be executed. The contractor assesses each of these factors when producing the cost plan.

There is one further risk in LCCs and that is employment cost. In most contracts the soft service element, such as cleaning, is market-tested for 5 or 7 years so the labour cost risk is effectively shortened. However, in the case of replacement, which requires labour to carry out the works required any time in the future, the cost is not market-tested and therefore there is a labour cost risk extending to 25–30 years. The risk associated is considered to be moderated by the assumption that the actual replacement cost of the materials is the significant element. However, predicting labour cost over a 25–30 year period is still difficult.

The standard contract for schools recognises the risk due to a change of law or due to vandalism and provides for it to be shared between the client and the project company. If there is a change of law which requires changes in the configuration of the building then a capital cost is incurred which is passed on to the client. Nevertheless, such laws are reasonably manageable and foreseeable as there is a 4–5 year gestation period before proposed legislation becomes law.

Taxation

SPVs are entities working in the UK where UK taxation rules should be applied. Models are normally designed using the best available tax information and should be subjected to any taxation change within the project period. Tax change is a risk that needs to be identified and agreed upon. Normally, if the changes within the general taxation regime affect everyone then it is the SPV's risk and they have to deal with it. If the change is considered to be a discriminatory change in the tax laws affecting only SPVs, or even more specifically SPVs running school projects then that could be classified as a council change and there would be compensation for the SPV.

Inflation

Inflation is contract specific. Typically the client would agree to pay a unitary charge indexed at a factor that follows the RPI which is the standard inflation in the UK. This means that the unitary charge will be increased annually but the client will have to choose whether they are prepared to take the full indexation risk, in which case the rate goes up according to RPI, or they can choose to share the inflation increase with the SPV. Some contractors work on two-thirds inflation of Imported Price Indices (IPI) which is a series of economic indicators that measure change in prices of goods and raw material imported into the UK. It is only a small proportion of the unitary charge that is affected by inflation. Inflation mostly affects FM costs, although the FM company usually takes this into account when obtaining funding to allow recovery of the difference between RPI and actual wage costs.

In most PFI contracts the inflation rate is fixed for the project duration. This means most of the risk is weighted towards one party. The client may choose to pay part of the inflation which distributes the inflation risk to both parties. This is one aspect of the financial risk the FM contractor faces. In the main, a fixed inflation rate is set that is higher than the current inflation rate for the project's duration. If the inflation rate is 2.5% the FM contractor may assume an inflation rate of 3.5% which means a gain of 1% is achieved if inflation remains at its current levels. If inflation increases to 4.3%, for example, the contactor would lose 0.8%.

Payment mechanism

School project payments used to be based on the number of pupils but this may not always be the case. For example, the standard version 3 of the Scottish School Contract is based on the gross service units (GSU) for the whole of the project. Contractors are not paid on the basis of how many pupils there are but on a utility charge depending on whether there are 100 pupils or 1000 pupils. The pupil numbers are simply used to break the utility charge down and calculate how much a particular classroom is worth. However, the construction companies and consortiums do not take volume risk. They are paid the full amount and any deductions are based on availability targets set out in the performance specifications. Therefore, the demand risk is assumed by the public sector.

Payment deductions

The level of deductions appears to be low in relation to the overall contract, but much higher relative to FM contracts. In general, the contractor will take certain design risks, but the FM contractor will take the majority of the risk. The FM contractor does however have rights against the main contractor in the case of a design failure that causes a loss. There is rather a misconception in terms of how much damage a particular risk can cause. Any compensation should be assessed in relation to the financial impact on the FM contractor rather than to the overall contract. It is believed that there are many early projects where the payment mechanism was not negotiated in this way. The key is to ensure that performance measurement is linked to the relevant players and that there is a good performance measurement system for measuring these. In advance of the procurement the authority should carefully test to ensure that the repair mechanism is properly calibrated. As nearly all projects will suffer reductions, they will always appear to be relatively low compared with the value of the overall contract.

Legal aspects

There are some legal aspects associated with the financial models. The financial models are contractual documents (including the software with all the formulae used to calculate changes). In principle if changes are instructed, the public sector will ask for the model to be used to calculate the associated costs of the change. Those changes are entered into the financial model and a new price is calculated whilst keeping the repayment schedules, debt, dividends and IRR the same.

It is important to detail in the contract who is responsible for the development of the financial model. There have been many projects where errors are discovered later in the process impacting on the overall costs of the project. Such errors, if discovered late in the project, or worse, after financial close, would cause major problems and ultimately affect the financial position of the key players. Often project funders ask for the financial model to be audited prior to the signing of the contract. This cost, typically in the range of £50 000–100 000, is usually borne by the contractor.

12.7   An Example of a PFI Financial Model for Schools

The construction industry needs to be more involved either in development or in applications of PFI strategies and processes. The number of PFI

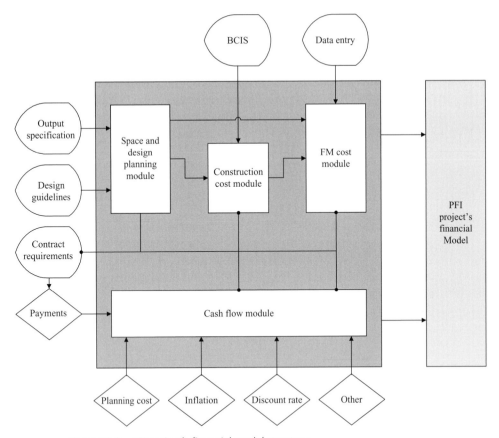

Figure 12.2   PFI project's financial model structure.

projects is growing in the UK and worldwide. The bidding process, the as-sociated costs and timespan are often seen as constraints by construction organisations when bidding for PFI/PPP projects. Developing an automated, or semi-automated, financial model for use at the early stages of a PFI project could provide a tool for decision making before any major cost is committed to the project.

This section outlines the development of a computer-based model to enable project teams to forecast and assess the cash flow of PFI projects at the tendering stage. It would also give the private sector the ability to predict the project's profitability and assist in negotiation with their sub-contractors as well as the project client. At the same time, the model could help the public sector to establish the potential scope and to develop the reference project.

The model structure as shown in Figure 12.2 contains four modules linked to each other. The strategy is to provide the required data at the beginning for use within each module and transfer to other modules. For example, the cash flow module will assess the (cash out) items from the construction cost module, the FM cost module, and from other data entries for other cost items such as initial cost (to be provided by the project company), and the cost of financing (to be provided by the bank through the SPV). Other data entries, such as interest rate and inflation rate, are entered by the user individually. Against that income data entries are provided by the user based on the client budget or assumptions to be evaluated. The final results should lead to the calculation of the project profit and the project cash flow following any repayments of the loans the project company commits to.

The construction cost module developed using data selected from the Building Cost Information Service (BCIS) on past schools projects is used to calculate construction cost depending on the output of the space planning module. The FM cost is based on the output of the space module and unit cost data entered by the user according to the Building Maintenance Index (BMI) cost categories. This calculated cost is adjusted for inflation using indices published by BCIS for indexing construction cost and tenders in the UK. These costs are exported to the cash flow module to calculate the project cash flow.

The model is applicable to school projects where the school net size is calculated according to the type of school and number of students. The proposed model provides the flexibility to adjust or amend the calculated size according to the project's specific needs and compare it with recommended school building size in the 'Building Bulletins: Area Guidelines for Schools' published to assist clients developing design briefs with the necessary detail and ensuring the priorities of the school are clearly expressed and can be carried through the design (DfES, 2004a,b).

12.8   Conclusion

A review of the literature and current practices in PPP financial modelling has revealed a need to examine the causes associated with the high cost of bidding. In the UK this has had the impact of seeing single-tender bidding starting to appear, in effect turning the process into a negotiated contract which may give single consortia too much power (Cartlidge, 2006). This highlights the need for a tool to enable clients to assess the project's affordability before they commit non-compensated funds. At the same time the private sector needs a tool to assist them in making the right decision(s) on a project before they spend a large amount of resources and to enable them to assess whether they are capable of competing for the project and if it is compatible with their bidding strategies.

This chapter outlines the issues that need to be considered when developing a financial model for PPP projects and the underpinning research and development associated. Whilst research in this particular area is scarce, the different models and methodologies developed for traditional building projects could still be used to support such an effort. The model outlined in this chapter was developed to support school PPP projects, but the structure and strategy could be applied for the development of PPP projects in other sectors.

References

Ahadzi, M. and Bowles, G. (2004) Public-private partnership and contract negotiations: and empirical study. *Construction Management and Economics*, 22, 967–978.

Akintoye, A. (2000) Analysis of factors influencing project cost estimatig practice. *Construction Management and Economics*, 18, 77–89.

Akintoye, A., Beck, M., Hardcastle, C. *et al.* (2001) *Framework for Risk Assessment and Management of Private Finance Initiative Projects*. Glasgow Caledonian University.

Al-Hajj, A. and Horner, M.W. (1998) Modelling the running cost of buildings. *Construction Management and Economics*, 16, 459–470.

Al-Sharif, F. and Kaka, A. (2004) PFI/PPP topic coverage in construction journals. *Proceedings of 20th Annual ARCOM Conference*, Heriot-Watt University. Association of Researchers in Construction Management, Vol. 1, pp. 711–719.

Ashworth, A. (2004) *Cost Studies of Buildings*, 4th edn. Pearson Education Limited, London.

Bing, L., Akintoye, A., Edwards, P.J. and Hardcastle, C. (2005) The allocation of risk in PPP/PFI construction projects in the UK. *International Journal of Project Management*, 23, 25–35.

Boussabaine, A. and Kirkham, R. (2004) *Whole Life-Cycle Costing: Risk and Risk Responses*. Blackwell Publishing, Oxford.

Cartlidge, D. (2006) *Public Private Partnerships in Construction*. Taylor & Francis, England.

Chan, S.L. and Park, M. (2005) Project cost estimation using principal component regression. *Construction Management and Economics*, 23, 295–304.

Cheetham, D., Kaka A.P. and Humphreys, G. (1997) Development and implementation of a system of financial planning and control for a medium sized building contractor. *Journal of Financial Management of Property and Construction*, 2(1), 5–34.

Daniel, H. (2002) Financial modeling strategies. In: *Financial Modeling Strategies for PFI/PPP Projects*, March 4–5, London. SMI.

DfES (2004a) *Building bulletin 99: Briefing Framework for Primary School Projects*. Department for Education and Skills, London.

DfES (2004b) *Building bulletin 98: Briefing Framework for Secondary School Projects*. Department for Education and Skills, London.

Flanagan, R. and Norman, G. (1983) *Life-Cycle Costing for Construction*. Royal Institute of Chartered Surveyors (RICS), London.

Ferry, D., Brandom, P.S. and Ferry, J.D. (1999) *Cost Planning of Buildings*, 7th edn. Blackwell Publishing, Oxford.

Fortune, C.J. (1999) *Factors Affecting the Selection of Building Project Price Forecasting Tools*. Unpublished PhD Thesis, Heriot-Watt University.

Fox, G. (2002) Sources of Finance for PPP/PFI projects: strategically tailoring the model to your project. In: *Financial Modelling Strategies for PFI /PPP Projects*, March 4–5, London. SMI.

Handley, M. (2002) Legal implications for the financial model post financial close. In: *Financial Modelling Strategies for PFI/PPP Projects*, March 4–5, London. SMI.

Kaka, A.P. (1994) Contractors' financial budgeting using computer simulation. *Construction Management and Economics* 12(2), 113–124.

Kaka, A.P. (2001) The case for re-engineering contract payment mechanisms. *Proceedings of 17th Annual ARCOM Conference*, 5–7 September 2001, University of Salford. Association of Researchers in Construction Management, Vol. 1, pp. 371–379.

Kaka A.P. and Cheetham D. (1997) The effect of some tendering and payment strategies to cash flow forecasting. *Proceedings of the 13th Annual Conference of the Association of Researchers in Construction Management*, pp. 574–583.

Kaka A.P. and Fortune C. (2002) Net cash flow modelling and the development of an alternative third generation approach. *Proceedings of 18th Annual ARCOM Conference*, Northumbria University, Newcastle-upon-Tyne, Sept 2002, Vol 1, pp. 73–83.

Kaka A.P. and Lewis J. (2003) Development of a company-level dynamic cash flow forecasting model (DYCAFF). *Construction Management and Economics*, 21, 693–705.

Kenley, R. (1999) Cash farming in building and construction: a stochastic analysis. *Construction Management and Economics*, 17(3), 393–401.

Kenley, R. and Wilson, O. (1986) A construction project cash-flow model: an idiographic approach. *Construction Management and Economics*, 4(3), 213–232.

Kishk, M. (2005) On the mathematical modeling of whole-life costs. *Proceedings of 21st Annual ARCOM Conference*, 7–9 September 2005, SOAS, University of London. Association of Researchers in Construction Management, Vol. 1, pp. 239–248.

Kishk, M., Al-Hajj, A., Pollock, R. *et al.* (2003) *Whole Life Costing in Construction: A State of the Art Review*. RICS Research Papers.

Kwakey, A. (1996) *Understanding Tendering and Estimating*. Gower Publishing, England.

MacMillan, J. (2002) Financial modeling techniques for bid evaluation: using modeling techniques to maximise value from bids In: *Financial Modelling Strategies for PFI/PPP Projects*, March 4–5, London. SMI

NAO (2001) *Modernising Construction*. National Audit Office, London.

Navon, R. (1995) Resource-based model for automatic cash-flow forecasting. *Construction Management and Economics*, 13(6), 501–510.

Odeyinka, H.A., Lowe, J.G. and Kaka, A. (2002) A construction cost flow risk assessment model. *Proceedings of 18th Annual ARCOM Conference*, 2–4 September 2002, University of Northumbria. Association of Researchers in Construction Management, Vol. 1, pp. 3–12.

RICS (1986)*A Guide to Life Cycle Costing for Construction*. Royal Institute of Chartered Surveyors, London.

RICS (1995) *The Private Finance Initiative: The Essential Guide*. Royal Institution of Chartered Surveyors, London.

RICS (1999) *Surveyors' Construction Handbook*. Royal Institute of Chartered Surveyors, London.

RICS (2006) *Construction Whole Life Costing*. Available online from http://www.hostref.com/NXT/GATEWAY.DLL/Environment/news?f=templates&fn=default.htm&vid=rics] Royal Institute of Chartered Surveyors. (Accessed on 25/05/2006)

Rintala, K. (2004) *The Economic Efficiency of Accommodation Service PFI Projects*. VTT Technical Research Center of Finland, Finland.

Seeley, I.H. (1996) *Building Economics*, 4th edn. Macmillan Press Ltd, London.

Skitmore, M. (1989) *Contract Bidding in Construction: Strategic Management and Modelling*. Longman Scientific and Technical, London.

Tas, E. and Yaman, H. (2005) A building cost estimation model based on cost significant work packages. *Engineering, Construction and Architectural Management*, 12(3), 251–263.

13

Application of Real Options in PPP Infrastructure Projects: Opportunities and Challenges

Charles Y.J. Cheah and Michael J. Garvin

13.1 Introduction

Much has been written and reported on private participation in infrastructure projects globally. Private participation comes in different modes – the most commonly cited ones are PPPs, PFIs and build, operate, transfer (BOT). Obviously, the structure and issues associated with these different forms are many, but generally speaking, the two common characteristics among them are: (a) private finance is at risk and (b) the arrangements are long term. Since the theme of this chapter is related to project evaluation techniques and risk management, our intention is not to differentiate or distinguish between the various private participation arrangements and structures. Likewise, since the issues discussed are applicable across PPP, PFI, BOT or other structures, we make no distinctions among these schemes and collectively they are referred to simply as 'PPP projects'.

PPP projects often have longer tenure and require more integration of services from project participants. Consequently, they are perceived as more risky than those delivered through 'conventional' modes (such as design–bid–build and design and build). It is important for all project participants, be it the sponsors, contractors, lenders or investors, to assess and manage the risks involved in a proactive manner. Firms continuously make decisions whether or not to invest in these risky projects. Specifically, many large-scale projects require huge initial outlays in exchange for an uncertain stream of future payoffs. The decision becomes more challenging when a project has a great deal of uncertainty *ex ante* regarding its value. Traditionally, the economic feasibility of a project is analysed using discounted cash flow (DCF) techniques. In fact, many standard corporate finance and engineering economics texts recommend the net present value (NPV) method as a better investment criterion than other conventional approaches, such as the internal rate of return (IRR) and the payback rule (Park and Sharp-Bette, 1990; Brealey *et al.*, 2005).

In the application of the NPV method[1], two important input parameters are the discount rate and the series of future cash flows. Unfortunately, an accurate discount rate is elusive, since in theory it should be the expected rate of return, whereas expectation for a non-traded asset is hard to measure (Garvin and Cheah, 2004). The assessment of alternative risk management strategies is sometimes founded on NPV analysis. By quantifying and comparing the impact of alternative risk management strategies on the present values of cash flows, an optimal choice is made. The NPV method, however, tends to ignore managerial flexibility in the project management process (Trigeorgis, 1999). NPV analysis works quite well when the risks during the lifespan of an asset remain relatively stable. Luehrman (1997) and Myers (1984), for example, suggested that traditional valuation methods are adequate for investment decisions regarding assets-in-place. In such cases, ongoing operations generate relatively safe cash flows and are held for this reason, not for less tangible strategic purposes. Traditional valuation also works well for typical engineering investments, such as equipment replacement, where the main benefit is cost reduction.

However, future cash flows could change drastically and are also difficult to predict. Investments often create future growth opportunities and cash flows would increase when an expansion is made (e.g. follow-on development if product or service demand is favourable). Likewise, these investments typically have contingency possibilities – the project may be contracted in scale, delayed or even abandoned when the economic environment turns out to be less favourable. In effect, the risk of subsequent cash flows can change as development proceeds or new information is received. By computing an expectation of future cash flows (i.e. in a mathematical sense, an 'average' scenario), the NPV method does not capture these flexibilities well since it implicitly excludes any possibility of altering decisions when circumstances dictate. But, this is not reality: managers often adapt their decision making to the latest development of a project. In such cases, the NPV method understates the value of a project. Among others, Amram and Kulatilaka (1999), Trigeorgis (1999), Dixit and Pindyck (1994) and Myers (1984) have all pointed to this shortcoming.

In fact, many PPP infrastructure projects possess such option-like features, so quantifying the value of such options can be quite significant to the timing of investments. These considerations have led to the recommendation of real options (RO) and contingent claim analyses, which are natural extensions of financial option pricing theory to real-life assets and projects. By definition an option is a right, but not an obligation, to exercise a certain action in the face of uncertainty. The notion of managerial and operating flexibility described earlier obviously matches the definition of an option. The fundamental pricing theory was first developed to value options on financial assets. The theory was subsequently expanded and applied to value options on real assets (e.g. development of a piece of vacant land; exploration of an oil field); hence the term 'real' options.

RO analysis is clearly not a new concept and has been applied to sectors and industries such as oil and gas (Paddock *et al.*, 1988), pharmaceutical (Rogers *et al.*, 2002), manufacturing (Kogut, 1991), airline (Stonier, 1999), mining (Moel and Tufano, 1999), R&D (Schwartz and Moon, 1999) and real estate (Grenadier, 1995) – just to name a few. Readers may also refer to Trigeorgis (1999), Amram and Kulatilaka (1999), Copeland and Antikarov (2001) for general expositions on the subject.

13.2   Infrastructure Project Flexibility as Real Options

Infrastructure projects are ripe with flexibility. Development often proceeds in a series of stages that aim to better define project scope and discover unknown information. Moreover, flexibility is often incorporated as an intuitive managerial approach to deal more effectively with uncertainty. Preliminary planning and feasibility studies, such as environmental impact studies, geotechnical surveys and traffic volume analyses, can reveal information that may alter further investment and development decisions. Flexible design permits infrastructure projects to more readily adapt to changing conditions, such as an increase or decrease in expected demand for the project's output. Staged construction can afford decision makers the opportunity to gain more information as market conditions become more certain. In short, flexibility can effectively reduce lifecycle costs by allowing a timelier and less costly response to a dynamic environment. Flexibility adds value, but it also comes at a cost in terms of money, time and complexity. Obviously, this added value should be weighed against its cost. Generally speaking, the value of an option itself is greater when the uncertainty of the asset value in question is higher. In addition, an option should probably be executed (or 'exercised') when the value of the asset exceeds its strike price (which is often the cost of asset development or a floor on the asset's value).

Part Two

13.2.1   Types of options

In general, the typical types of options found in PPP projects are:

- Call option: an option to secure/procure an underlying asset when the asset value exceeds a certain threshold known as the 'strike price'. Some common examples include: capacity expansion; procurement option; splitting of projects into two phases, whereby execution of the second phase would be contingent on the success of the first.
- Put option: an option to sell an underlying asset at a strike price when the asset value is lower than the strike threshold. Effectively, a put option provides a floor to the asset value (analogous to insurance). Examples include: capacity contraction; shutdown or sale of assets; abandonment option; guarantees granted by government or other parties.
- Switching option: this refers to the flexibility to alter the *modus operandi* of any given business/facility, so as to adopt the path that would derive the largest payoff. Examples include: switching between operation/construction modes; switching use of fuel source for power plants.
- Timing option: an option to defer a specific action such as developing a piece of vacant land or commencing construction. In such cases, the timing

option is technically similar to the call option since it entails an action of 'buying in'.

- Compound option: this is a combination of two or more types of options. For example, when a project is split into multiple phases, the execution of subsequent phases is contingent upon the success of the initial phases. In total, this represents a series of call options. When evaluating a compound option, interactions among two or more types of options impose greater complexity as the total value is usually not a simple sum of the parts (Trigeorgis, 1993). Furthermore, the value is affected by correlations of the asset movements that underlie the different types of options.

- Learning option: this is a more subjective type of option as it treats the entire project or business venture as a learning ground or part of a long-term strategic plan. For example, a pilot project in a politically unstable or less developed country is a learning option for a corporation to explore future business opportunities in the country.

Obviously project components do not automatically appear with 'real option' labels. It is therefore important for project stakeholders to identify scenarios or settings that give rise to flexibility and options. A short case study should help to illustrate some of the points mentioned.

Case Study 13.1   Texas High-Speed Rail Project

Project background

The Texas cities of Dallas, Fort Worth, Houston, Austin and San Antonio, through an unusual combination of geography and demographics, represent a promising market in the US for an inter-city high-speed passenger rail network. They are situated on a rough triangle with 250-mile long legs (Figure 13.1). Residents and visitors travel frequently between these cities. The Texas High-Speed Rail Corporation (THSRC) was formed by the Texas TGV ('Train à Grande Vitesse') Consortium, a group of private investors, to construct and operate a rail network. In 1991, the consortium was awarded an exclusive franchise for such a system by the State of Texas. In one early study, the volume of total traffic was projected to reach about 20 million person-trips per year by 2000, of which 12 million would be by air. Most of the air travel within the state was to and from Dallas–Fort Worth, a major air traffic hub, and most of it consisted of business trips. Although the project was eventually cancelled in August 1994, the original plan is worth studying here.

The THSRC project was planned to have five phases, each characterised by a different set of activities, uncertainties and funding needs. A preliminary phase of relatively low expenditure would last until sometime in 1993. A 2-year development phase would then mark the beginning of large expenditures and involve some irreversible investment. The construction phase would then last for 5 years and involve heavy expenditure in many locations. Next, a start-up phase was planned to begin in 1998 and would last for 2 years as the systems swung into full operation, so that THSRC could make adjustments to operating policies and systems. Finally, the high-speed rail system was expected eventually to arrive at a steady state characterised by stable operations and a new equilibrium in the Texas inter-city transportation market.

THSRC planned to operate trains along the eastern leg of the triangle beginning in 1998, and hoped to offer door-to-door travel times comparable to existing air service at prices competitive with airfare. By the end of 1999, services would then commence along the western corridor from San Antonio to Austin and Dallas–Fort Worth. A southern corridor, connecting Houston and San Antonio, would be added later if sufficient demand materialised.

Figure 13.1   Route alignment of the 'Texas Triangle'.

One expected problem was that the high-speed line between Dallas–Fort Worth and Houston would not be subsidised, so it was be difficult for THRSC to project whether it could match competing airfares and realise high enough load factors to produce an adequate return on investment.

In addition to the rail transport services, THRSC could provide several other potential services such as parking, package delivery and advertising. Furthermore, some of THSRC's right-of-way and station sites could create opportunities for real estate development, though none of THSRC's projections included costs or revenues from such projects. THSRC also planned to lay fibreoptic cables along its right-of-way.

Embedded options

Abandonment option throughout the five phases

Since THSRC would implement the project in phases, THSRC could make projections of the system's prospects based upon information collected in each stage. If the future prospects were favourable, then the project could be continued. Otherwise, the project could be abandoned to avoid larger losses and this is exactly what happened. The phased strategy gave THSRC a series of abandonment options during each of the five phases, and each abandonment option has a real economic value.

Expansion option

THSRC planned to only operate the eastern and western legs of the triangle initially. From its standpoint, construction of the southern corridor was an expansion option. The uncertainty of the southern leg's traffic volume was obviously a key factor, which led to an uncertain asset value but a positive option value. Once the present value of the southern corridor's revenues was greater than the present value of its cost, it would make sense to construct the southern corridor (i.e. once the value of the asset, the expected revenues from the southern corridor, exceeded the strike price, the southern corridor's construction cost, then development should proceed). Otherwise, the decision could be deferred.

Growth option from transporting air passengers

Clearly, the high-speed line between Dallas–Fort Worth and Houston would be competing against the airlines. Given this, THSRC could consider two alternatives. First, it could compete head-on with the airlines. As mentioned, this line would not be subsidised, so achieving an adequate rate of return could be challenging since a substitute transport service was available. Second, THSRC might cooperate with the airlines. By cooperating, the high-speed rail line could transport air passengers from outlying parts of Texas to the Dallas–Fort Worth hub which seems more feasible. The viability of the eastern and western lines would then highly depend upon the income negotiated with the airlines for delivering air passengers and the cost of cooperating with the airlines. In this strategic arrangement, cooperating with the airlines would represent a type of growth option to THSRC.

Call options from other revenue sources

The potential services from parking, package delivery, advertisement, real estate development and so on could form ancillary sources of revenue and could generate stable, if modest, cash flows. Laying fibreoptic cables along its right-of-way would involve low incremental costs during the system's construction phase, and create some immediate cash flow when the resulting telecommunications capacity was sold or leased. All these potential opportunities provided some call options for THSRC to capture additional revenue.

13.3   Real Options Literature Related to Architecture, Engineering, Construction and Infrastructure Projects

Although more commonly applied to other industries, RO remains a fairly novel concept to a large contingent of the architecture, engineering, construction and infrastructure (AECI) community. While the frequent presentation of the RO evaluation methodology as a black box has hindered its transfer into the community, some sceptics have stated their belief that the technical assumptions of the RO methodology do not suit the context of AECI projects. This is unfortunate since our previous discussions have illustrated the multiple flexibilities that exist naturally or are intentionally structured throughout the lifecycle of complex AECI projects. Still, attempts have been made to introduce this concept to the AECI community. Some of the RO articles related to the AECI context are briefly reviewed below.

13.3.1   Literature that is conceptual in nature

Tiong (1995) suggested that one of the most influential factors during the tendering stage is risk. When each party brings its own set of expectations and risk management strategies to the negotiation table, the outcome could well be a zero-sum game if all parties simply attempt to transfer unwanted risks to their counterparts. To infuse a spirit of positive thinking, it is important not to overlook the possibility of value enhancement. Value can be created by structuring operating options and flexibilities during the course of design and execution of a project. Cheah (2004) highlighted the importance of balancing risk and value in a PPP project and suggested that the RO methodology could play a valuable role in quantifying the value of these flexibilities. Without proper consideration or evaluation of these elements, the matching of risk and value cannot be obtained in a structured manner.

Leviäkangas and Lähesmaa (2002) commented that traditional cost–benefit analysis, which is often applied to investments in physical road infrastructure, failed to capture all the benefits or costs related to intelligent transport systems (ITS). While exploring the use of alternative evaluation methods, they confirmed that two different and independent RO approaches produced positive option values for ITS investments when compared to traditional capital investments. Despite being an experimental demonstration, the authors implied that RO stands as one of the two promising tools for evaluating ITS investments (the other one being multicriteria analysis).

Ford *et al.* (2002) derived the value of a flexible design strategy in a project by using a decision-tree analysis combined with a traditional cash flow analysis. In their case, two toll road project design strategies were developed – basic and engineered. Each of these design strategies carried different implications on the reduction of construction cost and also the procurement schedule. The authors showed that a flexible strategy that could use either the basic *or* the engineered design strategy, depending on the evolution of construction cost (which is uncertain *ex ante*), was preferred over either design strategy applied rigidly. The extra value derived by having a flexible approach is due to the existence of an option to choose a more beneficial design strategy when the construction cost moved up or down.

13.3.2   Literature that relates to discrete-time evaluation

In general, RO models fall into two categories: discrete-time and continuous-time. So far, discrete-time models have been more widely adopted in RO studies that are related to AECI projects.

Garvin and Cheah (2004) used a binomial model with a single time-step to evaluate a hypothetical deferment option in the Dulles Greenway project in Northern Virginia, US. Within 6 months of opening, the project was in financial distress since the actual traffic volume was far below forecasts. Since investment decisions normally hinge upon the strength of traffic forecasts and the linkage between the project and general economic conditions, Garvin and Cheah suggested that the project's developers should actually have considered the potential alternative of deferring the project. Deferment could allow acquisition of better information and observation of economic growth in the outlying regions. Analysis of a 5-year deferment option indicated that the value of this flexibility could be as high as $111.8m, as compared to a 'static' NPV value of *negative* $86.3m estimated based on the same set of operating cash flows.

Ho and Liang (2002) developed a five-step model to evaluate the financial viability of BOT projects (which they called a 'BOT-OV' model). They adopted discrete-time approximations to model the stochastic processes of two log-normally distributed variables, project value and construction cost, and subsequently solved for the BOT equity value using a lattice model. They argued that determination of equity value is more realistic this way, since traditional capital budgeting methods cannot account appropriately for the asymmetric payoff under bankruptcy risk. In addition, Ho and Liang asserted that the model can also assess the value of a government debt guarantee and its impact on the project's financial viability.

Zhao and Tseng (2003) emphasised that due to economic-based irreversibility, the expansion of a constructed facility may require enhancing the

Part Two

foundation (such as a building's columns) at an additional cost. This strategy, however, allows for future expansion if the demand for the facility so warrants. They demonstrated the significance of the flexibility trade-off between construction of a foundation strengthened for potential future demand and a foundation designed for expected demand for a public parking garage. They used a trinomial lattice to model the parking demand, and stochastic dynamic programming to determine the optimal expansion process. The case study proved that failure to capture flexibility turned out to be uneconomical.

In a detailed study of the Dabhol Power Project in India, Cheah and Liu (2005) first identified a number of RO that existed in the contractual arrangements among the project parties. By constructing a series of binomial models with multiple time steps, an expansion option, an abandonment option, an extension option and a switching option were evaluated. The value of these options ranged from 3.4–26% of the base case 'static' NPV. Clearly, the true value of the project to each party could be grossly undervalued if these options were not taken into consideration.

Other published works that fall into this category follow:

- In a real estate development venture, delay in construction completion will directly affect the financing costs and rental revenue and may even dictate the success or failure of the entire venture. The 'time to build' concept refers to the optimal rate at which construction should proceed based on the prevailing market conditions for the completed facility (Majd and Pindyck, 1987). Sing (2002) constructed a 'time to build' model for a large-scale construction project.
- Ng *et al.* (2004) set up a model to value a price cap contract and determine an optimal exercising policy in construction material procurement.
- Ng and Björnsson (2004) explained the similarities and subtle differences between RO and decision analysis (DA). They showed that in a complete market, DA and RO give the same valuation regardless of the choice of the utility function of the investor. The values however would differ in an incomplete market.
- In many PPP projects, government subsidies and guarantees can be viewed as a form of options (Mason and Baldwin, 1988). In Cheah and Liu (2006), the authors evaluated a revenue guarantee as a put option and suggested that a 'repayment' obligation (i.e. allowing profit sharing by the government if revenue exceeds certain thresholds) should be similarly demanded from the private sponsor in order to compensate the government for the value given away by the government when issuing the guarantee.
- Similarly, Chiara *et. al.* (2007) developed a novel valuation procedure for a government-sponsored revenue guarantee, which treated this option as a simple multiple-exercise RO. The technique extended the least-squares Monte Carlo method (LSM) introduced by Longstaff and Schwartz (2001) and constructed a stochastic dynamic programming model of the guaranteed party's decision process. The advantage of the technique over prevailing methods is its flexibility in estimating the guarantee's value. Specifically, the technique: allows the guarantor to grant as many exercise rights as they can afford/wish; and permits the guaranteed party to determine

when to exercise these rights *ex ante*, i.e. as information is revealed during the operating period of the project.

- Yiu and Tam (2006) made an interesting proposition that under-pricing in bidding strategy could be a result of the contractor's rational belief in its ability to defer and switch modes of construction in the face of cost uncertainties. The value of such options would compensate for having a more competitive pricing strategy in bidding.

- Mattar and Cheah (2006) introduced the notion of 'private' risk, which differs from the usual classification of a 'market' or a 'unique' risk in finance theory, and studied the effects of private risks in RO problems. They commented that although unique risk is potentially diversifiable, some investors in a PPP project might have difficulty in doing so. This could be due to large agency costs (because investors as insiders always have more information than the market), or simply because the investors are in a better position to retain and manage this risk in their portfolio.

13.4   Modelling Issues and Concerns

Although useful in principle, the many methods for modelling RO and their applications across different settings have created confusion. Moreover, this confusion is amplified by different assumptions that underlie the modelling techniques. Assumptions are often taken for granted, misunderstood or un-realistic. Not surprisingly, critics are quick to jump to the conclusion that the entire RO field is mathematically elegant, but hardly useful in practice.

In some ways, the critics' concerns are not unfounded. In 2000, Bain & Co., a management consulting firm, conducted a survey of 451 senior executives across more than 30 industries concerning their use of 25 management tools. Only 9% used RO. In another 2002 survey of 205 *Fortune* 1,000 chief financial officers conducted by a Colorado State university professor, only 11.4% responded that they used RO, which compared to 85.1% for sensitivity analysis, 66.8% for scenario analysis and a whopping 96% for NPV analysis (Teach, 2003).

In a critical review of the challenges to practical implementation of RO, Lander and Pinches (1998) attributed the lack of acceptance of RO in corporate decision making to three primary reasons:

- The types of models used are not well understood by corporate managers and practitioners. Many managers, practitioners or even academics may not have the required mathematical skills to set up a RO model comfortably and knowledgeably.
- Many of the required modelling assumptions are often and consistently violated in practical applications.
- The necessary additional assumptions required for mathematical tractability limit the scope of applicability of RO.

Borison (2003) provided additional insights with his classification of RO approaches into five different categories. He then reviewed the underlying assumptions and mechanics of each, before using a simple example to contrast

Part Two

the value of options obtained from all five approaches. He concluded that the differences in valuation were not minor, and that they lead to differences in value-maximising strategies. Borison, therefore, concluded that although there is a consensus about the conceptual appeal underlying RO, practitioners are left in 'troubling circumstances' with a good chance that either they will select the wrong approach or they will choose the correct approach but apply inappropriate assumptions.

In view of the above concerns, we provide a general review of the mechanics of common RO modelling approaches and their associated assumptions, before stating our opinion of RO's applicability to the AECI context.

13.4.1   Single- or multi-factor models

One of the first choices that a modeller must make is how to represent and model the value inherent in the project since the evolution of the asset's/project's value through time will ultimately dictate the value of any option linked to the asset/project. As our previous literature indicated, researchers have used both single-factor and multi-factor models of asset value. Obviously, additional factors increase the complexity of the model, but it might also improve its representation of reality. Copeland and Tufano (2004) have suggested that very often the underlying value of a real asset is driven by one key variable. On the other hand, many financial analysts of complex infrastructure projects are likely to agree that it is difficult to filter the factors that influence a project's value down to a single variable. Thus, this situation introduces one of the many trade-offs faced in RO modelling. A single-factor model permits the use of familiar and simpler techniques such as the binomial tree method whereas a multi-factor model will generally require the use of Monte Carlo simulation techniques (or more sophisticated lattice models). Until recently, the use of Monte Carlo simulation was only possible on the simplest form of options (i.e. European); however, recent advances have made its use more practical for other forms of options. Still, the implications here are quite clear: a single-factor model is likely to be simpler to implement but it may be less credible while a multi-factor model is likely to be more difficult to implement but it may be more realistic.

13.4.2   Continuous-time models

Continuous-time models represent one or more risk variables as stochastic processes. For PPP infrastructure projects, an analyst may select the value of the underlying facility as the variable of interest – a single-factor model. The value of the facility is often derived from the present value of net cash flows of the completed project or specific operating assets. The analyst may also model cash flow components at a more detailed level, so the value of the underlying project/asset is further decomposed into variables such as prices (e.g. tariffs), costs (e.g. fuel costs, labour and capital costs), and volume/quantity (e.g. traffic demand) – a multi-factor model. Understandably, many continuous-time models are restricted to just one or two risk variables due to computational complexity.

The change in value of the variable of interest (denoted as dV) can be modelled by the generalised Weiner process, a fundamental stochastic process with the following form:

$$dV = \mu dt + \sigma dz \qquad (1)$$

On the right-hand-side of equation 1, the first term is the expected drift, or a 'slope' parameter for a graph plotted with V on the vertical axis versus time on the horizontal axis. This 'slope' parameter sets the average rate of change in long-term value. Short-term fluctuation in value evolving around the slope, usually referred to as 'volatility', is modelled by the second term, $\sigma\, dz$. By itself, dz represents the basic Wiener process:

$$dz = \varepsilon\sqrt{dt} \qquad (2)$$

where ε is a random variable with a standardised normal distribution $\phi\,(0,1)$. Essentially, this is the Brownian motion commonly known in physics. The σ parameter, then, is a scaling factor or variance rate that controls the extent of volatility.

Further modifications can be introduced to equation 1 when the drift and variance rates become functions of the underlying variable and time. Known as the Ito process, this is mathematically represented by:

$$dV = \mu(V, t)dt + \sigma(V, t)dz \qquad (3)$$

Equation 3 is a more practical model since in many circumstances the drift and variance rates of the variable of interest are unlikely to remain constant over time. A favourite form of equation 3 is the Geometric Brownian Motion which treats both the drift and the variance terms as being proportional to the variable V:

$$dV = \mu V dt + \sigma V dz \qquad (4)$$

For RO modelling, however, a word of caution is warranted. A property inherent within this form of mathematical representation is that the variable V is log-normally distributed. Before applying this process to the selected variable, one should question whether it is correct to assume that the variable approximately follows a log-normal distribution. For example, if the traffic volume of a toll road, at some point in time, does not follow a log-normal distribution, then its stochastic evolvement should not be modelled with equation 4. Other forms of stochastic processes exist to broaden the regime of stochastic process modelling under different scenarios due to peculiar characteristics or distribution of the underlying. These include the mean-reverting process and the Poisson jump process. Detailed mechanics of these stochastic processes can be found in Dixit and Pindyck (1994), Hull (2005), McDonald (2002) and Wilmott (1998).

In continuous-time models, solving for the value of a RO contingent on variables such as V often requires the derivation of a partial differential equation (PDE). The procedure then transforms from an intuitive consideration of strategic issues into mathematical manipulation where the PDE is solved

Part Two

subject to a set of boundary conditions related to the features of the option. In the best case, a closed-form solution exists for the PDE, thus allowing it to be solved analytically. The Black-Scholes equation used to evaluate a European option is a good example. Pindyck (1991) provides another example of an analytical solution for the simplest version of an investment timing problem.

When the PDE does not have an analytical solution, one has to resort to numerical methods. Finite-difference schemes, both implicit and explicit forms, are commonly used. Mathematically, the two schemes differ in terms of robustness, convergence and computational efficiency. A point worth mentioning, however, is that some researchers, such as Brennan and Schwartz (1978) and Hull and White (1990), have developed and published algorithms that will simplify the valuation procedure using finite difference methods.

Another popular numerical procedure is the Monte Carlo simulation technique that simulates thousands of sample paths for a variable based on its stochastic process. For example, by taking random draws of ε in equation 2, sample paths for variable V can be simulated by substituting equation 2 into equation 3 with a predetermined time-step Δt. Evaluation is especially simple for a European option, where exercise of the option is only feasible at maturity, as illustrated in McDonald (2002).

RO, however, are often more akin to American options (or even compound options) in which optimal option exercise can happen before maturity. In other words, unlike the evaluation of a European option that resembles a forward induction procedure, values of RO in these situations often become path- and state-dependent, thus requiring a backward induction or dynamic programming procedure.[2] As with finite difference schemes, research has worked towards incorporating the features of backward induction and dynamic programming in Monte Carlo simulation (Barraquand and Martineau 1995; Broadie and Glasserman 1997).

Chiara (2006) recently developed a novel approach that models BOT revenue risk mitigation contracts as multiple-exercise RO. According to his approach, revenue guarantees may be structured into risk mitigation contracts and valued by two methods, the multi-least squares Monte Carlo method and the multi-exercise boundary method. Effectively, the two methods combine Monte Carlo simulation and dynamic programming techniques to price the multi-exercise options.

13.4.3   Discrete-time models

As pointed out in Section 13.3.2, most literature in the AECI context uses discrete-time models. Common forms of discrete-time models include the binomial model (single-factor) or the trinomial model and the lattice model (multi-factor). Many of these models have been developed with an intention to provide an approximation to continuous-time models 'in the limit'. For example, for a binomial model, Cox *et al.* (1979) recommended values for the up and down movements of the underlying variable V (and the associated probability expressions) so that its volatility matches that of the stochastic process given in equation 4 in the limiting situation. Boyle (1988) further expanded this approach for a trinomial and a lattice model while recommending appropriate jump magnitudes and probabilities. Thus, the underlying assumptions for this first category of discrete-time models and their continuous-time counterparts are basically the same. They are conceptually equivalent – but they are solved using different mathematical approaches.

Discrete-time models need *not* be mere approximations of their continuous-time counterparts, hence giving rise to what we consider as the second category of discrete-time models. This class of models essentially represents 'economically corrected' versions of decision-tree analysis so that the problems of payoff structure, risk characteristics and non-constant discount rates can be overcome (Trigeorgis, 1999). This is done by converting the real situation into a 'risk-neutral' one. More importantly, due to its resemblance to traditional decision-tree analysis, managerial flexibility can be modelled more explicitly in the tree structure. Consequently, modelling RO in this way is more intuitive. When simply imposing stochastic processes on underlying variables cannot represent a real option scenario, this approach is especially valuable.

The choice between a continuous-time and a discrete-time model depends on the particular context of the application. If one considers an equity investment in a power plant project or an oil field exploration project, utilising a continuous-time model is indeed a reasonable approach. Market prices for the output of these projects are given in the spot and future/forward markets, so historical data can be used to determine the most appropriate stochastic process to model the underlying asset's value. For example, Wey (1993) found that crude oil prices are likely to follow a mean-reverting process for a long time horizon, and evaluation of a RO contingent on oil prices can be modelled with this consideration in mind. The same argument, however, would not hold for a toll road project. Assuming that traffic volume (and hence toll revenues) follows a geometric Brownian motion or a mean-reverting process is quite a stretch. Perhaps, a stochastic process that incorporates multi-stage growth with jumps would better represent the evolvement of traffic volume, but its mathematical complexity might not warrant the effort required considering all the uncertainty assumptions. For this reason, it might be wiser to look for ways to represent the traffic growth scenario in terms of a decision tree, which is far more intuitive, but use the risk-neutral evaluation technique so that it is 'economically corrected' for non-constant discount rates due to asymmetry and flexibility.

13.4.4   Risk neutrality

The concept of risk neutrality features so prevalently in financial and RO pricing models that it warrants a discussion. The flexibility embedded in decision making and the asymmetry in option payoffs imply that the discount rate will change as the option progresses towards expiration (because the risk profile changes when an action is taken, i.e. flexible alternatives are executed). Assumption of risk neutrality overcomes this problem by transforming the actual setting into a risk-neutral world, which has the following characteristics:

Part Two

- Risk preference and the expected return of the underlying asset do not enter into the equation, and all assets will appreciate at the risk-free rate in this world. Consequently, a *single* risk-free rate can be used to discount cash flows in all periods.
- Even after transforming the problem from the real to a risk-neutral world, volatility of the underlying asset itself stays the same; therefore uncertainty of the project is still captured in the evaluation process.

The above outcomes are achieved by transforming probability measures from the actual function into a 'risk-neutral' function. Technically, this change of measure is an illustration of the Girsanov's Theorem (Mikosch, 2000).

The main assumption behind risk neutrality is that the market is complete, so that a tracking portfolio can be found among traded assets, which allows replication of the cash flow from the option (a form of the 'no arbitrage' principle). In the world of finance and asset pricing, two main types of risks exist – market and private. When a project is dominated by market risks, traded assets can usually be found in the market to construct a tracking portfolio. The market is said to be complete and the risk-neutral pricing method can be adopted (Luenberger, 1998). When a project is dominated by private risks, or its initial value and/or market prices of the underlying asset are unknown, a risk-neutral analogy is often difficult to establish since investors in this case do not have complete market securities and information to track or hedge the cash flows. Many infrastructure projects (and other 'real' assets) fall into this category. Not surprisingly then, risk neutrality is one dimension of RO analysis that critics deem inappropriate. Still, does this imply that the entire field and its concepts should fall apart?

Fortunately, as Smith and Nau (1995) have proven, when the market is partially complete and so long as investors' preferences are restricted, their utility functions would take on specific forms. Certainty equivalents of risky cash flows can then be derived using the investors' probabilities for private uncertainties. This conversion procedure virtually creates an equivalent replicating trading strategy as in the case of a complete market and allows the usual evaluation of project cash flows (which include both market and private risks) to proceed using a single risk-free rate. Alternatively, the lack of replicating traded portfolios to hedge project cash flows can also be overcome if a priced asset that is reasonably correlated with the project cash flows can be identified. In this case, the 'projection price' of the project payoff can be derived through correlation pricing (Luenberger, 2001) – this is indeed the technique used in Ng et al.'s (2004) example. In principle, cash flows from infrastructure projects should still be able to be reasonably tracked by traded proxies in the market, albeit with some errors, despite the specificity of projects and the infrequency of real asset trading.

Further, the assumption of risk neutrality is not *absolutely necessary* in RO modelling when traded proxies do not exist or are difficult to identify for real assets. This situation is particularly true when the underlying value of an asset is expected or is known to follow a stochastic process that is substantially different from those of traded assets/securities. In such cases, the analyst could choose to simulate the discount rate, so that it responds to changes in the asset's risk profile over time. In other words, the analyst would abandon the need for the use of a constant discount rate and the presumption of a risk-neutral world. Clearly, however, such a decision introduces an additional dimension of extreme complexity. Thus, analysts find themselves again in the position of weighing model complexity against model accuracy or credibility.

After all, risk-neutral pricing in RO models does not make assumptions that are any stronger than those of the NPV technique (University of Maryland Roundtable, 2003; Brealey & Myers, 2000). Even in NPV, the estimation of an appropriate risk-adjusted discount rate is by itself subjective (Garvin and Cheah, 2004). Indeed, who can verify that a calculated NPV will turn out to be the true value after the fact?

13.4.5   Are real options suitable for the AECI context?

RO as a tool

The *motivation* behind the usage of RO as a tool is something that many critics have ignored. We would argue that this is an important aspect in judging the usefulness of RO. If RO is used for *pricing* (i.e. to define, for example, an exact value of a guarantee and specify this explicitly in a contract), then obviously a higher level of due diligence and care in modelling assumptions and techniques is quite necessary. If NPV is taken as the standard pricing technique for valuation (assuming that NPV can indeed be used to value a guarantee), then RO should serve as its *substitute* since one can only adopt a single price for an asset.

However, if RO is used to facilitate *decision making* (i.e. a 'go or no-go' decision for project investment or a timing decision to defer the project), the precision of RO values is far less significant. If two sets of modelling assumptions/techniques result in deferment option values of $100m and $200m respectively, then even this seemingly large difference should not matter, since both values send the same signal to the manager to defer the project. In fact, smart managers usually rely on several techniques to arrive at a final judgement. In this case, NPV and RO are *complementary*, since their results can be evaluated and compared to make decisions. Sensitivity analysis can be further implemented to each of these processes to assess the significance of the assumptions and their impact on the results.

RO as a strategic concept

Valuation issues aside, people intuitively seek to 'keep options open' in the face of uncertainties in corporate and project management. Bowman and Hurry (1993) developed an option-theoretic perspective for organisational strategic management, which integrates resource allocation, sense making, organisational learning and strategic positioning. Similarly, Lessard and Miller (2000) suggested that creating options to allow for a greater range of responses in line with uncertain future outcomes is critical. They commented that many successful projects are not selected *a priori*, but rather are shaped

Part Two

and structured along the way; projects are shaped in episodes to gradually transform the initial hypothesis, make progress on critical issues, and solidify initial coalitions of players to achieve temporary and final commitment.

The appeal of RO as a concept in strategic management is affirmed by industry examples like American Express, which adopted an RO approach to planning for its operations in Asia (Leander, 2005). In the context of infrastructure development, RO thinking can generate the premeditated design of project flexibility, as exemplified by Case Study 13.1.

13.5   An Integrated Real Options–Risk Management Process for PPP Infrastructure Projects

RO as a strategic concept can be further extended to risk management. Traditionally, a standard risk management process follows three steps: risk identification, risk analysis or measurement and risk mitigation. Risk mitigation approaches can be further divided into risk avoidance, risk reduction, risk shifting or transfer, and risk retention. Although the four risk mitigation strategies can help to manage uncertainty, they presume a certain degree of losses as a starting point. Clearly, this mindset may limit a manager's ability to recognise and exploit opportunities to increase project value. Accordingly, it is necessary to develop a risk management framework which considers the flexibility of management.

Figure 13.2 shows a revised risk management process which incorporates RO concepts. The process includes risk identification, risk analysis and subsequent consideration of various risk mitigation strategies. After identifying and analysing the threat of major risks, risk retention is considered. Parties who are willing to assume total or partial risks will adopt an 'option

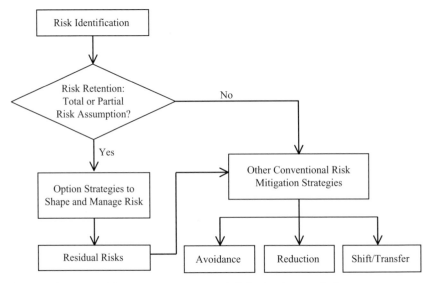

Figure 13.2 A new risk management process in infrastructure projects.

mindset' to shape and manage the risks. This is conceptually similar to Lessard and Miller's 'layering model' (Lessard and Miller, 2000: 88). For example, flexibility can be built into contracts by introducing special clauses which can be used to alter the timing and sequence of activities to achieve reduced risks (Floricel and Miller, 2000). Other potential option design strategies include creating flexibility to switch between different construction modes or technologies and negotiating for ancillary sources of revenue during the initial phase of project planning.

Naturally, however, options can only be designed to shape and mitigate certain risks – there are other residual risks that cannot be handled. Thus, managers should resort to the remaining conventional risk mitigation strategies: avoidance, reduction and shifting/transfer. As a whole, the design of option strategies/flexibilities and conventional risk mitigation strategies are complementary in nature. Moreover, an integrated framework provides a more comprehensive means for handling project uncertainties.

13.6   Summary

We hope that this chapter has achieved its purpose of distinguishing RO myths and realities while also illustrating its potential application to PPP infrastructure projects. Accordingly, we have chosen not to present a full length case study. Instead, we have provided a very comprehensive list of references which comprise many full length case studies, and we have discussed many salient issues that should allow the reader to review this literature in a more informed manner.

To summarise, flexibilities either exist inherently in stages of a project's lifecycle or they are intentionally structured into a project's design or execution process by a shrewd planner who recognises the potential need to react to real world uncertainties. Some will choose to ignore these flexibilities and continue to rely upon traditional evaluation methods with which they feel more comfortable. Others might explore the possibility of quantifying such flexibilities using RO techniques; this group probably accepts that the assumptions/methods are imperfect, but they also appreciate the utility of establishing a project/asset value that is likely to be closer to its 'true' worth. The key is to understand the limitations, implications and the trade-offs of the modelling assumptions/methods in order to establish a suitable level of confidence with the figures derived. Furthermore, as discussed, the level of precision required differs when one is using the figures for pricing or decision-making purposes. Most importantly, the analyst must know how to communicate the rationale of the assumptions/techniques employed to decision or policy makers. By doing so, obstacles in PPP projects, such as negotiating concession agreements, attracting equity investors and securing long-term financing, may be either eliminated or, at least, reduced.

Given the widely accepted conceptual appeal of the methodology, we do not advise AECI professionals to abandon RO techniques altogether. Only through use and experience will the community begin to understand the strengths and weaknesses of this tool, and thereby modify the quantification

Part Two

processes to suit particular purposes and circumstances. Indeed, RO analysis is a flexible tool and not a formula that should be applied in a rigid or uninformed manner.

Notes

1. The concepts of DCF and present value still play an important role in some of the RO modelling techniques, especially where spreadsheets are used to compute values of the underlying asset or exercise price. Here, the NPV method refers to 'static' or 'straight' NPV analysis without consideration of any managerial or operating flexibility.
2. Dynamic programming is a general tool for breaking a whole sequence of decisions into just two components: the immediate decision and a valuation function that encapsulates the consequences of all subsequent decisions (Dixit and Pindyck, 1994). When the planning horizon is finite, the very last decision at its end has nothing following it, so the valuation function in earlier time-steps can be determined through backward induction by first solving for the optimal decision in the final time-step and working backwards.

References

Amram, M. and Kulatilaka, N. (1999) *Real Options: Managing Strategic Investment in an Uncertain World*. Harvard Business School Press, Boston.

Barraquand, J. and Martineau, D. (1995) Numerical Valuation of High Dimensional Multivariate American Securities. *Journal of Financial and Quantitative Analysis*, 30(3), 383–405.

Borison, A. (2003) Real Options Analysis: Where are the Emperor's Clothes? *Proceedings of the 7th Annual Conference on Real Options*, 10–12 July, Washington, DC.

Bowman, E.H. and Hurry, D. (1993) Strategy through the option lens: an integrated view of resource investments and the incremental-choice process. *Academy of Management Review*, 18(4), 760–782.

Boyle, P.P. (1988) A lattice framework for option pricing with two state variables. *Journal of Financial and Quantitative Analysis*, 23(1), 1–12.

Brealey, R. and Myers, S. (2000) *Principles of Corporate Finance*, 6th edn. McGraw-Hill/Irwin, New York.

Brealey, R.A., Myers, S.C. and Allen, F. (2005) *Principles of Corporate Finance*, 8th edn. McGraw-Hill/Irwin, New York.

Brennan, M. and Schwartz, E. (1978) Finite difference methods and jump processes arising in the pricing of contingent claims: a synthesis. *Journal of Financial and Quantitative Analysis*, 13(9), 461–474.

Broadie, M. and Glasserman, P. (1997) *Monte Carlo Methods for Pricing High-Dimensional American Options: An Overview*. Working Paper, Columbia University.

Cheah, C.Y.J. (2004) Public-private partnerships in infrastructure development: on value, risk and negotiation. *Proceedings of the CIB W107 Globalization and Construction Symposium*, 17–19 November, Bangkok, Thailand, pp. 735–746.

Cheah, C.Y.J. and Liu, J. (2005) Real option evaluation of complex infrastructure projects: the case of Dabhol Power Project in India. *Journal of Financial Management of Property and Construction*, 10(1), 55–68.

Cheah, C.Y.J. and Liu, J. (2006) Valuing governmental support in infrastructure projects as real options using Monte Carlo simulation. *Construction Management and Economics*, 24(5), 545–554.

Chiara, N. (2006) *Real Option Methods for Improving Economic Risk Management in Infrastructure Project Finance*. PhD Dissertation, Columbia University, New York.

Chiara, N., Garvin, M.J. and Vecer, J. (2007) Valuing simple multiple-exercise real options in infrastructure projects. *Journal of Infrastructure Systems*, 13(2), 97–104.

Copeland, T.E. and Antikarov, V. (2001) *Real Options: A Practitioner's Guide*. Texere LLC, New York.

Copeland, T. and Tufano, P. (2004) A real-world way to manage real options. *Harvard Business Review*, 84(3), 90–99.

Cox, J., Ross, S. and Rubinstein, M. (1979) Option pricing: a simplified approach. *Journal of Financial Economics*, 7(3), 229–263.

Dixit, A.K. and Pindyck, R.S. (1994) *Investment under Uncertainty*. Princeton University Press, Princeton.

Floricel, S. and Miller, R. (2000) Strategic systems and templates. In: *The Strategic Management of Large Engineering Projects*, Miller R. and Lessard, D. (eds). MIT Press, Cambridge, MA., pp. 113–130.

Ford, D., Lander, D. and Voyer, J. (2002) A real option approach to valuing strategic flexibility in uncertain construction projects. *Construction Management and Economics*, 20(4), 343–351.

Garvin, M.J. and Cheah, C.Y.J. (2004) Valuation techniques for infrastructure investment decisions. *Construction Management and Economics*, 22(4), 373–383.

Grenadier, S. (1995) Valuing lease contracts: a real-options approach. *Journal of Financial Economics*, 38(3), 297–331.

Ho, S. and Liang, Y. (2002) An option pricing-based model for evaluating the financial viability of privatized infrastructure projects. *Construction Management and Economics*, 20(2), 143–156.

Hull, J. (2005) *Options, Futures, and Other Derivatives*, 6th edn. Prentice Hall, Upper Saddle River, NJ.

Hull, J. and White, A. (1990) Valuing derivative securities using the explicit finite difference method. *Journal of Financial and Quantitative Analysis*, 25(3), 87–100.

Kogut, B. (1991) Joint ventures and the option to expand and acquire. *Management Science*, 37(1), 19–33.

Lander, D.M. and Pinches, G.E. (1998) Challenges to practical implementation of modeling and valuing real options. *Quarterly Review of Economics and Finance*, 38(4), 537–567.

Leander, T. (2005) View from Asia. *CFO*, 21(14), 29.

Lessard, D. and Miller, R. (2000) Mapping and facing the landscape of risk. In: *The Strategic Management of Large Engineering Projects*, Miller, R. and Lessard, D. (eds). MIT Press, Cambridge, MA., pp. 75–92.

Leviäkangas, P. and Lähesmaa, J. (2002) Profitability evaluation of intelligent transport system investments. *ASCE Journal of Transportation Engineering*, 128(3), 276–286.

Longstaff, F.A., and Schwartz, E.S. (2001) Valuing American options by simulation: a simple least-squares approach. *Review of Financial Studies*, 14(1), 113–147.

Luehrman, T. (1997) What's it worth? A general manager's guide to valuation. *Harvard Business Review*, 75(3), 132–142.

Luenberger, D.G. (1998) *Investment Science*. Oxford University Press, New York.

Part Two

Luenberger, D.G. (2001) Projection pricing. *Journal of Optimization Theory and Applications*, 109(1), 1–25.

Majd, S. and Pindyck, R.S. (1987) Time to build, option value, and investment decisions. *Journal of Financial Economics*, 18(1), 7–27.

Mason, S.P. and Baldwin, C.Y. (1988) Evaluation of government subsidies to large-scale energy projects. *Advances in Futures and Options Research*, 3, 169–181.

Mattar, M. and Cheah, C.Y.J. (2006) Valuing large engineering projects under uncertainty – private risk effect and real options. *Construction Management and Economics*, 24(8), 847–860.

McDonald, R.L. (2002) *Derivatives Markets*. Addison-Wesley, Boston, MA.

Mikosch, T. (2000) *Elementary Stochastic Calculus – With Finance in View*. World Scientific, Singapore.

Moel, A. and Tufano, P. (1999) Bidding for the Antamina Mine. In: *Project Flexibility, Agency, and Competition: New Developments in the Theory and Application of Real Option Analysis*, Brennan, M.J. and Trigeorgis, L. (eds). Oxford University Press, New York, pp. 128–150.

Myers, S.C. (1984) Finance theory and financial strategy. *Interfaces*, 14(1), 126–137.

Ng, F.P. and Björnsson, H.C. (2004) Using real option and decision analysis to evaluate investments in the architecture, construction and engineering industry. *Construction Management and Economics*, 22(5), 471–482.

Ng, F.P., Björnsson, H.C., and Chiu, S.S. (2004) Valuing a price cap contract for material procurement as a real option. *Construction Management and Economics*, 22(2), 141–150.

Paddock, J.L., Siegel, D.R. and Smith, J.L. (1988) Option valuation of claims on real assets: the case of offshore petroleum leases. *Quarterly Journal of Economics*, 103(August), 479–508.

Park, C.S. and Sharp-Bette, G.P. (1990) *Advanced Engineering Economics*. John Wiley & Sons, New York.

Pindyck, R. (1991) Irreversibility, uncertainty, and investment. *Journal of Economic Literature*, 29(3), 1110–1148.

Rogers, M.J., Gupta, A. and Maranas, C.D. (2002) Real options based analysis of optimal pharmaceutical research and development portfolios. *Industrial and Engineering Chemistry Research*, 41(25), 6607–6620.

Schwartz, E.S. and Moon, M. (1999) Evaluating research and development investments. In: *Project Flexibility, Agency, and Competition: New Developments in the Theory and Application of Real Option Analysis*, Brennan, M.J. and Trigeorgis, L. (eds). Oxford University Press, New York, pp. 85–106.

Sing, T.F. (2002) Time to build options in construction processes. *Construction Management and Economics*, 20(2), 119–130.

Smith, J.E. and Nau, R.F. (1995) Valuing risky projects: option pricing theory and decision analysis. *Management Science*, 41(5), 795–816.

Stonier, J. (1999) Airline long-term planning under uncertainty: the benefits of asset flexibility created through product commonality and manufacturer lead time reductions. In: *Real Options and Business Strategy: Applications to Decision Making*, Trigeorgis, L. (ed). Risk Books, London.

Teach, E. (2003) Will real options take root? *CFO*, 19(9), 73–74, 76.

Tiong, R.L.K. (1995) Risks and guarantees in BOT tender. *ASCE Journal of Construction Engineering and Management*, 121(2), 183–188.

Trigeorgis, L. (1993) The nature of option interactions and the valuation of investments with multiple real options. *Journal of Financial and Quantitative Analysis*, 28(1), 1–20.

Part Two

Trigeorgis, L. (1999) *Real Options: Managerial Flexibility and Strategy in Resource Allocation*. MIT Press, Cambridge, MA.

University of Maryland Roundtable on Real Options and Corporate Finance (2003) *Journal of Applied Corporate Finance*, 15(2), 8–23.

Wey, L. (1993) *Effects of Mean-Reversion on the Valuation of Offshore Oil Reserves and Optimal Investment Rules*. Unpublished Undergraduate Thesis, Massachusetts Institute of Technology, Cambridge, MA.

Wilmott, P. (1998) *Derivatives: The Theory and Practice of Financial Engineering*. John Wiley, Chichester.

Yiu, C.Y. and Tam, C.S. (2006) Rational under-pricing in bidding strategy: a real options model. *Construction Management and Economics*, 24(5), 475–484.

Zhao, T. and Tseng, C. (2003) Valuing flexibility in infrastructure expansion. *ASCE Journal of Infrastructure Systems*, 9(3), 89–97.

Part Two

14

Financial Implications of Power Purchase Agreement Clauses in Revenue Stream of Independent Power Producers in Nepal

Raju B. Shrestha and Stephen Ogunlana

14.1   Introduction

PPPs have a critical role to play in the implementation of development projects in developing countries. Traditionally, the responsibility of constructing and operating infrastructure facilities had rested with the government, but with the growth of economy outstripping infrastructure supply (Gupta and Sravat, 1998) and the need for providing infrastructure facilities to keep up the pace of development, the burden of infrastructure development has been shifted to the private sector through concession contracts such as build, own, transfer (BOT). In the electricity sector, the participation of the private sector is in the form of independent power producers (IPPs) with long-term power purchase agreements (PPAs) selling electricity in bulk to a utility which in turn sells it to end-users.

PPAs are considered the most important contract underlying the construction and operation of a power plant usually drawn at the implementation phase of IPP projects, as there can be no project if PPA is not reached. It is also an extremely complex and politically sensitive issue, since it is the PPA that ultimately governs the price of electricity delivered to end-users. The other project agreements including those covering engineering, procurement, construction, lending, operations and maintenance, can be negotiated only after the PPA is concluded (Crow, 2001).

In developing countries, the purchaser is, in almost all of the cases, the state-owned enterprise, often with a monopoly on generation, transmission and distribution of electric power, while the project developers may be foreign investors, local investors, or a joint venture between local and foreign investors. The PPA reached between the state-owned enterprise and the project sponsor is the contractual arrangement for sharing the risks and responsibilities between the contracting parties for a term, which may span up to 25 or 30 years depending on the contract.

In Nepal, over the past 10 years, IPPs, with local and foreign investment, have played a major role in installing new hydropower facilities in the country, contributing more than 25% of the total generation capacity. With the introduction of the Electricity Act 1992 following the adoption of Hydropower Policy in 1992, a comprehensive legal framework for the development of hydropower was put in place for private participation in hydropower projects in Nepal (Nepal, 2001). The majority of the hydropower projects are owned by the national utility, the Nepal Electricity Authority (NEA), an undertaking of His Majesty's Government of Nepal. However, with the recent policies of the government to attract private investments in the hydropower sector, some major projects have now been developed by private investors, namely the Butwal Power Company (BPC), Himal Power Ltd. and Bhote Koshi Power Company. These private power producers generated 144 MW of power and 840 GWh of energy in 2003, which is 35% of the total generation in the country (Nepal Electricity Authority, 2004). A critical issue in designing PPAs is to create a level playing field for the players to secure successful and sustainable IPPs and PPAs. However comparative studies have shown that discriminatory clauses and unequal treatment of IPPs are present in the key issues of PPAs.

Before projects are undertaken, they are normally evaluated using financial modelling to ensure that the project's cash flows are adequate both to service project debt in a timely manner and to provide an acceptable rate of return to equity investors. In the absence of required data, the common approach of the analysis is to make hypothetical assumptions and estimates to accommodate the methodology in the exercises. Assumptions made in IPPs can be fairly accurate as power purchase guarantees specify the type and quantum of energy that is to be purchased from the IPP and the price for each megawatt of electricity delivered to the utility. This allows financial modelling to be used for evaluating major clauses of PPAs with financial implications.

In designing PPAs, a key contract in private power projects, the principal issues are (Shrestha and Ogunlana, 2006):

- Power purchase guarantees
- *Force majeure* guarantees
- Financial and foreign exchange guarantees
- Operation risks
- Dispute resolution and insurance issues

As specified in the PPAs, the power purchase guarantees establish the type and quantum of energy that is to be purchased from the IPP and the price for each megawatt of electricity delivered to the utility. These specifications in PPAs will have significant impact on the returns from the projects to the investors.

The specifications in purchase guarantees of power with financial implications are:

- Take or pay clauses. The clause obligates the purchaser to pay for the deemed energy irrespective of whether the purchaser is able to dispatch the energy or not or else pay an amount equal to the cash value of the difference between the contractually specified energy to be purchased and energy actually purchased.

- Guarantees of purchase of interim energy. A guarantee is provided by the utility to purchase the energy before the required commercial operation date (RCOD).
- Supply guarantees of minimum energy. Guarantees provided by the project sponsors on supply of the minimum contracted energy.
- Guarantees of purchase of excess energy. Guarantees provided by the utility to purchase the energy in excess of the contracted energy.
- Allowance of third party sales. Allowance given to the project sponsors to sell energy to a third party.

14.2   Financial Analysis

For the purpose of analysing the clauses with financial implications, a study of financial analyses with hypothetical cash flows based on the contract volumes, prices and clauses (as specified in the existing PPAs) were developed. The analyses have been carried out for projects based on the availability of the relevant data from the agreements and its relevance for the study. Some assumptions that reflect the actual pattern of investment have been made. The projects were selected from a pool of 25 projects that have signed PPAs with the utility on the basis of the type of investment, size and the year of commissioning, to reflect a broad spectrum of agreements.

14.2.1   Projects reviewed

- **Mardi Khola (3.1 MW).** The project company, Gandaki Hydropower Development Company, will build, own and operate the power plant, which is located in western Nepal. A local investor, who has the licence to operate the plant for 35 years, owns the project company and has reached a PPA with the utility for 25 years but has yet to reach a financial closure. The PPA was signed in September 2003 (Nepal Electricity Authority and Bavarian Hydropower Nepal, 2003).
- **Chilime (20MW).** The project company, Chilime Hydropower Company limited, has built, owns and operates the power plant, located in central Nepal. The project company is controlled by local investors in joint venture with the utility and signed a PPA with the utility in June 1997 (Nepal Electricity Authority and Chilime Hydropower Company, 1997). The project has been in operation since 2002.
- **Lower Nyadi (4.5 MW).** The project company, Bavarian Hydropower Nepal, owns and operates the plant. Foreign investors control the project company. The PPA was signed on 1 December 2003 (Nepal Electricity Authority and Gandaki Hydropower Development Co., 2003) and RCOD was 15 May 2005 (Nepal Electricity Authority and Gorkha Hydropower Nepal, 1999); however the project is delayed and not yet in operation.

- **Daram Khola (5.0 MW)**. The project company, Gorkha Hydropower, owns and operates the plant. Local investors control the project company. The PPA was signed in December 1999; but the project is delayed and not yet in operation.

The above projects represent a broad spectrum of the agreements reached between the utility and IPPs (Nepal Electricity Authority, 2005) with a mix of projects with foreign and local investors, of different capacities, and commissioned at different periods of time.

14.2.2   Relevant data for analyses

Private investors and sponsors make their investments based on the likely returns and the risks and uncertainties prevalent in the project. Financial analyses of typical real projects are conducted from the project sponsor's point of view for profitability and financial liquidity with the clauses in the corresponding PPAs and assumed parameters. The analyses were carried out for the IPP hydropower projects using seasonal energy prices as offered in PPAs. The cash flows were developed using the assumed parameters as shown in Table 14.1, for the analyses:

- Analysis period: 25 years from the commercial operation date, the term of most PPAs entered between NEA and the developer. The project is assumed to take 3 years to construct.
- Reference date: the reference date for costs, exchange rate and discounting is the end of year of project commissioning year.
- Exchange rate: an exchange rate prevailing at the reference date.
- Investment cost: the total investment required for the project including customs, local tax and VAT determined is financial cost as it is based on unit costs prevailing currently in the market and therefore adopted for the financial analysis. The project, as mentioned above, is assumed to require 3 years to complete. The percentage project cost disbursement is assumed to be 25%, 50% and 25% of the total project cost in the first, second and third year respectively for most projects, but can be different if stated. The interest during construction (IDC) has been estimated to be a rate of 10%.
- Operation and maintenance costs: annual operation and maintenance costs of the plant in the first year of commercial operation following completion of the project have been assumed to be 1.5% of the total project cost with an escalation rate of 5%.
- Insurance premium: annual insurance premium of the plant is assumed to be 0.5% of the total project cost.
- Debt/equity (D/E) ratio: the project is assumed to be developed with 70% debt and 30% equity.
- Conditions on long-term loan: the long-term debt is assumed to carry an annual interest rate of 10% with 10 years of repayment period following project completion. Interest during construction will be capitalised. The outstanding debt at the end of project completion will be amortised annually over the 10-year loan repayment period.

Part Two

Table 14.1   Assumptions for the cash flow projections.

	Lower Nyadi	Chilime	Mardi	Daram
Installed capacity MW	4.5	20	3.1	5.0
Year of commissioning	2006	2002	2006	1999
Analysis period	25 years	25 years	25 years	25 years
Dry season energy MWh	9402.48	48995	6756	14040
Wet season energy MWh	23972.67	62941	17115	22010
Total energy	33375.15	111936.5	23871	36050
Project cost (Rs. millions)				
Project construction cost	588.89	2497	413.113	758.70
Total equity 30%	176.67	749	123.934	227.61
Total debt 70%	412.223	1748	289.18	531.09
Interest during construction	64.44	273	45.20	83.01
Total cost	653.33	2770.615	458.32	841.72
Dry season energy price	5.687	4.081	5.68	4.505
Wet season energy price	4.014	4.081	4.01	3.18
Escalation	—	8%	—	6%
Escalation period years	—	8	—	4
Insurance	0.5% of total project cost	0.5% of total project cost	0.5% of total project cost	0.5% of total project cost
Operation and maintenance costs	1.5% of project cost	1.5% of project cost	1.5% of project cost	1.5% of project cost
Escalation on maintenance costs	5%	5%	5%	5%
Conditions on long-term loan	10% interest	10% interest	10% interest	10% interest
Loan repayment	10 years from project completion	10 years from project completion	10 years from project completion	10 years from project completion
Discount rate	10%	10%	10%	10%
Royalty				
Depreciation	4% straight line method	4% straight line method	4% straight line method	4% straight line method
Tax rate	20%	20%	20%	20%
Bonus and welfare	2% of net profit	2% of net profit	2% of net profit	2% of net profit

- Discount rate: the discount rate of 10% has been used to calculate the NPV and benefit:cost ratio of the project as well as to compare the calculated IRR on equity investment.
- Energy price: the dry and the wet season energy selling price has been assumed to be the prevailing rates of NRs.5.52 and NRs.3.90 with the escalation rates provided in PPAs.
- Royalty: according to Hydropower Development Policy, 2001, for hydropower plants ranging between 1 and 10 MW, the government imposes, for the first 15 years from the date of commercial operation, NRs.100 per year for each installed kW as capacity royalty and 1.75% of energy

revenue as energy royalty. From the 16th year onwards, the capacity royalty increases to NRs. 1000 per year for each installed kW and the energy royalty increases to 10% of energy revenue. The given capacity royalty rates are, however, for the base year of 2001. Thereafter, each year it is increased by 5%.

- Depreciation: the depreciation rate applied is 4% per annum and the straight-line method per annum is used.
- Tax rate: as stipulated in the Income Tax Act 2008, the applicable corporate tax rate for enterprises undertaking electricity generation is 20%.
- Bonus and welfare: the bonus and welfare fund has been assumed to be 2% of the net profit.

A simplistic project based financial analysis from the utility's point of view is also presented with the benefits from the sale of energy from the project to consumers and expenses for dispatching the energy.

For the utility, the clause can be unfavourable when it is not able to dispatch the contracted energy to retail customers. This scenario is more likely during monsoon when the generation capability of the utility is at its maximum.

14.3   Analyses of Clauses

14.3.1   Take or pay clauses

Rationales for take or pay clauses in PPAs

A take or pay contract obligates the purchaser of the project's output or services to pay for the output or services whether or not the purchaser takes delivery. In PPAs, the clause obligates the purchaser to pay for the deemed energy irrespective of whether the purchaser is able to dispatch the energy or not or else pay an amount equal to the cash value of the difference between the contractually specified energy to be purchased and energy actually purchased. The obligation makes the utility assume the demand/market risk; this is not illogical as, invariably, the utility controls the system and is responsible for the demand forecast. The capital-intensive and site-specific nature of the project, unpredictable output and high construction risks also make the take or pay clause appropriate in PPAs. Lenders rely upon the take or pay contracts for repayment of the loans.

Financial analysis considering take or pay clauses in PPAs

A financial analysis carried out considering the sale of only the contract energy under take or pay clauses without any additional benefits from the sales of excess energy, interim energy or any other benefits the clauses of the PPAs allow, show that the NPVs of the project to the IPPs (sponsor's perspective) are moderately high and the real rate of return exceeds 20% in all but one case (Tables 14.2, 14.3). Under the take or pay clause, the revenue stream is steady and predictable as long as the IPPs do not default on commitments

Table 14.2 Financial cashflow of Lower Nyadi Project from sponsor's point of view (with contract energy only). Figures given in millions of NRs., 2005 level.

Year	Equity	Revenue	Operating exps.[a]	Operating CF	Depreciation[b]	Interest on debt	Capital repayment	Profit before tax	Corporation tax	Profit after tax	Bonuses	Cash flow	DSCR
-2	44.17											(44.17)	
-1	88.33											(88.33)	
0	44.17											(44.17)	
1		149.72	21.33	128.385	26.13	47.67	29.91	54.59	10.92	43.67	0.87	39.02	1.66
2		149.72	21.80	127.920	26.13	44.68	32.90	57.11	11.42	45.69	0.91	38.01	1.65
3		149.72	22.29	127.431	26.13	41.39	36.19	59.91	11.98	47.93	0.96	36.92	1.64
4		149.72	22.80	126.918	26.13	37.77	39.81	63.02	12.60	50.42	1.01	35.73	1.64
5		149.72	23.34	126.379	26.13	33.79	43.79	66.46	13.29	53.17	1.06	34.45	1.63
6		149.72	23.90	125.814	26.13	29.41	48.17	70.27	14.05	56.22	1.12	33.06	1.62
7		149.72	24.50	125.220	26.13	24.59	52.98	74.50	14.90	59.60	1.19	31.55	1.61
8		149.72	25.12	124.597	26.13	19.29	58.28	79.17	15.83	63.34	1.27	29.92	1.61
9		149.72	25.78	123.942	26.13	13.46	64.11	84.35	16.87	67.48	1.35	28.15	1.60
10		149.72	82.31	67.407	26.13	7.05	70.52	34.22	6.84	27.38	0.55	-17.56	0.87
11		149.72	27.19		26.13			96.40	19.28	77.12	1.54	101.71	
12		149.72	27.94		26.13			95.64	19.13	76.51	1.53	101.12	
13		149.72	28.74		26.13			94.85	18.97	75.88	1.52	100.49	
14		149.72	29.58		26.13			94.01	18.80	75.21	1.50	99.84	
15		149.72	30.45		26.13			93.13	18.63	74.51	1.49	99.15	
16		149.72	31.37		26.13			92.21	18.44	73.77	1.48	98.43	
17		149.72	32.34		26.13			91.24	18.25	73.00	1.46	97.67	
18		149.72	33.36		26.13			90.23	18.05	72.18	1.44	96.87	
19		149.72	34.42		26.13			89.16	17.83	71.33	1.43	96.04	
20		149.72	35.54		26.13			88.04	17.61	70.43	1.41	95.16	
21		149.72	36.72		26.13			86.87	17.37	69.49	1.39	94.24	
22		149.72	37.95		26.13			85.63	17.13	68.51	1.37	93.27	
23		149.72	39.25		26.13			84.34	16.87	67.47	1.35	92.25	
24		149.72	40.61		26.13			82.97	16.59	66.38	1.33	91.19	
25		149.72	42.04		26.13			81.55	16.31	65.24	1.30	90.06	

NPV @10% = 215.41
IRR = 20.12%

[a] Operating expenses include operation maintenance, insurance and royalties.
[b] Depreciation is not included as cash expense but used for tax purposes.

Table 14.3 Comparison of NPVs and IRRs of various projects.

Projects	Lower Nyadi	Chilime	Daram	Mardi
NPV at time of commission (NRs. millions)	215.41	1964.02	91.87	166.66
IRR at time of commission %	20.12	25.40	13.25	21.13

due to faulty hydrology, bad management, or any other reasons, and as long as the utility honours the PPA.

The utility, on the other hand, is seen to be losing from the financial point of view based on the revenue it receives from the retail customers at existing average selling rate of 6.28 Rs./kWh and the expenses it bears in power purchase, transmission, distribution and loss costs. It will continue to bear losses even with the present trend of rate increase on energy prices. The NPV of the projects from the utility's perspective as seen in Tables 14.4 and 14.5 are negative for various projects. The negative NPV of the projects is due to the selling price (retail price) being lower than the cost of purchase, transmission, distribution and other associated costs. The NPV of the projects will be even lower if the utility is not able to dispatch the projects as it will still have to pay for the energy not dispatched under the take or pay clauses. This scenario, of not being able to dispatch energy, is possible due to the seasonal nature of the ROR power plants, which are able to produce electricity at full capacity during the wet season and partial production during the dry season when the demand for electricity is at its peak. Moreover, the ROR and PROR types of plant dominate the utility's own generation capacity. The only attraction of the PPA for the utility is not having to raise the funds for financing the generation capacity for the benefit to the consumers today, albeit at the expense of tomorrow's consumers. This undermines the ability of the utility to fulfil its commitment for payment under long-term contracts. The only recourse for the utility will be either to raise energy prices at the cost of the consumers, which is not always possible due to political reasons, or to revisit the PPAs for adjustments to clauses.

The lenders relying upon the take or pay clauses for repayment will be satisfied as long as the revenue is sufficient for the payment of fixed and variable operating costs and to service project debt. Given the price offered for energy in the existing PPAs, Table 14.2 shows that there is no difficulty in debt servicing for the project studied. The project is able to service debt with the cash flow from the sales of contracted energy under take or pay clause with a ratio of over 1.5. However, the agreement reached by utilities in their short-sighted need to benefit the consumers today at the expense of tomorrow's consumers, may not be considered as an unconditional obligation to pay, as the utility may not be able to honour its obligation due to the losses it has to bear and may go bankrupt as mentioned above.

Even though the utility has employed the take or pay clause in all the PPAs to date, they are not favourable to it because of the losses it has to bear from the clause. However, the value it has received in exchange is that it has been able to provide service to its customer. The alternatives the utility has in

Part Two

Table 14.4   Financial cashflow of Lower Nyadi Project from utility's point of view.

Assumed escalation (from past trends)		
Revenue rate	6.28 Rs/KWh	3.00%
Transmission cost	0.37 Rs/KWh	2.00%
Distribution costs	1.25 Rs/KWh	5.00%
Loss cost	1.08 Rs/KWh	2.00%

(Millions of NRs.)

Ref. year	Total revenue	Power purchase cost	Transmiss. cost	Distribution cost	Loss cost	Total cost	Net cash flow
1	209.60	149.72	12.35	41.72	36.05	239.83	(30.24)
2	215.88	149.72	12.60	43.80	36.77	242.89	(27.00)
3	222.36	149.72	12.85	46.00	37.50	246.06	(23.70)
4	229.03	149.72	13.10	48.29	38.25	249.37	(20.34)
5	235.90	149.72	13.37	50.71	39.02	252.81	(16.91)
6	242.98	149.72	13.63	53.25	39.80	256.39	(13.42)
7	250.27	149.72	13.91	55.91	40.59	260.13	(9.86)
8	257.78	149.72	14.18	58.70	41.40	264.01	(6.23)
9	265.51	149.72	14.47	61.64	42.23	268.06	(2.55)
10	273.48	149.72	14.76	64.72	43.08	272.27	1.20
11	281.68	149.72	15.05	67.96	43.94	276.67	5.01
12	290.13	149.72	15.35	71.35	44.82	281.24	8.89
13	298.83	149.72	15.66	74.92	45.71	286.02	12.82
14	307.80	149.72	15.97	78.67	46.63	290.99	16.81
15	317.03	149.72	16.29	82.60	47.56	296.17	20.86
16	326.54	149.72	16.62	86.73	48.51	301.58	24.96
17	336.34	149.72	16.95	91.07	49.48	307.22	29.12
18	346.43	149.72	17.29	95.62	50.47	313.10	33.33
19	356.82	149.72	17.64	100.40	51.48	319.24	37.58
20	367.53	149.72	17.99	105.42	52.51	325.64	41.89
21	378.55	149.72	18.35	110.69	53.56	332.32	46.23
22	389.91	149.72	18.72	116.23	54.63	339.30	50.61
23	401.61	149.72	19.09	122.04	55.73	346.57	55.03
24	413.66	149.72	19.47	128.14	56.84	354.17	59.48
25	426.07	149.72	19.86	134.55	57.98	362.11	63.96

NPV @ 10% = (30.32)
IRR = 8%

Table 14.5   Comparison of NPVs and IRRs of various projects – utility's point of view.

Projects	Lower Nyadi	Chilime	Daram	Mardi
NPV at time of commission	−30.32	−2169.34	−210.44	−64.03
IRR %				18%

addition to the take or pay clause in future negotiations are: take if offered clause or take and pay clause where the utility will not have to pay for the energy that it is not able to dispatch which will significantly increase the NPVs of the projects from utility's point of view.

14.3.2   Interim energy

Rationales for purchase of interim energy (energy before RCOD)

The clause guarantees purchase by the utility of interim energy or energy produced by the IPP before the required commercial operation date. This guarantee allows sponsors to enter into construction contracts with an incentive for speedy completion of the project. This in turn minimises the risk of completion of the project facility for the sponsors. Concurrently, the purchaser will be able to take advantage of the project sponsors to complete the project before the required date and make the services available to its customers. However, the purchasers will have to make arrangements and changes to its system to be able to accept and purchase metered energy. There may also be a potential conflict between the requirements of the system, which varies from time to time, depending on the expansion scheme and the interests of the project sponsors.

Financial analysis considering 'purchase of interim energy (energy before RCOD)'

The impact of purchase guarantees of interim energy in PPAs on the financial NPVs of the projects largely depends upon the season the interim energy is produced as most PPAs guarantee purchase of dry season interim energy only. The energy produced in wet season will have little impact on the NPV, as it is valued much less than dry season energy. Most PPAs do not even have guarantees of purchase of wet season interim energy.

14.3.3   Supply guarantee of minimum energy

Rationales for supply guarantee of minimum energy

The guarantee of supply of minimum quantum of energy allocates the supply risk along with the hydrological risks to the project sponsors. The supply of minimum energy, often called the contract energy, is usually based on the hydrology of the river and type of power plant. Failure of supply of minimum contracted energy doubly penalises the project company as penalty clauses are normally imposed on top of the losses from reduced revenues.

Financial analysis considering supply guarantee of minimum energy

Predictions of future hydrological conditions used in hydropower feasibility studies have been shown to be extremely unreliable. Of 63 hydropower

Part Two

Table 14.6   Effect of reduction in dry season energy on the NPV Lower Nyadi.

Reduction	0%	30%	40%	50%
Dry season energy	9402.48	6581.74	5641.49	4701.24
NPV	215.41	149.15	84.97	20.79
IRR %	20.12	17.52	14.30	11.06

dams reviewed by WCD (World Commission on Dams) 55% generated less power than predicted (Clarke, 2000). A hypothetical case of sustained reduction in energy for any reason e.g. faulty studies will, due to the penalty clauses in PPAs, result in reduced revenues and penalty for the energy not supplied at prevailing rates (see Table 14.6).

A sensitivity analysis carried out to determine the impact of reduction of dry season energy by 30–50% on the financial NPVs of a project shows how it will significantly affect the outcome of the project. The IRR of the project is reduced from 22.57% to 17.52% with 30% reduction in 4 months of dry season energy and IRR is halved when there is a 50% reduction in dry season energy.

The risk of failure to supply the minimum contracted energy also increases when there is a potential for upstream developments that can interfere with the flow patterns of the project. This problem is further compounded when the area is beyond the control of the host government. For example, the catchment area of Bhotekoshi, one of the reviewed projects, lies in Tibet and any upstream development there will affect the flow patterns of the project.

It can be summed up from the above discussion that the guarantee of supply of minimum energy is crucial to the utility. For the sponsors, the guarantee of supply of minimum energy makes them assume the hydrological risk and from the financial analyses it can be seen that sustained reduction of energy can have significant implication on the project.

14.3.4   Purchase guarantee of excess energy

Rationales for purchase guarantee of excess energy

The heavy wet monsoon of the region makes it possible to produce energy in excess of the committed contract energy, and with favourable hydrological conditions, production of energy in excess of the committed energy is also possible during the dry season. If there is a purchase guarantee of excess energy, sponsors can plan to install a plant with higher capacity as long as such an installation does not come in conflict with other clauses of the PPA (Table 14.7). This clause is also an incentive for the IPPs to operate efficiently. However, there may be a potential conflict between the requirements of the system of the utility, which may vary from time to time, and the interest of the sponsors. Purchasers with their own generating capacities with maximum production capabilities during the wet season will not always be able to dispatch the produced energy. The need for the excess energy will depend largely upon the load demand pattern of the utility.

Table 14.7 Projected cash flow from Lower Nyadi's IPP investors' point of view (with 15% excess energy only). Figures given in millions of NRs., 2005 level.

Year	Eq.	Rev.	Operating exps.[a]	Operating[b]	Depreciation[a]	Interest on debt	Capital repayment	Profit before tax	Corporation tax	Profit after tax	Bonus	Cash flow	DSCR
−2	44.17											−44.17	
−1	88.33											−88.33	
0	44.17											−44.17	
1		160.95	21.53	139.42	26.13	47.67	29.91	65.62	13.12	52.49	1.05	47.67	1.80
2		160.95	22.00	138.95	26.13	44.68	32.90	68.14	13.63	54.52	1.09	46.66	1.79
3		160.95	22.48	138.46	26.13	41.39	36.19	70.95	14.19	56.76	1.14	45.57	1.78
4		160.95	23.00	137.95	26.13	37.77	39.81	74.05	14.81	59.24	1.18	44.38	1.78
5		160.95	23.54	137.41	26.13	33.79	43.79	77.49	15.50	61.99	1.24	43.10	1.77
6		160.95	24.10	136.85	26.13	29.41	48.17	81.31	16.26	65.05	1.30	41.71	1.76
7		160.95	24.70	136.25	26.13	24.59	52.98	85.53	17.11	68.42	1.37	40.20	1.76
8		160.95	25.32	135.63	26.13	19.29	58.28	90.20	18.04	72.16	1.44	38.57	1.75
9		160.95	25.97	134.97	26.13	13.46	64.11	95.38	19.08	76.30	1.53	36.80	1.74
10		160.95	82.51	78.44	26.13	7.05	70.52	45.25	9.05	36.20	0.72	−8.91	1.01
11		160.95	27.38	133.56	26.13			107.43	21.49	85.95	1.72	110.36	
12		160.95	28.14	132.81	26.13			106.67	21.33	85.34	1.71	109.77	
13		160.95	28.94	132.01	26.13			105.88	21.18	84.70	1.69	109.14	
14		160.95	29.77	131.18	26.13			105.04	21.01	84.03	1.68	108.49	
15		160.95	30.65	130.30	26.13			104.17	20.83	83.33	1.67	107.80	
16		160.95	31.57	129.38	26.13			103.24	20.65	82.60	1.65	107.08	
17		160.95	32.54	128.41	26.13			102.28	20.46	81.82	1.64	106.32	
18		160.95	33.55	127.39	26.13			101.26	20.25	81.01	1.62	105.52	
19		160.95	34.62	126.33	26.13			100.19	20.04	80.16	1.60	104.69	
20		160.95	35.74	125.21	26.13			99.07	19.81	79.26	1.59	103.81	
21		160.95	36.92	124.03	26.13			97.90	19.58	78.32	1.57	102.89	
22		160.95	38.15	122.80	26.13			96.66	19.33	77.33	1.55	101.92	
23		160.95	39.45	121.50	26.13			95.37	19.07	76.29	1.53	100.90	
24		160.95	40.81	120.14	26.13			94.01	18.80	75.21	1.50	99.83	
25		160.95	42.24	118.71	26.13			92.58	18.52	74.06	1.48	98.71	

NPV @10% = 274.39
IRR = 22.93%

[a] Operating expenses include operation maintenance, insurance and royalties.
[b] Depreciation is not included as cash expense but used for tax purposes.

Part Two

Table 14.8 Financial analysis with purchase of excess energy.

Effects of purchase of excess energy in Financial NPVs – Lower Nyadi				
			Excess energy	
	Contract energy	5%	10%	15%
Energy	33 375.15	1669.00	3337.52	5006.27
NPVs	231.11	250.51	269.90	274.39
IRRs %	21.61	22.57	23.53	23.93

Effects of purchase of excess energy in Financial NPVs – Chilime				
			Excess energy	
	Contract energy	5%	10%	15%
Energy	111 936.50	117 533.33	123 130.15	128 726.98
NPVs	1964.02	2137.84	2285.57	2433.3
IRRs %	25.40	26.68	27.76	28.84
NPVs Utility	−2169.34	−2424.63	−2679.89	−2935.16

Effects of purchase of excess energy in Financial NPVs – Mardi				
			Excess energy	
	Contract energy	5%	10%	15%
Energy	23 871.00	1194	2387.1	3580.65
NPVs	166.66	179.23	188.65	198.08
IRRs %	21.13	22.96	23.63	24.31
NPVs	64.03	−38.68	−55.19	−71.71

Financial analysis considering purchase guarantee of excess energy

As in the case of interim energy, the price of excess dry season energy is usually higher than the wet season energy or, as in most cases, there are purchase guarantees of only dry season excess energy. A hypothetical position with the effect of purchase of dry season excess energy at 50% the prevailing rates as stipulated in most PPAs are as shown in Table 14.8. The IRR of Lower Nyadi Project increases from 21.61% to 22.57% with a sale of mere 5% excess energy in the dry season (4 months) and increases to nearly 24% with a sale of 15% excess energy in dry season. Similarly, the IRR of Chilime increases from 25.4% to 26.68% with a sale of 5% of excess energy and increase to 28.84% from sale of 15% of excess energy during the dry season. For projects, like Khimti and Bhotekoshi, with purchase rates of excess energy at the prevailing rates, the increase in the IRR will be even more significant.

As can be seen from the financial analyses, purchase of excess energy is a big incentive for the sponsors. On the other hand, the utility is not in favour of purchase of excess energy because of the losses it has to bear from the purchase of the excess energy. However, past practices show that the clause has been incorporated in past PPAs and has ranged from purchase of all excess energy to agreements with no purchase of excess energy. The value received by the utility in exchange for the inclusion of the clause is the added service to its customer.

14.3.5   Guarantee of third party sales

Rationales for 'guarantee of third party sales'

If the purchaser is not able to dispatch all of the energy produced by the IPP, it might be beneficial to both the parties to include a guarantee provided by the utility allowing third party sales of the energy. The revenues earned from the sales of energy to a third party can be set off against amounts due to the sponsors from the utility under the take or pay clauses. Provisions will have to be made by the utility to wheel the energy to the consumers in such third party sales, that is to allow the IPP to use its transmission grid for supplying energy to a third party. The allowance of third party purchase will also give comfort to the lenders when the credibility of the purchaser is questionable (Kerf *et al.*, 1998).

Financial analysis considering guarantee of third party sales

A financial analysis with a hypothetical case of selling the energy that the utility cannot dispatch to its consumers to a third party is presented in Table 14.9. The table shows the impact it will have in the NPVs of the project from the utility's point of view. If 25% of wet season energy of Chilime project, which has a provision of third party sales, can be sold to a third party and the revenue from it set off against the utility's payment, it will alleviate the financial losses borne by the utility besides the extra revenue it will earn from the wheeling charges. Table 14.10 shows how proceeds from third party sales of 25% of the wet season energy from a project can increase the NPV.

Financial analyses have indicated that the sale of excess energy the utility is not able to dispatch to third party can create a win–win situation for all the concerned parties. The past PPAs indicate that the clause has been incorporated in some agreements and with some limitations in others. However, actual sales have not materialised due to limitations in the clause in pricing.

14.4   Summary

From the above discussions, it can be concluded that the issue of power purchase guarantees have a direct bearing on financial viability of the project as well as in defining the quantum of energy to be traded between the parties in the form of contract energy, excess energy, and interim energy.

14.4.1   Take or pay clause

For the sponsors along with the lenders, it can be concluded that take or pay clauses have been a successful instrument in ensuring revenue streams. They

Table 14.9   Financial cashflow of Chilime Project from utility's point of view.

(With third party sales of 25% of wet season energy)					assumed escalation	
Revenue rate	6.28 Rs/KWh				3.00%	
Transmission cost	0.37 Rs/KWh				2.00%	
Distribution costs	1.25 Rs/KWh				5.00%	
Loss cost	1.08 Rs/KWh				2.00%	

(millions of NRs.)

Ref. year	Total revenue	Power purchase cost	Transmission cost	Distribution cost	Loss cost	Total cost	Net Cash Flow
1	702.961	457	41.417	139.921	120.891	759	(56)
2	724.050	493	42.245	146.917	123.309	806	(82)
3	745.772	533	43.090	154.262	125.775	856	(110)
4	768.145	576	43.952	161.976	128.291	910	(142)
5	791.189	622	44.831	170.074	130.857	967	(176)
6	814.925	671	45.727	178.578	133.474	1029	(214)
7	839.372	725	46.642	187.507	136.143	1095	(256)
8	864.554	783	47.575	196.882	138.866	1166	(302)
9	890.490	846	48.526	206.726	141.644	1243	(352)
10	917.205	846	49.497	217.063	144.476	1257	(339)
11	944.721	846	50.486	227.916	147.366	1271	(327)
12	973.063	846	51.496	239.312	150.313	1287	(314)
13	1002.255	846	52.526	251.277	153.320	1303	(300)
14	1032.322	846	53.577	263.841	156.386	1319	(287)
15	1063.292	846	54.648	277.033	159.514	1337	(274)
16	1095.191	846	55.741	290.885	162.704	1355	(260)
17	1128.046	846	56.856	305.429	165.958	1374	(246)
18	1161.888	846	57.993	320.701	169.277	1394	(232)
19	1196.744	846	59.153	336.736	172.663	1414	(217)
20	1232.647	846	60.336	353.572	176.116	1436	(203)
21	1269.626	846	61.543	371.251	179.638	1458	(188)
22	1307.715	846	62.774	389.814	183.231	1481	(174)
23	1346.946	846	64.029	409.304	186.896	1506	(159)
24	1387.355	846	65.310	429.770	190.634	1531	(144)
25	1428.975	846	66.616	451.258	194.446	1558	(129)

NPV @ 10% = (1826.92)

Table 14.10   Effects of third party sales to financial NPV of utility.

	Energy MWh		NPVs
	Dry season	Wet season	
Contract energy	48 995	83 922	−2169.34
With third party sales (of 25% wet season energy)	48 995	62 942	−1826.92

Part Two

value the clause and are seen to favour including it in PPAs. The utility that has employed the take or pay clause in all the PPAs to date, but it is seen to be making losses due to the clause. The alternative it has in future negotiations are: take if offered clause or take and pay clause where the utility will not have to pay for the energy that it is not able to dispatch. However, the utility will have to be more transparent in the load forecast and generation expansion plan so that the IPPs can plan their projects accordingly, minimising the possibility of the utility rejecting the energy produced by IPPs.

14.4.2   Interim energy

Projects can benefit significantly by including purchase guarantees of interim energy. To capitalise on the sale of interim energy, the project has to plan on early dry season commission of the project, when the demand of energy is highest. For the utility, the purchase of interim energy allows utility to provide service sooner.

14.4.3   Supply guarantee of minimum energy

The guarantee of supply of minimum energy is crucial to the utility for uninterrupted services to its customer. For the sponsors, the guarantee of supply of minimum energy makes them assume the hydrological risk. From the financial analyses, it can be seen that sustained reduction of energy can have significant implication in the project. Most of the projects are given a 25% waiver on the declared energy. Guaranteeing supply of minimum energy has to be done with the consideration of the authenticity of the hydrological studies and the reliability of the power plant. The 25% waiver of supply can be beneficial to cushion shortfalls of energy due to hydrological reasons or plant failure.

14.4.4   Purchase of excess energy

Purchase guarantees of excess energy, as can be seen from the financial analyses, are a big incentive for the IPPs. The presence of the clause can even allow IPPs to plan for projects with higher capacities to be able to produce energy above the declared availability, maximising the use of the country's resources. Dry and wet season excess energy has been purchased at rates ranging from 50–100% of the prevailing rates from various IPPs. Inclusion of the purchase guarantees of excess energy will have a significant positive impact on the outcome of the project.

14.4.5   Third party sales guarantees

In spite of a 'third party sales clause' being prevalent in some PPAs of the past, it has not been practised to date. The sale of excess or energy the utility is not able to dispatch to a third party can create win–win situation for all

Part Two

the concerned parties. The revenues earned from the sales of energy to a third party can be set off against amounts due to the sponsors from the utility under the take or pay clauses. The allowance of third party purchase will also give comfort to the lenders.

References

Clarke, C. (2000) *Cross-check survey, Final report*. Prepared for the World Commission on Dams (WCD), http//www.dams.org.

Crow R.T. (2001) *Foreign Direct Investment in New Electricity Generating Capacity in Developing Asia: Stakeholders, Risks, and the Search for a New Paradigm*. http://APARC.stanford.edu.

Gupta J.P. and Sravat A.K. (1998) Development and project financing of private power projects in developing countries: a case study of India. *International Journal of Project Management*, 16(2), 99–105.

Kerf, M., Gray, R.D., Irwin, T. *et al*. (1998) *Concessions for Infrastructure: a Guide to their Design and Award*. Technical paper no. 399, The World Bank.

Nepal (2001) *The Hydropower Development Policy*. Kathamndu, Government Printer.

Nepal Electricity Authority and Bavarian Hydropower Nepal (2003) *Power Purchase Agreement between Nepal Electricity Authority (NEA) and Bavarian Hydropower Nepal (P.) Ltd concerning Lower Nyadi Hydroelectric Project*.

Nepal Electricity Authority and Chilime Hydropower Company (1997) *Power Purchase Agreement between Nepal Electricity Authority (NEA) and Chilime Hydropower Co. Ltd concerning Chilime Hydroelectric Project*.

Nepal Electricity Authority and Gandaki Hydropower Development Co. (2003) *Power Purchase Agreement between Nepal Electricity Authority (NEA) and Gandaki Hydropower Development Co.(P.) Ltd concerning Mardikhola Hydroelectric Project*.

Nepal Electricity Authority and Gorkha Hydropower Nepal (1999) *Power Purchase Agreement between Nepal Electricity Authority (NEA) and Gorkha Hydropower Nepal (P.) Ltd concerning Daram Khola Hydroelectric Project*.

Nepal Electricity Authority (2004) *Annual Report: Fiscal year 2003/04 – A Year in Review*. Kathmandu.

Nepal Electricity Authority (2005) *Annual Report: Fiscal year 2004/05 – A Year in Review*. Kathmandu.

Shrestha, R.B. and Ogunlana, S. (2006) Comparative study of power purchase agreements in the Nepalese environment. *Journal of Financial Management of Property and Construction*, 11(3), 133–147.

Part Two

15

Government Policy on PPP Financial Issues: Bid Compensation and Financial Renegotiation

S. Ping Ho[1]

15.1 Introduction

Private participation has been recognised as an important approach for providing public works and services (Walker and Smith, 1995; Henk, 1998). Whereas BOT, PFI and DBFO etc. are popular variations, PPPs can be considered the most general term for schemes of this kind. According to a report by US Federal Highway Administration (2005), between 1985 and 2004 there were approximately 1120 major PPP projects funded and completed worldwide at a total cost of US$450bn. PPPs are now major initiatives supplying public works in the UK and have become increasingly popular in Asia. In 1999 Japan passed the PFI Law supporting the use of PPPs. Other Asian countries have adopted PPPs including Hong Kong, Taiwan, Thailand, China, Singapore, Korea and the Philippines. In 2000 Taiwan enacted *The Act for Promotion of Private Participation in Infrastructure Projects* and began to aggressively promote the use of PPPs. Up to April 2005 there have been 280 PPP projects funded in Taiwan, with US$25bn invested by private parties. Started in January 2007, the Taiwan High Speed Railway, a US$18.4bn project, is the largest PPP project in Taiwan and one of the largest PPP projects in the world. At 508 m in height, the Taipei 101 building is currently the tallest building in the world and was also funded by a PPP.

Because PPPs involve special relationships between public and private parties and complex financing issues, the administration of PPP projects has been a challenging task. Too often serious problems occur in PPPs, mainly due to bad administration policies. In practice there are various guidelines for managing PPP projects; however these guidelines cannot be universal and need to be modified to fit the specific environment of a country. When there are opportunities to participate in policy making, decisions should be based on solid economics instead of intuition-based superficial reasoning. The purpose of this chapter is to introduce two game-theory models for PPP administration to address two key issues: bid compensation and financial renegotiation.

Bid compensation is a stipend or honorarium paid by the owner to the unsuccessful bidders to compensate them for the cost of bid preparation and is often used for projects with high preparation costs. Many project owners, especially government authorities, are keen to know whether they should offer bid compensation and how much to offer. Owners may waste money in bid compensation if it is not effective, and if governments are not aware of its ineffectiveness, they lose their chance to adopt other approaches for improving the quality of bids received. In this chapter, a model will be introduced that examines how bidders react to bid compensation and the associated policy implications.

Financial renegotiation problems play a more crucial role in the success of a PPP project. Financial renegotiation is the financial subsidy negotiated after a contract has been signed when conditions change unfavourably and significantly. The importance of financial renegotiation policy goes beyond whether governments should renegotiate with private parties. The greater concern is that the government may bail out a distressed project and re-negotiate with the developer in PPPs which can cause serious opportunism problems in project administration. The issue for PPPs is then how to reduce the probability of future renegotiation and the opportunism due to the pos-sibility of renegotiation. A second model will be discussed that investigates how the government and project developers behave in various renegotiation situations when a PPP project is in distress, and what impact government rescue has on procurement and management polices. This model may help to provide theoretic foundations to policy makers to assist in the development of effective PPP procurement and management policies. The model may also offer researchers a framework to understand the behavioural dynamics of the parties involved in PPPs.

It is worth noting that compared to survey- or case-based research, the models introduced here have the advantage of considering contracting and administration problems without the limitation of specific practical or study environments (Ho, 2006b). In other words, environmental differences can be factored into an analytical model with some degree of simplification and the model generalised, providing policy makers or governments with a more rigorous framework for crafting administration policies or PPP guidelines suitable for their environments.

15.2   Game Theory

Game theory can be defined as 'the study of mathematical models of con-flict and cooperation between intelligent rational decision-makers' (Myerson, 1991). Game theory by far is one of the most important analytical tools in studying the economic behaviour of individuals, corporations and societies. As more and more problems are analysed and understood by applying game theory, the theory itself continues to advance.

Game theory has been applied to construction management in two areas. Ho (2001) applied game theory to analyse information asymmetry during the procurement of a BOT project and its implication for project financing

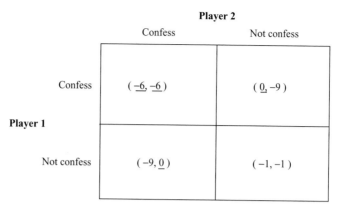

Figure 15.1 Prisoner's dilemma.

and government policy. Ho and Liu (2004) developed a game theory model to analyse the behavioural dynamics of builders and owners in construction claims. In PPPs, conflicts and strategic interactions between public and private parties are common, and thus game theory can be a useful tool for analysing these problems.

There are two basic types of games: static and dynamic. In a static game players act simultaneously, meaning each player makes a decision without knowing the decisions made by others. The bid compensation issues discussed in Section 15.3 are modelled on static games. Conversely, in a dynamic game players act sequentially. The financial renegotiation model proposed in Section 15.4 is a dynamic game where private parties and the government take turns in making decisions after observing the other party's action. The players of a game are assumed to be rational, which is one of the most important assumptions in economic theory, and it is assumed that the players will always try to maximise their payoffs.

A well-known example of a static game is the 'prisoner's dilemma' shown in Figure 15.1. Two suspects are arrested and held in separate cells. If both of them confess they will be sentenced to jail for 6 years. If neither confesses each will be sentenced for only 1 year. However, if one of them confesses and the other does not, then the honest one will be rewarded by being released (in jail for 0 years) and the other will be punished by 9 years in jail. In each cell of the table the first number represents player 1's payoff and the second one represents player 2's.

In a dynamic game players move *sequentially* instead of *simultaneously*. It is more intuitive to represent a dynamic game by a tree-like structure, also called the 'extensive form'. The concept of dynamic games can be illustrated by the following simplified *Market Entry* example. A new firm, New Inc., wants to enter a market to compete with a monopoly firm, Old Inc. The monopoly firm does not want the new firm to enter the market because new entry will reduce the old firm's profits. Old Inc. threatens New Inc. with a price war if New Inc. enters the market. Figure 15.2 shows the extensive form of the market entry game. The game tree shows (1) New Inc.'s options

Part Two

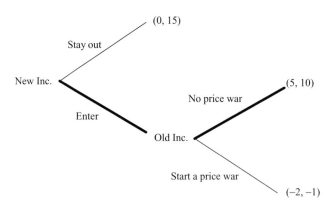

Figure 15.2 Simplified market entry game.

to enter the market or not, and then Old Inc.'s options to start a price war or not, and (2) the payoff of each decision combination.

The '*Nash equilibrium*' describes the action that will be chosen by each player and is one of the most important concepts in game theory. In a Nash equilibrium, each player's strategy should be the best response to the other player's strategy, and no player wants to deviate from the equilibrium solution. Thus, the equilibrium or solution is 'strategically stable' or 'self-enforcing' (Gibbons, 1992). Conversely, a non-equilibrium solution is unstable since at least one of the players can be better off by deviating from the non-equilibrium solution. In the prisoner's dilemma only the (confess, confess) solution, where both players choose to confess, satisfies the stability test or requirement of Nash equilibrium. Although the (not confess, not confess) solution seems better for both players this solution is unstable since either player can obtain extra benefit by deviating from this solution.

In the simplified dynamic market entry game, an intuitive conjecture is that New Inc. will 'stay out' because Old Inc. threatens to 'start a price war' if New Inc. plays 'enter'. However, Figure 15.2 shows that the threat to start a price war is *not credible* because Old Inc. can only be worse off by starting a price war. Conversely, New Inc. knows the pretence of the threat, and therefore will maximise the payoff by playing 'enter'. As a result, the Nash equilibrium is (enter, no price war), a strategically stable solution. This simplified game did not consider that there might be other companies trying to enter the market if the old company did not maintain their reputation regarding the credibility of the threat. A dynamic game can be solved by maximising each player's payoff *backward recursively* along the game tree (Gibbons, 1992); a technique that will be applied in solving the financial renegotiation game in PPPs.

In the following analysis, a certain degree of simplification and abstraction was necessary to obtain insightful results. The insights and qualitative implications from a model are often more important than the exact game solutions obtained. Therefore, it is not necessary to go through every detailed derivation in this chapter to understand the insights obtained from the models. Readers may choose to forego the equations and focus on the qualitative implications and insights implied by the equations.

Part Two

15.3   Is Bid Compensation Effective in PPP Tendering?

15.3.1   Bid compensation myth

For projects with high bid preparation costs it is often suggested that the owner should consider paying compensation to unsuccessful bidders. According to DBIA (1995), 'the provision of reasonable compensation will encourage the more sought-after design–build teams to apply and, if short-listed, to make an *extra effort* in the preparation of their proposal'. As bid preparation cost depends on project scale, delivery method and other factors, the cost of preparing a proposal is often relatively high in some types of schemes, including PPPs. Therefore, the government's bid compensation policy is important to practitioners and worth further investigation.

Before Ho (2005), the bid compensation strategy for PPP projects was not formally modelled in literature. It is in a project owner's own interest to understand whether they can stimulate high-quality inputs from bidders during bid preparation and under what conditions. Whilst the argument for using bid compensation may be intuitively sound, there is no theoretical basis or empirical evidence for such argument. Therefore, it is crucial to study under what conditions bid compensation is effective and how much compensation is adequate with respect to different bidding situations. Based on game theory analysis and numeric trials, a bid compensation model has been developed. The model provides a quantitative framework and qualitative implications on bid compensation policy. This model may help project owners to develop bid compensation strategies for specific competition situations and project characteristics.

A paradox exists in this model: on one hand the model solves the equilibrium conditions for effective bid compensation; on the other, it is shown that offering bid compensation is not very effective and thus not recommended in most cases. This conclusion is partly confirmed by Connolly (2006) who stated that they had 'found payment of bid compensation on large international construction projects to be counterproductive in several sectors'.

15.3.2   Bid compensation model

Assumptions and model set-up

To perform a game theory study, it is essential to make necessary simplifications so one can focus on the issues and obtain insightful results. The assumptions made in this model are summarised as follows. Note that these assumptions can be relaxed for more general purposes.

- Average bidders: bidders are assumed to be equally good in terms of their technical and managerial capabilities. Since PPPs focus on quality issues, the pre-qualification process imposed during procurement reduces the variation of the quality of bidders. As a result it is not unreasonable to make the 'average bidders' assumption.

272	Policy, Finance & Management for PPPs

- Bid compensation for the second best bidder: it is assumed that bid compensation will be offered to the second best bidder.
- Two levels of effort: it is assumed that there are two levels of effort in preparing a proposal, high and average, denoted by 'H' and 'A' respectively. 'A' effort is defined as the level of effort that does not incur extra cost to improve quality. Conversely, 'H' effort is defined as the level of effort that will incur extra cost, denoted as 'E', to improve the quality of a proposal, where the improvement is detectable by an effective proposal evaluation system, such as the evaluation criteria and respective weights specified in the request for proposal.
- Fixed amount of bid compensation, 'S': a fixed amount of bid compensation can be expressed as a percentage of the average profit, denoted as 'P', assumed during the procurement by an average bidder.
- Absorption of extra cost, 'E': for convenience, it is assumed that 'E' will not be included in the bid price so that the high-effort bidder will win the contract under the price-quality competition. This assumption simplifies the trade-off between quality improvement and bid price increase.

Two-bidder game

In this game there are only two qualified bidders. The possible payoffs for each bidder in the game are shown in a normal form in Figure 15.3. If both bidders choose H, denoted by (H, H), both bidders will have 50% probability of winning the contract and, at the same time, have 50% probability of losing the contract but being rewarded with the bid compensation, S. As a result, the expected payoffs for the bidders in (H, H) solution are $(S/2 + P/2 - E, S/2 + P/2 - E)$. The computation of the expected payoff is based on the assumption of the average bidder. Similarly, if the bidders choose (A, A), the expected payoffs will be $(S/2 + P/2, S/2 + P/2)$. If the bidders choose (H, A), bidder 1 will have 100% probability of winning the contract, and the expected payoffs are $(P - E, S)$. Similarly, if the bidders choose (A, H), the expected payoffs will be $(S, P - E)$. Payoffs of an n-bidder game can be obtained by the same reasoning.

		Bidder 2 H	A
Bidder 1	H	(S/2+P/2-E, S/2+P/2-E)	(P-E, S)
	A	(S, P-E)	(S/2+P/2, S/2+P/2)

Figure 15.3 Two-bidder game.

Since the payoffs in each equilibrium are expressed as functions of S, P and E, instead of a particular number, the model will focus on the conditions for each possible Nash equilibrium of the game. Here, the approach to solving for Nash equilibrium is to find conditions that ensure the stability or self-enforcing requirement of Nash equilibrium.

First, check the payoffs of (H, H) solution. For bidder 1 or 2 not to deviate from this solution:

$$S/2 + P/2 - E > S \rightarrow S < P - 2E \qquad (1)$$

Therefore, condition 1 guarantees (H, H) to be a Nash equilibrium. For bidder 1 or 2 not to deviate from (A, A), condition 2 must be satisfied.

$$S/2 + P/2 > P - E \rightarrow S > P - 2E \qquad (2)$$

Condition 2 guarantees (A, A) to be a Nash equilibrium. Note that the condition 'S = P − 2E' will be ignored since the condition can become 1 or 2 by adding or subtracting an infinitely small positive number. Thus, since S must satisfy either condition 1 or condition 2, either (H, H) or (A, A) must be a unique Nash equilibrium. For bidder 1 not to deviate from H to A: P − E > S/2 + P/2; i.e. S < P − 2E. For bidder 2 not to deviate from A to H, S > S/2 + P/2 − E; i.e., S > P − 2E. Since S cannot be greater than and less than P − 2E at the same time, the (H, A) solution cannot exist. Similarly, the (A, H) solution cannot exist either. This also confirms the previous conclusion that either (H, H) or (A, A) must be a unique Nash equilibrium.

Bid compensation is designed to serve as an incentive to induce bidders to make more effort to win a contract. Therefore, the concerns of bid compensation strategy should focus on whether S can induce more effort and how effective it is. According to the equilibrium solutions, the bid compensation decision should depend on the magnitude of P − 2E or the relative magnitude of E compared to P. If E is relatively small such that P > 2E, then P − 2E will be positive and condition 1 will be satisfied even when S = 0. This means that bid compensation is not an incentive for high effort when the extra cost of high effort is relatively low. Moreover, surprisingly S can be damaging when S is high enough that S > P − 2E.

On the other hand, if E is relatively large so that P − 2E is negative, then condition 2 will always be satisfied since S cannot be negative. In this case, (A, A) will be a unique Nash equilibrium. In other words, when E is relatively large, it is not in the bidder's interest to incur extra cost to improve the quality of their proposal, and therefore, S cannot provide any incentives for high effort.

To summarise, when E is relatively low, it is in the bidder's interest to make an increased effort even if there is no bid compensation. When E is relatively high, the bidder will be better off by making an average effort. In other words, bid compensation cannot promote extra effort in a two-bidder game, and ironically, bid compensation may discourage high effort if the compensation is too high. Thus, in the two-bidder procurement, the owner should *not* use bid compensation as an incentive to induce high effort.

Part Two

Bidder 3		H		A	
Bidder 2		H	A	H	A
Bidder 1 H		(S/3+P/3-E, S/3+P/3-E, S/3+P/3-E)	(S/2+P/2-E, 0, S/2+P/2-E)	(S/2+P/2-E, S/2+P/2-E, 0)	(P-E, S/2, S/2)
Bidder 1 A		(0, S/2+P/2-E, S/2+P/2-E)	(S/2, S/2, P-E)	(S/2, P-E, S/2)	(S/3+P/3, S/3+P/3, S/3+P/3)

Figure 15.4   Three-bidder game.

Three-bidder game

Figure 15.4 shows all the combinations of actions and their respective payoffs in a three-bidder game. Similar to the two-bidder game, the Nash equilibrium can be solved by ensuring the stability of the solution. Details of the derivation and associated equations may be found in Ho, 2005.

There are four possible equilibria: (H, H, H), (A, A, A), (2H + 1A) and (1H + 2A) where the last two equilibria are 'mix strategy Nash equilibria'. According to the concept of 'mix strategy', 2H + 1A means that each bidder randomises actions between H and A with certain probabilities, and the probability of choosing H in 2H + 1A is higher than that in 1H + 2A. From this perspective, the difference between 2H + 1A and 1H + 2A is not very distinctive. In other words, one should not consider, for example, 2H + 1A, to be two bidders playing H and one bidder playing A. Instead, one should consider each bidder to be playing H with higher probability. Similarly, 1H + 2A means that the bidder has lower probability of playing H, compared to 2H + 1A.

The effectiveness of bid compensation

The equilibrium conditions for a three-bidder game are numerically illustrated and shown in Figure 15.5, where 'P' is arbitrarily assumed as 10% for numerical computation purpose and 'E' varies to represent different costs for higher efforts.

The '*' in Figure 15.5 indicates that zero compensation is the best strategy, i.e. bid compensation is ineffective in terms of stimulating extra effort. According to the numerical results, Figure 15.5 shows that bid compensation can promote higher effort only when E is within the range of P/3 < E < P/2 where zero compensation is not necessarily the best strategy. The question is whether it is beneficial to the owner to incur the cost of bid compensation when P/3 < E < P/2. The answer to this question lies in the concept and definition of the mixed strategy Nash equilibrium, 2H + 1A, as explained previously. Since 2H + 1A indicates that each bidder will play H with significantly higher

Equilibrium E; P=10%	3H	2H+1A	1H+2A	3A
E < P/3 e.g. E=2%	S < 14% *	N/A	N/A	14% < S
P/3 < E < P/2 e.g. E=4%	2% < S < 8%	S < 2%	N/A	8% < S
P/2 < E < (2/3)P e.g. E=5.5%	N/A	N/A	S < 3.5% *	3.5% < S
(2/3)P < E e.g. E=7%	N/A	N/A	N/A	Always *

Figure 15.5   Compensation impacts on a three-bidder game.

probability, 2H + 1A may already be good enough, knowing that only one bidder out of three is needed to actually play H. As a result, if the 2H + 1A equilibrium is good enough, the use of bid compensation in a three-bidder game is not recommended.

15.3.3   Nash equilibrium of N-bidder game

Mixed strategy Nash equilibrium

As mentioned earlier, in a mixed strategy players randomly select actions H and A to confound other players. From a more dynamic perspective, every player observes which strategy works and changes their strategy if the one used did not perform as well as others. This strategy-adjusting process continues until the proportion of players in the population who play a particular strategy is equal to the mixed strategy Nash equilibrium probability. A mixed strategy can occur when there are multiple pure strategy equilibria or when there is no pure strategy equilibrium. In fact, a pure strategy equilibrium can be considered a mixed strategy equilibrium with 100% probability of playing the pure strategy. Therefore, the major concern in mixed strategy equilibrium is the probability of playing each strategy.

   In the bid compensation problem, one issue is how to compute the probabilities for choosing actions H and A. The Sale Competition Game, shown in Figure 15.6, illustrates how mixed strategy probabilities are solved. Suppose two stores are considering whether they should have a winter sale. If both stores run the sale the payoffs would be $300 for each because of intensive price competition. If none of the stores has a sale the payoffs would be $500 for each. If only one store has a sale the payoffs would be $700 and $400 for the sale store and the regular store respectively. So there are two pure strategy equilibria in the Sale game: (Sale, No Sale) and (No Sale, Sale) where no player has an incentive to change. However, it is difficult to explain why there is a player who would always choose 'No Sale'. In fact, there is a better equilibrium, the mixed strategy equilibrium where each store will randomise

	Sale (w/p λ)	No Sale (w/p 1 − λ)
Store 2		
Sale	(300, 300)	(700, 400)
Store 1		
No Sale	(400, 700)	(500, 500)

Figure 15.6   Sale competition game.

'Sale' and 'No Sale' with certain probabilities. The probabilities can be solved by following the definition of mixed strategy Nash equilibrium. According to Gibbons (1992), in a two-player game, each player's mixed strategy is a best response to the other player's mixed strategy. In the sale game suppose 'λ' is the probability that store 2 has a sale and λ is known by store 1, store 1's expected payoffs are $(\lambda)300 + (1 - \lambda)700$ from playing 'Sale' and $(\lambda)400 + (1 - \lambda)500$ from playing 'No Sale'. If $\lambda > 2/3$ then store 1's best response is to play 'No Sale'. If $\lambda < 2/3$ then store 1's best response is to play 'Sale'. If $\lambda = 2/3$ then store 1's best response is to play either strategy with any probabilities. When $\lambda = 2/3$ store 1 can choose any mixed strategies as a best response to store 2's mix strategy. In this regard, half of the equilibrium definition is satisfied. Logically, if store 2's best response is to play any mixed strategies the equilibrium definition 'each player's mixed strategy is a best response to the other player's mixed strategy' will be satisfied. Thus, the mathematical requirement for the mix strategy Nash equilibrium is that each player's mix strategy probabilities will make the other player indifferent between potential strategies. Since the Sale game is symmetrical, i.e. the payoff patterns for store 1 and 2 are identical, the mixed strategy probability for store 1 to choose 'Sale' is also 2/3. Thus, the mixed strategy Nash equilibrium of the Sale game is that each store will choose 'Sale' with a probability of 2/3 and 'No Sale' with a probability of 1/3.

Mixed strategy Nash equilibrium in the N-bidder game

A numerical method, such as trial-and-error, is needed for solving probability. For an n-bidder game of symmetrical payoffs, the mixed strategy probability, q^*, can be obtained by solving equation (3):

$$(1 - q^*)^{n-1}(P - E) + \sum_{i=2}^{n}\left[(q^*)^{i-1}(1 - q^*)^{n-i}C_{i-1}^{n-1}\left(\frac{S}{i} + \frac{P}{i} - E\right)\right]$$

$$= (1 - q^*)^{n-1}\left(\frac{S}{n} + \frac{P}{n}\right) + q^*(1 - q^*)^{n-2}(n - 1)\left(\frac{S}{n-1}\right) \qquad (3)$$

where C_{i-1}^{n-1} is the number of combinations of $n - 1$ things choosing $i - 1$.

Equilibrium E ; P=10%	4H	3H+1A	2H+2A	1H+3A	4A
E < P/4 e.g. E=2%	S < 22%	N/A	N/A	N/A	S > 22%
P/4 < E < P/3 e.g. E=3%	2% < S <18%	0 < S < 2% S=0, q=0.829 S=1%, q=0.914	N/A	N/A	S > 18%
P/3 < E < P/2 e.g. E=4%	6% < S <14%	2% < S < 6% S=2%, q=0.697 S=4%, q=0.854	S < 2% S=0, q=0.578 S=1%, q=0.632	N/A	S > 14%
P/2<E<(3/5)P e.g. E=5.5%	N/A	6.5% < S <8% S=6.5%, q=0.550 S=7.5%, q=0.661	3% < S < 6.5% S=3%, q=0.341 S=5.5%, q=0.457	S < 3% S=0, q=0.296 S=1%, q=0.306	S > 8%
(3/5)P<E<(3/4)P e.g. E=6.5%	N/A	N/A	N/A	S < 4% S=0, q=0.140 S=2%, q=0.102	S > 4%
(3/4)P < E	N/A	N/A	N/A	N/A	ALWAYS

Figure 15.7   Mixed strategy probabilities in a four-bidder game.

The left hand side (LHS) of equation 3 is the bidder's expected payoff by choosing H, given that each of the competing bidders plays H and 'A' with probabilities q and $1 - q$, respectively. The first term of LHS is the bidder's expected payoff when all competitors play A. The second term of LHS sums up the bidder's expected payoff with $(n - i)$ competitors playing A and $(i - 1)$ competitors playing H, with C_{i-1}^{n-1} different combinations for each i. The right hand side (RHS) is the bidder's expected payoff by choosing A. The first term of RHS is the bidder's payoff when all competitors play A. The second term is the bidder's payoff when there is only a competitor playing H. When there are at least two competitors playing H, the bidder's expected payoff would be zero. A computer programme was developed to solve equation 3. Figure 15.7 shows some mixed strategy probabilities with respect to various S. For example, when E is equal to 5.5% and in the range of P/2 < E < (3/5)P, the probability of choosing H without compensation, q^* will be 0.296. If the compensation is designated to cover all extra cost; i.e., S = 5.5%, then q^* will be equal to 0.457. On the other hand, when E is smaller, e.g., E = 4%, q^* will be equal to 0.578 without compensation, significantly larger than the aforementioned probability with E = 5.5%.

Optimal bid compensation decisions

As stated earlier, it is assumed that the owner's evaluation criteria are effective so a higher quality proposal can be identified and awarded the contract. As a

Part Two

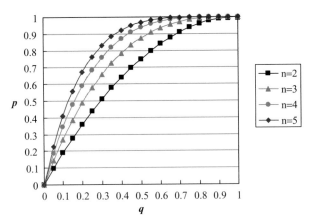

Figure 15.8   Probability p versus probability q for different numbers of bidders.

result, the owner will only need *one* high effort bidder during procurement. The major concern of the owner will be the probability that there is at least one bidder with effort H, which is computed by equation 4. This equation shows that the probability of having at least one H bidder, p, expressed as a function of S, can be computed by one minus the probability of not having a single H bidder.

$$p(S) = 1 - [1 - q^*(S)]^n \qquad (4)$$

In a three-bidder game, when 'E' equals 4% as shown in Figure 15.5, according to equation 3, q^* will be equal to 0.8 for S = 0. Equation 4 shows $p(S = 0) = 0.992$, which confirms previous conjecture that the 2H + 1A mixed strategy is good enough in a thee-bidder game and bid compensation should not be used in a three-bidder game.

Figure 15.8 shows the values of p with respect to different values of q^* in a two-, three-, four-, and five-bidder procurement. For example, if the owner wants p to be 0.97, the requirements for q in the cases of two, three, four and five bidders are approximately 0.83, 0.7, 0.6, and 0.5, respectively. Although q^* will be equal to 1 only in the nH equilibrium, Figure 15.2 shows that when there are at least three bidders, the mixed strategy equilibrium tends to become a satisfactory solution. The examples in Figure 15.7 show that when n = 4 and E = 4%, p will be equal to 0.968 for S = 0 even though q^* only equals 0.578. As a result, bid compensation is not necessary in this case. For another case, when E grows to 5.5%, p will be equal to 0.754 for S = 0, and thus bid compensation becomes more effective. In this case, p will be increased from 0.754 to 0.913 with S = 5.5%. However, the owner may not be better off by offering S = 5.5% in exchange for a higher p.

The issue now is how to determine whether a certain amount of bid compensation, S, is appropriate. It is argued from the economic perspective that an appropriate S should be justified by the marginal benefit obtained through the increase of p. Therefore, it is suggested that the owner should determine the magnitude of bid compensation according to the objective function in equation (5):

$$B = \underset{S}{Max}\{[p(S) - p(S = 0)](\Gamma_H - \Gamma_A) - S\} \qquad (5)$$

where Γ_H and Γ_A are the net values of a project to the owner with effort H and effort A, respectively. $\Gamma_H - \Gamma_A$, the marginal benefit due to higher effort, can be expressed as a percentage of total cost, so to be consistent with the expressions of E and S. For previous example, when n = 4 and E = 5.5%, if $\Gamma_H - \Gamma_A$ equals 20%, B will be maximised when S = 0 according to the equation 5. Thus, in this case, it is not in the owner's interest to use bid compensation to promote higher effort.

Equation 5 implies the owner has to award the project to a bidder even when all bidders invest limited effort A. This is true when it is very costly to repeat a procurement exercise for a project. So, for a large-scale or complex project, the implicit assumption in equation 5 should be reasonable. However, if it is possible to repeat the procurement exercise until a better H bidder appears, the cost–benefit analysis must be evaluated differently. Specifically, the owner's cost of project procurement and the expected rounds of procurement should be considered.

15.3.4   Bid compensation policy in PPP procurement

The bid compensation policy is based on analysis of games with two, three, four and n (n > 4) bidders. The bid compensation model provides the owner or government a theoretical framework for bid compensation decisions. Although the equilibrium conditions for effective bid compensation are solved, it does not mean that the model supports the use of bid compensation. Four important policy implications on PPP bid compensation are concluded:

1. Inappropriate use of bid compensation could discourage high effort.
2. Bid compensation strategies can be regarded as a problem of three-dimensions: the number of bidders, the complexity of project and the project profitability. Project complexity can be characterised by the amount of extra effort needed for improvement, denoted as 'E' in this model. Project profitability, denoted as 'P', is the expected profit before compensation.
3. Bid compensation is not desirable when the cost of extra effort, E, is very small or large compared to the expected profit margin before compensation. More specifically, bid compensation is not recommended for two- or three-bidder procurement because of the ineffectiveness of compensation, no matter how simple or complex the project. When there are four or more bidders, bid compensation becomes more effective in promoting higher effort.
4. It is not necessarily better to use bid compensation even when the bid compensation becomes more effective in stimulating higher effort. In fact, the final decisions of whether to use bid compensation and the amount of compensation should be judged by the marginal cost–benefit analysis as indicated in equation 5.

It is worth noting that in PPP projects it is not unusual for the number of bidders to be limited to two or three. In this case the owner or government should not use bid compensation as an incentive. For projects with minimum complexity and small contract profit margin, such as highways or factory plants, the use of bid compensation is not recommended either, even when there are more than three bidders. Bid compensation should only be considered when there are at least four bidders and the costs for high effort are moderate, not too high compared to the profit margin.

An incentive mechanism that is more effective than offering bid compensation may be required. The extra effort invested in a bid by the contractor does not necessarily lead to improvement in bid quality since those extra efforts may not be consistent with the project owner's needs. From this perspective, bid compensation is a passive approach without proactive participation by the project owner. One alternative to the design–build type or PPP delivery system is the design competition in which the owner pays the bidders to develop individual concepts to the point where the documents become the technical scoping brief. 'Variations of the method have the owner choosing the concept that is best in the owner's view, and all contractors bidding that one as the basis' (Connolly, 2006).

15.4   Financial Renegotiation and its Associated Problems

Financial renegotiation may happen when project cost, market demand or other market conditions become significantly unfavourable. The fact that a government may rescue a project and renegotiate with the developer causes major problems in project procurement and management. The dilemma faced by the government is that although financial renegotiation is not considered an option in advance, it is often desirable when a project goes awry. Such time inconsistency creates serious problems in project administration. Here, a game theory-based model is proposed to analyse procurement and management policies from the perspective of renegotiation.

15.4.1   Problems caused by financial renegotiation

Opportunistic bidding behaviour during project procurement

Opportunistic bidders, in their proposal, will intentionally understate the possible risks involved or overstate the project profitability to outperform other bidders. In their pilot study, Ho and Liu (2004) developed a game theory claims decision model (CDM) for analysing the behavioural dynamics of builders and owners in construction claims and the implications on opportunistic bidding. Their model shows that if a builder can make an effective construction claim the builder will have an incentive to bid opportunistically. Following their logic, if a request for renegotiation is always granted, developers would have an incentive to bid optimistically to win the project. An overly optimistic proposal can have a higher chance of winning because some crucial and developer-specific information regarding the project can be difficult to verify and, as a result, can be stated untruthfully in the proposal. For example, the developer's cost and profit structures, the project's commercial and technical risk and the risk impacts may not be fully revealed in the developer's proposal. Because of the information asymmetry in PPPs, opportunistic bidding may succeed during procurement. If developers have an incentive to bid opportunistically due to the *ex ante* expectation of *ex post* renegotiation, the effectiveness of project procurement and contract management can be significantly influenced. Since this logic between government rescue and project administration effectiveness is not straightforward, the importance of financial renegotiation problems is underemphasised.

Principal–agent problem

This occurs where the principal is played by government and the agent is played by the developers. This problem is also considered a 'moral hazard' problem, which only occurs after the contract is signed. In his repossession game example, Rasmusen (2001) shows that if renegotiation is expected, the agent may choose inefficient actions that will reduce overall or social efficiency but increases the agent's payoff. In PPPs, after signing the concession, moral hazard problems may also occur if renegotiation is expected. As developers are frequently the major contractors of PPP projects, they may not be concerned about cost overruns because they may benefit from such overspending.

In short, if the government always bails out a financially distressed project, renegotiation will be expected by developers and such expectation can cause opportunism problems. Unfortunately, government is often tempted to bail out distressed projects because of the *ex post* renegotiation benefits to government and/or the society.

Part Two

15.5   Financial Renegotiation Game and its Equilibrium

The behavioural dynamics of the renegotiation, or government rescue, plays a central role in PPP administration when information asymmetry exists. Here, game theory is applied to analyse when the government renegotiates with the developer and the impact of such renegotiation on the project. While this study is motivated by real world cases from various countries, the goal of this model is to provide a framework that is not restricted to a particular environment. In other words, the model is expected to consider various environments characterised by the parameters of the model.

15.5.1   Model set-up

The game theory framework for analysing a PPP investment shown in Figure 15.9 is a dynamic game expressed in an extensive form. Suppose

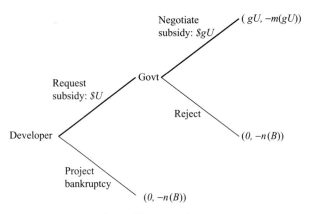

Figure 15.9   Renegotiation game's equilibrium path.

a PPP contract does not specify any government rescue or subsidy in the face of financial crisis. Neither does the law prohibit the government from bailing out the PPP project by providing a debt guarantee or extending the concession period. Suppose also that government is not encouraged to rescue a project without compelling and justifiable reasons. Cost overrun or operation losses caused by inefficient management or normal business risk are not considered just reasons for government rescue, whereas adverse events caused by unexpected or unusual equipment/material price escalation may be justified more easily. It should be reasonable to assume that if the government grants a subsidy to a project on the basis of unjustifiable reasons, the government may suffer the loss of public trust or suspicion of corruption.

The dynamic game, shown in Figure 15.9, starts from adverse situations where it is in the developer's or lending bank's best interests to bankrupt the project if the government does not rescue the project. Alternatively, the developer can also request government rescue and subsidise for the amount of U, even though the contract clause does not specify any possible future rescue from the government. Here U is defined as the present value of the net financial viability change, and is considered as the maximum possible requested subsidy. Note that U is not the actual subsidy amount. The actual subsidy is determined in the renegotiation process discussed later.

If the developer chooses project bankruptcy, the payoff will be $-\delta$. Here it is assumed $\delta \to 0$. If the situation calls for bankruptcy, the value of the *equity shares* held by the developer should approach zero before project bankruptcy. Consequently the developer, being an equity holder, will lose little if the distressed project is bankrupted. Thus, it is assumed that $\delta = 0$ in the model. Some may argue that δ is significant due to the loss of reputation. The loss of reputation occurs when the project is in distress, no matter where the developer chooses to request rescue subsidies or project bankruptcy. If δ is defined as bankruptcy payoff, then δ should not be regarded as the loss of reputation. Loss of reputation may discourage opportunistic behaviour. The effect of this, from the game theoretic perspective, is beyond the scope of this model.

On the other hand, if a PPP project is bankrupted, the payoff of government is $-n(B)$, where B is government's 'budget overspending' when a project is bankrupted and retendered, and n, a function of B, is the *political cost due to project retendering*. Generally, from either a financial or political perspective, it is costly for the government if a PPP project is bankrupted. Suppose that for a PPP project to proceed beyond procurement stage, the project must have been shown to provide facilities or services that can be justified economically. Then it is reasonable to assume that a bankrupted PPP project should be retendered to another new developer, unless, in rare occasions, the marginal subsidy for improving the project financial viability is greater than the net benefits from the facility/service. Logically, for the government to 'permanently' terminate a project without retendering, after spending millions or billions of dollars, would signify that the project was not worth undertaking in the beginning and that the government had made a serious mistake during project procurement. Therefore, in this game, it is assumed that retendering is desired by government if a project is going bankrupt.

Alternatively, as shown in Figure 15.9, the developer can negotiate a subsidy starting with the maximum amount $\$U$, where the subsidy can be in various forms such as debt guarantee or concession period extension. Typically the bank will not provide extra capital without a government debt guarantee or other subsidies. Because the debt guarantee is a liability to the government, but an asset to the developer, a debt guarantee is equivalent to a subsidy from government. Other forms of subsidy may include the extension of the concession period, more tax exemption for a certain number of years, or an extra loan or equity investment directly from government.

After the developer's request for subsidy, the game proceeds, as shown in Figure 15.9, to its sub-game: 'negotiate subsidy' or 'reject'. If the government rejects the developer's request, the project will be bankrupted and retendered and the payoff for both parties will be $(0, -n(B))$. If the government decides to negotiate a subsidy, expressed by the *rescuing subsidy ratio g*, a ratio between 0 and 1, the payoff to the developer and the government will be $(gU, -m(gU))$, respectively, where m is the *political cost due to the rescuing subsidy to a private party*. Note that although the political cost, m, is a function of budgeting spending, function m is different from function n, because in the two functions the budget spending goes to different parties. To rescue a PPP project and provide rescuing subsidy to the original PPP firm could bring serious criticism toward government. If the government lacks compelling reasons for the subsidy, the criticism will cause significant *political* cost depending on the magnitude of the subsidy. The differences between the two functions will be discussed in detail later. Here 'g' is not a constant and is used to model the process of 'offer' and 'counter-offer'. More details on negotiation modelling using g can be found in Ho and Liu (2004).

15.5.2 'Rescue' or 'no rescue' Nash equilibria of the rescue game

As mentioned previously, the financial renegotiation game tree derived above will be solved backward recursively and its Nash equilibrium solutions will be obtained. Since the values for the variables in the game's payoff matrix are undetermined, the payoff comparison and maximisation cannot be solved for a unique solution. However, the conditions for possible Nash equilibria of the game can be analysed. There are three candidates for the Nash equilibria: (1) the developer will 'request a subsidy', and the government will 'negotiate a subsidy', (2) the developer will 'request a subsidy' and the government will 'reject', and (3) the developer will choose 'project bankruptcy'.

The developer will 'request a subsidy' and the government will 'negotiate a subsidy'

Here, since government chooses to 'negotiate subsidy', this equilibrium is called the 'rescue' equilibrium in this model. Solving backward from the government's node first, if the payoff from negotiation is greater than that from rejection, i.e. $-m(gU) \geq -n(B)$, government will 'negotiate subsidy' with the developer. Therefore, the condition for negotiation or rescue can be rewritten as

$$m(gU) \leq n(B) \tag{6}$$

This condition is straightforward: the political cost of rescue should be less than or equal to the political cost for not rescuing the project. As indicated by the latter bold line in Figure 15.9, the payoff for the developer and the government will now be $(gU, -m(gU))$ respectively.

The next step is to solve Figure 15.9 backwards again from the developer's node, and obtain the final solution. Now the payoffs for 'request a subsidy' are $(gU, -m(gU))$, and the developer will request a subsidy if $gU \geq 0$. Since g and U will not be negative numbers, the condition for the developer to request subsidy will always be satisfied. In other words, it is always to the developer's benefit to negotiate a subsidy if equation 6 is satisfied.

Figure 15.9 also shows the equilibrium path expressed in bold lines that go through the game tree. Note that when the developer requests subsidy for U, the final settlement for the subsidy will be a portion of U, gU, which satisfies equation 6. From equation 6, it is known that so long as $n(B) - m(gU) \geq 0$, the rescue equilibrium will be the solution of the game, where no one can be better off by deviating from this equilibrium. The condition for this equilibrium needs to be refined for other reasons which will be discussed further in later sections.

The developer will 'request a subsidy' and the government will 'reject'

If equation 6 is not satisfied, 'reject' would be a preferable decision to the government, and the payoff matrix for both parties is $(0, -n(B))$. Now turn to the developer's node: it seems that the payoff of either 'request a subsidy' or 'project bankruptcy' is $0, and the developer is indifferent between the two actions. From the game tree, it is not obvious which action the developer will choose. However, if the developer recognises the existence of the cost incurred in the process of requesting subsidy, although it may be relatively small compared to other variables in the game tree, the developer should choose 'project bankruptcy' instead of requesting a subsidy. From this perspective, although the cost of requesting subsidy is suppressed in the game tree for clarity, the cost of requesting a subsidy should be recognised whenever there is a tie between 'request a subsidy' and 'project bankruptcy'. To summarise, if the developer knows government will 'reject' the subsidy request, the developer will choose 'project bankruptcy' instead of 'request a subsidy' in the first place, and this is the third possible equilibrium, 'project bankruptcy'. Thus, the second equilibrium solution cannot exist.

The developer chooses 'project bankruptcy'

Since the developer knows that government will choose to 'reject' the subsidy request, the developer will choose project bankruptcy in the first place. This shall be termed the 'no rescue' equilibrium. As argued above, the developer will choose project bankruptcy if, and only if, it is optimal for government to 'reject' the subsidy request. Therefore, the condition of this Nash equilibrium would be

$$m(gU) > n(B) \qquad (7)$$

In other words, for 'project bankruptcy' to be an equilibrium solution, it must be impossible to achieve the 'rescue' solution. Equation (7) can be rewritten as

$$n(B) - m(gU) < 0 \qquad (8)$$

To conclude, equations 6 and 8 are the 'rescue' and 'no rescue' equilibrium conditions respectively. Both equilibria depend solely on the knowledge of government's political cost for rejecting a subsidy and granting a subsidy. It is assumed that the PPP game is a game with *complete information*, where $n(B)$ and $m(gU)$ are common knowledge and both parties know the other party is equally rational and smart. From a practical perspective, it is not easy for both parties to quantify $n(B)$ and $m(gU)$, because it is difficult to measure political cost in terms of monetary units. Fortunately the game depicted above can still be analysed without knowing the exact functions for $n(B)$ and $m(gU)$, and game theory analysis can still lead to important qualitative and quantitative implications on PPP policies and decision making.

15.5.3   Modelling of game parameters

To perform this analysis it is necessary to examine the characteristics of the PPP project, especially its bankruptcy conditions and the political costs associated with bankruptcy.

Part Two

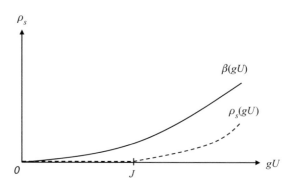

Figure 15.10   Political cost function of budgeting overspending, $\beta(gU)$, and political cost function of over subsidisation, $\rho_s(gU)$.

Political cost of rescuing a project by subsidy

If government negotiates the subsidy with the existing developer and rescues the project, the function of the political cost to government is modelled here as:

$$m(gU) = \begin{cases} \beta(gU) & \text{if } gU \leq J \\ \beta(gU) + \rho_s(gU) & \text{if } gU > J \end{cases} \tag{9}$$

where J is the amount of the subsidy that can be justified without criticism of over subsidisation, $\beta(gU)$ is the political cost of budget overspending, and $\rho_s(gU)$ is the political cost of over subsidisation. The subscript 's' of $\rho_s(gU)$ denotes subsidy.

The modelling of the political cost of subsidy in equation 9 is based on the most fundamental concept in economics that resources are scarce. If the government had unlimited funds to spend there would be no political cost for negotiated subsidy. Since the government only has limited budget there will be a political cost should the funds be allocated inappropriately. The more the subsidy, the higher the political cost. As a result, the political cost of subsidy should be an increasing function of the amount of subsidy, gU. In equation 9, the political cost is broken into two elements, namely, $\beta(gU)$ and $\rho_s(gU) \cdot \beta(gU)$, as illustrated in Figure 15.10, is an increasing function of gU, representing the political cost caused by budget overspending on subsidies, and is considered the 'basic' political cost.

In addition to the basic political cost, it is argued that for the subsidy exceeding a justifiable amount, further political costs, $\rho_s(gU)$, would be incurred to reflect a more serious resource misallocation. In the model 'J' is termed the 'justifiable subsidy', which is considered by the public an eligible claim for subsidy. 'J' can be measured by imagining the amount of 'claim' that could be granted to the developer had the case gone to court. For example, the damages due to *force majeure* might be considered justifiable. If the subsidy is less than the justifiable claim, the government will not be blamed for over subsidisation, and therefore $\rho_s(gU)$ will be considered zero when $gU \leq J$. However, when the subsidy is greater than J, the government will

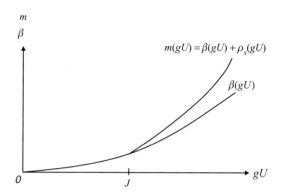

Figure 15.11 Political cost function of rescuing a project, m(gU).

be criticised for over subsidisation or suspected of corruption, and will suffer further political cost, $\rho_s(gU)$, in addition to the basic political cost, $\beta(gU)$. Figure 15.10 illustrates the function of the political cost of over subsidisation, $\rho_s(gU)$. The shapes of the functions in Figure 15.10 are for illustration purposes only. The functions need not to be continuous or convex. The only requirement is that these functions are strictly increasing. Figure 15.11 shows the function $m(gU)$ obtained by combining the curves in Figure 15.10 as defined in equation 9.

Political cost of retendering a project

A very common bankruptcy condition is the inability of the borrower to meet the repayment schedule. In PPPs, the lending bank will impose conditions to trigger bankruptcy and protect the loan should adverse events happen. The lenders could specify the upper limit of cost overrun during the project development. According to financial theory, rational lenders will prevent the *net value of the project to date* from being below the *current outstanding debt*. Since project value and cost may be volatile from time to time during the project lifecycle, to ensure the security of debt, lenders need to evaluate the project viability and debt security periodically in terms of project's gross value and required debt.

Assuming the lending bank can effectively monitor the project financial status, it may be inferred at the time of bankruptcy that the overall value of the project will be less than, but close to, the estimated total outstanding debt. As a result, under near bankruptcy conditions, it is unwise for the bank to continue providing additional capital because it is likely that the PPP firm will not be able to repay any further borrowing. Unless the government guarantees the repayment of the loan, or secures the additional debt by other means, the lending bank will deny further capital requests, even when such capital is still within project's original loan contract.

When a project is bankrupted, it will be considered 'sold' to government and retendered to another private developer given the earlier assumption that the project is still worth completing. The government may want to regain control of the project after earlier unsuccessful development because a PPP contract is usually related to public facilities or services and cannot

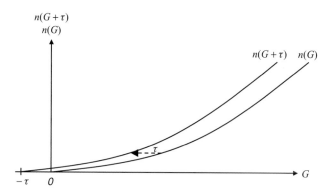

Figure 15.12   Function $n(G + \tau)$ with respect to G given a fixed τ.

be transferred directly to a new developer without a new contract. Consequently, the government would consider bankruptcy a costly replacement of the developer. Under normal situations, the bankrupted project acquired by the government will still be mainly financed by debt, and subsidies for securing the lending bank's new loan are essential to complete the project or continue the operation. As a result, when a project is bankrupted, the amount of budgeting overspending can be modelled as:

$$B = G + \tau \tag{10}$$

where G is the least required subsidy that can persuade the lending bank to support a distressed project, and τ is the opportunity cost for replacing developers, which may include the retendering cost and the cost of interruption due to the bankruptcy and retendering process.

Similar to the political cost of rescuing a project, the political cost of project retendering can be modelled by:

$$n(B) = \beta(B) \tag{11}$$

Substituting equation 10 into 11, equation 11 can be rewritten as:

$$n(G + \tau) = \beta(G + \tau) \tag{12}$$

Figure 15.12 shows functions $n(G)$ and $n(G + \tau)$, defined by equation 12, where given τ is fixed, the variable of horizontal axis will be G. Thus function $n(G + \tau)$ is depicted differently from $n(G)$, as shown in Figure 15.12, by shifting the original $n(G)$ to the left by τ.

Mathematical characteristics of the parameters in PPPs

- Characteristic 1: if the government intends to rescue a project, the project subsidy must be at least equal to G, i.e., $gU \geq G$.
- Characteristic 2: the developer replacing opportunity cost is always positive and significant, i.e., $\tau \gg 0$.

Part Two

- Characteristic 3: since not all losses due to financial viability change can be justified for subsidy during renegotiation, the range of 'J' can be modelled as

$$J \in [0, U] \tag{13}$$

The amount of justifiable subsidy depends on how the public may agree with the subsidy considering the developer's justifiable reasons. 'J' may also be quantitatively determined should the subsidy request be brought to court.

Characteristic 4: according to the NPV investment rule, 'G' may be defined by the equality: $G + NPV_t = 0$, meaning 'G' will revert the project NPV to zero. This characteristic comes from the requirement that 'G' should improve a project from negative NPV_t to zero NPV. Zero NPV indicates that the project has normal profit and is worth continuing for developers.

15.5.4   Refined Nash equilibrium

Previous sections concluded that equations 6 and 8 are the conditions for 'rescue' and 'no rescue' equilibria respectively. However, it is also noted that these conditions need to be refined. According to characteristic 1, to rescue a project the subsidy must be at least equal to 'G', i.e., $gU \geq G$. As a result, the condition for rescue equilibrium becomes:

$$m(gU) \leq n(B) \quad \text{where} \quad gU \geq G \tag{14}$$

Substituting equation 10 into 14, equation 14 can be rewritten as:

$$m(gU) \leq n(G + \tau) \quad \text{where} \quad gU \geq G \tag{15}$$

Since $m\,(gU)$ is an increasing function, gU must have an upper limit, below which the inequality in equation 15 is satisfied. The upper limit of gU can be obtained by solving $n(G + \tau) - m(gU) = 0$. Thus, the condition for rescue equilibrium can also be reorganised and expressed by the lower and upper limits of the subsidy as shown in equation (16):

$$gU \in \{x : G \leq x \leq m^{-1}[n(G + \tau)]\} \tag{16}$$

where $m^{-1}[n(G + \tau)]$ is the inverse function of m. Here equation 16 will be called the 'renegotiation offer zone'. Figure 15.13 shows the rescue equilibrium condition, equation 16, and the renegotiation offer zone, indicated by the grey bar in the x axis. Given any G in Figure 15.13, $n(G + \tau)$ will be determined first, and then $m^{-1}[n(G + \tau)]$ is obtained so that any gU between G and $m^{-1}[n(G + \tau)]$ will satisfy equation 15. In other words, the negotiation settlement will fall within the range between G and $m^{-1}[n(G + \tau)]$, expressed as $[G, m^{-1}[n(G + \tau)]]$.

15.6   Propositions and Rules

15.6.1   Propositions

This section presents propositions implied by the equilibrium of game model. Detailed proofs of the propositions may be found in Ho (2006a).

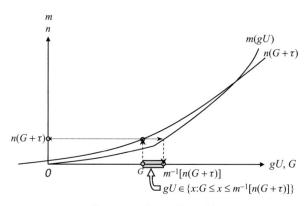

Figure 15.13   Renegotiation offer zone in 'rescue' equilibrium.

Proposition 1

Assume that the rescue renegotiation process follows the game tree in Figure 15.9, that g, U, J, G and τ are non-negative and common knowledge, and that m and n are non-negative increasing political cost functions and common knowledge. Given U, G, τ and functions m and n, if $m(gU) \leq n(G + \tau)$, where $gU \geq G$, the government will 'rescue' a distressed PPP project with a negotiated subsidy and the renegotiation offer zone is $gU \in \{x : G \leq x \leq m^{-1}[n(G + \tau)]\}$.

Proposition 1 is graphically illustrated in Figure 15.7, where the renegotiation offer zone is indicated.

Proposition 2

Suppose all assumptions in proposition 1 hold. Given U, τ and functions m and n, when there exists a S_α defined by $S_\alpha = m^{-1}[n(S_\alpha + \tau)]$ and $\forall x \leq S_\alpha : m(x) \leq n(x + \tau)$, the equilibrium must be to 'rescue' if $G \leq S_\alpha$ and must be 'no rescue' if $G > S_\alpha$. This is illustrated in Figure 15.14.

Proposition 3

Suppose all assumptions in proposition 1 hold. It must be true that the larger ρ_s function will yield a smaller S_α. Proposition 3 is illustrated in Figure 15.15, which shows that the steeper the function m, the smaller the S_α.

15.6.2   Rules due to the propositions

The propositions can be transferred into rules to assist policy makers analysing various renegotiation situations.

Part Two

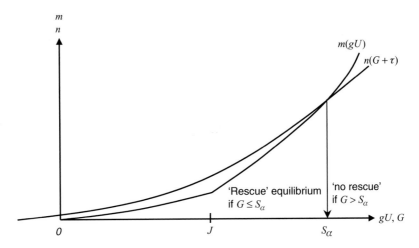

Figure 15.14   Illustration of proposition 2.

Rule 1: Equilibrium determination rule

The equilibrium determination point is S_α. The equilibrium is to 'rescue' if $G \leq S_\alpha$, and is 'no rescue' if $G > S_\alpha$.

Rule 2: S_α determination rule

S_α will depend negatively on ρ_s, and positively on τ and J. If ρ_s is small enough to be ignored, then S_α will approach ∞ and the equilibrium will always be to 'rescue'. A direct inference from this rule is that in a more dictatorial country the government will be more inclined to rescue a distressed project, justifiably or not, given that the project is still socially beneficial. Also, given other variables fixed, $\tau = 0$ will yield the smallest S_α, which will be J, and functions $m(x)$ and $n(x)$ will be on the same curve for all $x \leq S_\alpha = J$.

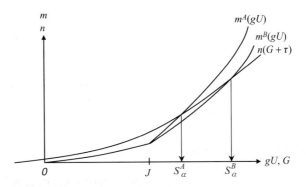

Figure 15.15   Illustration of proposition 3.

Part Two

Rule 3: renegotiation offer zone rule

If the equilibrium is to 'rescue', the renegotiation offer zone will be $gU \in \{x : G \leq x \leq m^{-1}[n(G + \tau)]\}$.

This solution is considered a Pareto optimal solution for both parties since both parties' payoffs will be improved compared to 'no rescue' solution. The difference between $m^{-1}[n(G + \tau)]$ and G is the surplus obtained by reaching the settlement. The remaining question is how this surplus will be divided. The division of the surplus may depend on each party's negotiation power and risk attitude (Binmore, 1992).

Rule 4: interval of renegotiation offer zone rule

If the equilibrium is to 'rescue', then the interval of the renegotiation offer zone will depend positively on τ. Particularly, when $\tau = 0$ the interval of the zone will be zero, the rescuing subsidy will reach at $gU = G$.

Literature has attributed the occurrence of renegotiation to the hold-up problem due to the opportunity cost of contract termination, e.g. in this model, the developer replacing cost, τ. This rule confirms that the larger the replacing cost, the more serious the hold-up problem and the wider the interval of the renegotiation offer zone. However, surprisingly, rule 4 shows that when there is no replacing cost, i.e., $\tau = 0$, the equilibrium still guarantees the occurrence of renegotiation given that the 'rescue' condition in rule 1 is met. The major reason is the existence of the least required retendering subsidy, G. Apparently, G becomes the new basic factor for the hold-up problem when the project is financed through the PPP scheme. By the definition of project distress, G must be positive and, therefore, the hold-up problem must exist.

15.7	Governing Principles and Policy Implications for Project Procurement and Management

Governing principles and administration policy implications can be obtained from the propositions, corollaries and rules derived from the model. Whilst the proposed model does not provide the approaches to quantifying the game parameters this pilot study focuses on the characteristics of the game parameters/functions and the relationship between these parameters. Particularly, the political cost functions m and n may be the most difficult to be quantitatively determined. Fortunately, useful insight can still be drawn without knowing the approach to quantifying parameters. The focus will be on which strategies can reduce the renegotiation problem and enhance the administration in PPPs. Suggested governing principles and administration policies for PPP projects are given as follows.

15.7.1   Governing principle 1

Be well prepared for renegotiation problems, as it is impossible to rule out the possibility of renegotiation and the 'rescue' equilibrium.

Part Two

Practically, S_α will be greater than 0 as S_α cannot be 0 unless $J = 0$ and $\tau = 0$. Thus, it is always possible that $G \leq S_\alpha$ given that G is uncertain; i.e. it is impossible to rule out the 'rescue' equilibrium. As a result, the government should be well prepared for the opportunism problems induced by the *ex ante* expectation of renegotiation as discussed previously. Policy implications from this principle include:

- In project procurement, while the developer's financial model is typically included in the proposal for reference, the government should recognise the possibility of opportunism problems and always have reasonable doubt about the proposal provided by developer.
- The government could devise a mechanism to ensure developers provide true information. For example, the government can establish a formal policy to disqualify a developer if they are shown to have the history of behaving opportunistically.

15.7.2   Governing principle 2

Although renegotiation is always possible, the probability of reaching 'rescue' equilibrium should be minimised and could be reduced by strategies that increase the political cost of over subsidisation, ρ_s, and reduce the developer replacing cost, τ, and the justifiable subsidy, J.

One way to reduce the opportunism problems is to minimise the probability of 'rescue' equilibrium and the developer's expectation of the probability. According to rule 1, the probability of 'rescue' can be reduced by having a smaller S_α, which can be achieved by strategies that increase ρ_s and reduce τ and J. Policy implications of this principle may include:

- Laws may regulate the renegotiation and negotiated subsidy, and such laws will increase ρ_s when the subsidy is not justifiable.
- A good monitoring or 'early warning' system can give the government enough lead time to prepare for replacing a developer with minimal impact, and hence, reduce τ.
- To reduce J, the government should pay attention to the quality of the contract in terms of content and implementation, e.g. the scope, risk allocation, documentation and contract management process.

15.7.3   Governing principle 3

During the renegotiation process, the government should try to settle the rescuing subsidy at G, the least required subsidy to retender a project, and spend more effort on determining G objectively and conveying such information to the developer, rather than on negotiation skills.

Since the 'rescue' equilibrium is a better solution for project developer, the government should try to settle the negotiation at G, the lower bound of the renegotiation offer zone. One policy implication may be that the government could regulate the negotiated subsidy using laws explicitly forbidding a subsidy being greater than G. The government should spend more efforts

Part Two

on determining G objectively and conveying such information to the project developer. For example, G can be assessed through a survey of major bankers on the least required retendering subsidy for a particular project. Therefore, it is suggested that a government should build an objective and transparent standard procedure for determining G.

15.7.4   Governing principle 4

The government should determine a fair justifiable subsidy, J, which corresponds to the developer's responsibilities and allocated risks specified in the contract.

Holliday *et al.* (1991) argue that because of the scale and complexity of BOT projects they are *often developer-led*, and it is extremely difficult to identify a clear client–contractor relationship. The 'developer-led' phenomenon implies information asymmetry and an opportunism problem in PPP projects where the developer may hide information and have an incentive to behave opportunistically. *J* is fair only when the allocation of risks and responsibilities is appropriate. As Ho and Liu (2004) and Rubin *et al.* (1983) argue a harsh contract will only encourage opportunistic behaviours. When the amount of *J* is brought to court or special committee, the court or committee will consider not only the contract clauses but also the fairness of the contract. Policy implications may include:

- The government can separate the developer from the builder/contractor in a PPP project to have a clearer client–contractor relationship.
- The government can assign third party experts to serve on the board of the project company to ensure proper monitoring and the collection of accurate information.
- The government can form a special committee consisting of outside experts to determine a fair *J* for the project.
- The government should spend more effort on appropriate risk allocation in the contract rather than developing harsh contract clauses.
- Risk assignment between the concessionaire and government should be made explicitly in the agreement. This could help to determine a fair *J* in the future.
- The government should carefully consider and specify when they may intercede:
  - The government could step in and temporarily take over a project when the project shows signs of potential distress according to the monitoring system.
  - By temporarily taking over a project the government may have more information regarding the project; who is responsible, how to minimise the impact and how much subsidy could be justified. Even if the distress is inevitable, the government will obtain more objective information regarding *J* and *G*, and will reduce τ due to longer lead time to respond and prepare for the retendering.

☐ The government should not intervene too hastily since the risk and responsibility may be partly transferred back to government if the step-in itself cannot be justified.

☐ The step-in decision should be made cautiously by government officials and outside experts following a standard procedure.

Case Study 15.1:   Taiwan High Speed Rail

Taiwan's first act to support the partial use of PPPs in transportation infrastructures was passed in 1994. In 2000, Taiwan enacted the Act for Promotion of Private Participation in Infrastructure Projects to support the use of PPPs in most public infrastructures and services. Up to April 2005 there had been about 280 PPP projects funded in Taiwan, with approximately US$25bn invested by private parties.

Background of Taiwan high speed rail

The Taiwan High Speed Rail (THSR) project is the country's first high speed rail system connecting major cities from north to south by running trains at up to 300 km/hour along the 345 km route. This project is the largest transportation infrastructure in Taiwan and also one of the largest projects in the world delivered through PPPs. This project was developed using a BOT scheme and within the 35-year concession period the awarded concessionaire must deliver the project in return for the operating profit from the rail system.

The procurement of the project officially began in January 1997 and was awarded to the Taiwan High Speed Rail Corporation (THSRC) in September 1997. After 10 months of negotiation the project concession agreement was signed in July 1998. Construction began in February 2000 and, after almost 7 years, was completed in January 2007 with a 14-month delay. The total cost of the project was $18.4bn, including $3.4bn committed by the government and $15bn invested by private parties, with $2bn of cost overruns. Major works completed by private investment included civil works, stations, track work, electrical and mechanical systems and financing. Items undertaken by government, called 'government-assisted items', were mostly related to the exercise of government authority, such as land acquisitions and project supervision.

The capital structure of the THSRC was originally targeted at 30% equity ratio and 70% debt ratio, later revised to 25%:75% respectively. While the total equity to be raised was about $4bn, 9 months after the contract was signed the THSRC had only $0.6bn of equity. The THSRC had substantial difficulty raising the rest of the equity according to the contracted schedule and was forced to renegotiate total equity down to $3.3bn. In fact, the THSR project encountered many major difficulties before its completion and most of these were related to financing.

Awarding of THSR project

Only two teams competed for the project: Taiwan High Speed Rail Alliance and China High Speed Rail Alliance, with the project being awarded to the Taiwan High Speed Rail Alliance. Since the technical concerns are limited due to the maturity of high speed rail technology, the competition focused on financial issues. In their financial proposal China High Speed Rail Alliance requested the government invest $4.6bn in addition to the government-assisted items, to make the project financially viable. Taiwan High Speed Rail Alliance requested zero additional government investment and further promised that the government might receive at least $3.2bn payback from the project operation revenue by the end of the concession period.

The government made several serious mistakes in the procurement phase of the THSR project. First and most critically, the government should not have adopted PPPs in the THSR, a project that could not be allowed to default, when the government had no experience in PPPs. According to the renegotiation model, opportunism will be most serious when the government cannot allow the project to fail and, thus, it is almost certain that the government will bail out the project at any cost. If there is sufficient incentive for opportunism, the developer's financial proposal will tend to be overly optimistic. In fact, after the awarding decision, the government was criticised for naively believing in the winner's financial proposal. However, it is difficult to differentiate whether a financial forecast is fair or too optimistic, particularly in PPPs, where creativeness and efficiency from private parties are emphasised. According to the model, instead of focusing on the figures in the financial plan, the government should focus more on eliminating the sources of opportunism.

Current practice in PPPs that involve construction inherently creates incentives for developers to behave opportunistically. For example, in the THSR, the construction phase of the project undertaken amounted to $3.3bn while the total equity invested by the firms was only $0.36bn. This type of stakeholder profit structure would make promoters greatly emphasise short-term construction profit instead of long-term operational profit. Given the existence of such incentives to behave opportunistically, the importance of reducing the possibilities for opportunism cannot be overstated.

The debt financing crisis

The first crisis faced by the THSRC was the inability to obtain debt financing of $10bn after signing the concession contracts. The developer did not utilise the international debt markets for financing partly because the Taiwan government was expected to subsidise the loan at an interest rate far below the market. However, since the THSR was the first PPP mega project in Taiwan, the banks had no faith in financing the project at a rate below fair market without 'full' debt guarantees from the government. Since the full debt guarantee was a significant liability to the government and was neither anticipated by the government nor specified during the procurement process, the provision of debt guarantees became a controversial issue and the government hesitated to offer the debt guarantee. In fact, the doubt from the public was that the project might have been financially non-viable if a fair market interest rate had been imposed. After several rounds of fruitless negotiation, the THSRC gave the government an ultimatum: if the government could not help to settle the debt financing negotiation by 31 July 1999, they would abandon the project. In response to the ultimatum the government offered full debt guarantees and signed the agreement in August 1999 with syndicate banks and the THSRC. Among the $10bn of debt financing, $8.6bn came from government-owned banking systems and only $1.4bn belonged to private commercial banks. The Prime Minister, Mr Hsiao, explicitly stated that 'the project is not allowed to fail' and the 'government will do everything to support the project'.

In this crisis, the political cost of not rescuing the project was the political cost of spending 3 more years, and the procurement cost, to replace the developer. Conversely, the political cost of rescuing the project was relatively low. The rescue was easily rationalised by the government's role in facilitating the transactions between the developer and the banks. Additionally, the statement made by the Prime Minister declared the importance of the project to the society. However, the attitude that 'the project (was) not allowed to fail' unfortunately gave the developer more advantage and opportunity to renegotiate later during the construction stage.

The equity raising crisis

According to the concession contract, the total amount of equity to be raised was $4bn and the timetable for equity raising was specified in the debt financing contract. The fulfilment of the timetable was a prerequisite for withdrawing funds from the loan credit facility. The THSRC only had $0.6bn equity in September 1999, 9 months after signing the concession contract. For the next 7 years, the THSR constantly had difficulties fulfilling the equity raising requirement. Their inability to raise sufficient equity caused the breach of contract by THSRC. Two major reasons contributed to this equity raising crisis. Firstly, at the time of initial equity raising, Taiwan's economy was still in the after shock of the 1997 East Asian financial crisis and the climate for taking a risk and investing in the unfamiliar high speed rail was very conservative. Secondly, the market had substantial doubts about the profitability of the project, suspecting that the THSRC's financial proposal was too optimistic. From the financial perspective, the return of initial equity will be much higher than that of later equity if the project is expected to be successful at the time of following equity raising. However, if there is substantial doubt about the profitability of the project, the low offering price will hurt the initial equity's profitability. The doubt about the project profitability could also be seen from the initial shareholders' reluctance to invest more equity later although they had the capacity to do so. As a result, a couple of rounds of renegotiation between the THSRC and banks took place and finally the banks had to accept THSRC's proposal to reduce the total equity amount from \$4bn to \$3.3bn.

The Taiwan government played a crucial role in bailing out THSRC in this crisis. The government was criticised for having government-owned/-controlled enterprises (GOEs) and non-profit organisations make substantial equity investment in the THSRC. The last equity investment of \$0.23bn by government-controlled non-profit organisations in September 2005, a very small amount compared to previous similar investments, caused the most serious criticism of unjustifiable aid and failure to monitor the project. During this crisis the government announced again that the 'government (was) determined to ensure the completion of the high speed rail'. However, soaring criticism and associated political costs forced Prime Minister Hsieh to publicly assure that 'government (would) make no further equity investment in the THSRC because it (was) against the will of the society and people'. It was later determined by the court that the September 2005 equity investment by a non-profit organisation was illegal. Currently the total equity of the THSRC is close to the revised target, \$3.3bn, with common stocks and preferred stocks at about 49% and 51% of total equity respectively. Total passive equity investment by GOEs and government-owned banks is about 23% of total equity, or 35% of total equity if considering investments from government-controlled non-profit organisations, while initial equity invested by the promoters is only about 28.5% of total shares.

Unlike the guarantee for debt financing, equity investment is an asset so the political cost of having GOEs make several rounds of equity investment in the early construction stage was relatively low and the government chose to continue to help the THSRC. However, equity investments in the later construction stage caused increasing criticism since the failure to raise equity when the project was near completion signified pessimistic profitability expectation and thus the equity investments were seen by the public as a government subsidy. From the perspective of renegotiation model, the political shock due to the 'September 2005 equity investment' could be considered the result of the sharp political cost increase when the subsidy passed the 'justifiable' amount even though the amount of that particular investment was relatively small.

The cost overrun crisis

One year before project completion, only 3 months after the government's 'September 2005 equity investment', the THSRC announced that the total cost overrun was estimated to be \$2bn due to the estimated 1-year schedule delay and other causes for cost overrun. Because of the serious political impact of previous unjustifiable government investment, the government had ruled out the possibility of providing any equity investment or liability guarantees. Moreover, for the first time the government formally announced that they would make plans to takeover the project if the THSRC could not raise either equity or debt to finance the additional capital need. Since it had been almost impossible for them to raise any additional equity, THSRC decided to supplement the capital gap through debt financing. It was a daunting task for the THSRC to obtain another \$2bn debt at this stage, as the debt ratio had just passed over the revised 75% at that time and that the market now had further doubts about the financial viability of the project because of the cost overruns.

THSRC finally obtained \$1.4bn debt financing by arranging a 'second mortgage financing' type loan, in which the THSRC used the concession rights on project-associated real estate development as collateral for the loan. This arrangement again brought government criticism. Since all the physical assets obtained during the project had been assigned as collateral during earlier debt financing, the rights on project-associated real estate development cannot independently exist if the THSRC defaults; therefore, it did not make too much sense to use the development rights as collateral. Moreover, in this arrangement the government had to officially agree to the collateral being assigned to the banks. The government was blamed for agreeing to the collateral assignment and urging the leading syndicate bank to accept such a deal. Nevertheless, the criticism was not as harsh as that from the earlier equity investment.

The cost overrun crisis almost became the final straw and made the government prepare to take over the project. From the perspective of the renegotiation model, any significant subsidy after the 'September 2005 equity investment' rendered the political cost of rescuing larger than that of taking over the project. Although the cost of taking over and retendering the project was supposed to be substantially large for a project nearing completion, the even higher political cost of providing more subsidies demonstrated how steep the slope of political cost associated with unjustifiable subsidy could be, as shown in Figure 15.5.

Lessons learned: the perspectives of the financial renegotiation model

- *Do not have a project that is not allowed to default*: from a societal perspective, projects that are too important to fail or too expensive to default are not good candidates for PPPs. Such projects will create more opportunities of opportunism than others. Unfortunately, the THSR project was too important and too expensive to default.

- *Do not focus too much on the bidder's financial proposal*: the greater the incentives and opportunities for opportunism, the lower the credibility of the bidder's financial proposal. Therefore, a more optimistic proposal requires more justification for positive figures. In the THSR project, the government failed to ask for justification for the attractive proposal.

- *Do not adopt PPPs too hastily when the government has limited experience and incomplete support systems*: incomplete support systems and a lack of experience are also a source of opportunism opportunities. Governments should limit the scope for using PPPs when they are initially introduced. Unfortunately, since the enacting of Taiwan's PPP law in 2000, the Taiwan government has aggressively promoted the use of PPPs for almost all public infrastructure projects.

- *Do not force local governments to use PPPs*: the Taiwan government set a yearly goal of signing $3.1bn of PPP projects for the promoting federal agency, the Public Construction Commission. This goal was then passed and allocated to local governments as an important criterion of their performance. Under such pressure, the local government would use PPPs on projects even where PPPs were not the best choice and would become very soft on contract negotiation.

- *Do consider separating the developer and contractors as much as possible*: although it is not always possible, the government should encourage the separation of the developer and contractor in the procurement process by, for example, giving such separation higher scores in bid evaluation. The separation of the developer and contractor will make the developer emphasise long-term profits and reduce the incentive for opportunism.

- *Do prepare in advance for project default*: advance preparation for project default and take over will reduce the cost of project retendering and hence renegotiation expectation and opportunism. In the THSR project, when the government announced their intention to take over the project if the THSRC could not obtain financing for cost overruns, the THSRC did not even try to renegotiate.

- *Do use professional help*: using professional consulting firms to provide support in evaluating financial proposals and negotiating contract terms will largely reduce the potential for developers to behave opportunistically and the possibility of awarding projects to opportunistic bidders.

- *Do know that the transaction costs of PPP projects are much higher than that of government projects*: the higher transaction costs for PPP projects may include the costs due to a more complex project procurement process and the higher capital costs compensating for fair market required returns on equity and debt. Lack of government funding should not be the major reason for adopting PPPs. The use of PPPs for a project should be justified by higher creativeness and efficiency due to private participation. For example, in the UK the use of PPPs for a project is required to meet the VFM criteria. Blindly promoting PPPs only because of the lack of government funding will generate more problems and difficulties in the future.

15.8   Conclusion

The cost of solving problems due to inferior project concept development or financial renegotiation is enormous. If these problems take place persistently, the subsequent high transaction costs due to the increased level of monitoring imposed by the public will make PPP an infeasible approach for providing public infrastructures and services. This chapter introduced two models as the theoretical foundations for PPP policies on two important financial issues: bid compensation and financial renegotiation; it is hoped that the use of these models could prevent these types of problems from occurring in the first place.

There is a paradox in the bid compensation problem; whilst the model solves the equilibrium conditions for effective bid compensation for practical reasons offering bid compensation is not generally recommended. So, a creative approach to stimulate quality inputs from bidders for PPP projects is required.

From the financial renegotiation model, governing principles and policies for PPP administration are inferred. The policy implications cover issues in project procurement and management, in addition to renegotiation itself. Although advances in public project procurement practice have reduced the opportunities for opportunism, opportunism will never cease to exist in the mind of every rational and economic individual. The exploitation of renegotiation possibility in a complex contract or PPPs is a typical behaviour of opportunism that poses many serious problems. The model is expected to help government authorities and policy makers establish more effective polices for PPP projects. The case study of the Taiwan High Speed Rail project shows how the renegotiation model can help to prevent or alleviate the opportunism problems.

Some simplified assumptions are made in these models so that useful insights can be drawn from real life complex situations. These insights could provide decision makers with useful concepts and directional principles despite the real situation being more complex. The insights and qualitative implications of an economic model are often more important than the solutions obtained. Furthermore, the two models can consider various project environments characterised by the parameters of the model. The validity of this model does not require the government and developer to explicitly 'use' game theory. The only requirement is that all players are rational decision makers.

Whilst there are many guidelines for PPP schemes, these guidelines cannot be global and must be re-examined to fit the specific environment. The models in this chapter may help to understand various problems and make appropriate modifications. Rigorous theories concerning government policy

Part Two

in PPPs are difficult to find. It is hoped that the pilot study introduced in this chapter may provide a theoretical foundation and analytical logic for making effective PPP administration policies and respective guidelines for different governments.

Note

1. S. Ping Ho is Associate Professor, Construction Management Program, Dept. of Civil Engineering, National Taiwan University, Taipei, TAIWAN.

References

Binmore, K. (1992) *Fun and Games: A Text on Game Theory*. D.C. Heath, Lexington.
Connolly, J.P. (2006) Discussion of 'Bid compensation decision model for projects with costly bid preparation' by S. Ping Ho. *Journal of Construction and Engineering Management*, 132(4), 430–431.
Design-Build Institute of America (DBIA) (1995) *Guidelines: Request for Design-Build Qualifications & Proposals for Competitive Selection for a Public or Institutional Facility*. Washington, DC.
Gibbons, R. (1992) *Game Theory for Applied Economists*. Princeton University Press, Princeton, NJ.
Henk, G. (1998) Privatization and the public/private partnership. *Journal of Management in Engineering*, 14(4), 28–29.
Ho, S.P. (2001) *Real Options and Game Theoretic Valuation, Financing and Tendering for Investments on Build-Operate-Transfer Projects*. Ph.D. Thesis, Department of Civil and Environmental Engineering, University of Illinois at Urbana-Champaign, Urbana.
Ho, S.P. (2005) Bid compensation decision model for projects with costly bid preparation. *Journal of Construction and Engineering Management*, 131(2), 151–159.
Ho, S.P. (2006a) Model for financial renegotiation in public-private partnership projects and its policy implications: game theoretic view. *Journal of Construction and Engineering Management*, 132(7), 678–688.
Ho, S.P. (2006b) Closure to 'Bid compensation decision model for projects with costly bid preparation' by S. Ping Ho. *Journal of Construction and Engineering Management*, 132(4), 430–431.
Ho, S.P. and Liu, L.Y. (2004) Analytical model for analyzing construction claims and opportunistic bidding. *Journal of Construction and Engineering Management*, 130(1), 94–104.
Holliday, I., Marcou, G. and Vickerman, R. (1991) *The Channel Tunnel: Public Policy, Regional Development and European Integration*. Belhaven Press, New York.
Myerson, R.B. (1991) *Game Theory: Analysis of Conflict*. Harvard University Press, Cambridge.
Rasmusen, E. (2001) *Games and Information*. Blackwell Publisher Inc., Malden.
Rubin, R.A., Guy, S.D., Maevis, A.C. and Fairweather, V. (1983) *Construction Claims*. Van Norstrand Reinhold, New York.
U.S. Dept. of Transportation, Federal Highway Administration (2005) *Synthesis of Public-Private Partnership Projects for Roads, Bridges and Tunnels From Around the World – 1985–2004*. Prepared by AECOM Consult, Inc.
Walker, C. and Smith, A.J. (1995) *Privatized Infrastructure – The BOT Approach*. Thomas Telford Inc., New York.

Part Three
PPP Management

16

Innovation in PPP

David Eaton and Rıfat Akbiyikli

It has always been the case that the private sector has provided goods and services to the public sector. A widespread feature of the last two decades has been the shift away from the in-house provision of services by the public sector towards the contracting out of services to be provided by the private sector. These services are a contribution and an addition to the provision of services by the government to the public, but the services are supplied by the private sector employees.

Private infrastructure provision is not a new idea. Bridges have been privately owned for centuries (Dupuit, 1844). Infrastructure concessions were first granted in France in the mid-seventeenth century (Winch, 2002). One of the first documented concessions was granted in 1782 in France (Walker and Smith, 1995). At this time the Perrier brothers founded a company that was granted licence to supply piped water in the Paris area for 15 years. The agreement did not survive the political changes that took place in conjunction with the French Revolution, as the city council cancelled the franchise (Walker and Smith, 1995; OECD, 2000).

The late 1700s also saw the concept of toll roads become increasingly common in the US, many of which were constructed with some federal assistance in the form of land grants or subsidies (Levy, 1996). From the beginning of the twentieth century various governments increasingly incorporated the procurement of assets with strategic policies for development and therefore preferred the use of their own fiscal and sovereign resources of finance (Walker and Smith, 1995; Winch, 2002).

Growing concern over state budgetary deficits and concerns over the inability of the public sector to manage complex infrastructure efficiently in an increasingly competitive environment, led to the reversal of the state ownership as a norm to provide infrastructure to the public at large (Vickerman, 2002a,b). Over the period 1970–1996, according to Debande (1999), large reductions in government investments were observed in OECD countries. Privatisation and public sector expenditure constraints had given rise to a substantial reduction in both private and public sector investment. Some commentators, for example Birnie (1998, 1999), state that the Maastricht criteria and European Monetary Union (EMU) have played a role in the implementation of PPP/PFI, as governments throughout Europe have been forced to take action to enable conversion to the single European currency.

In order to achieve infrastructure development and to reduce the associated government debt burden, the public authorities and the national governments sought to involve the private sector and private capital to implement design and build infrastructure projects and to provide infrastructure services previously in the domain of the public sector (Debande, 1999; de Lemos *et al.*, 2000; Heald and McLeod, 2002; Quiggin, 2002).

As a part of the above trend, the PFI was launched in 1992, as a legal framework for concessions in the UK to encourage private capital investment into the construction industry. In the PFI framework the public sector defines the output specification for the services to be purchased from the private sector with a predefined payment mechanism. The public purchases a service not an asset. After 1997, and the change of government from Conservative to Labour control, PFI gained momentum in the UK and it is expected to continue expanding as a procurement instrument in the future (Eaton and Akbiyikli, 2005).

PPP is about establishing arrangements, often a legally binding concession agreement, that will bring benefits to both sectors. The private sector needs to earn a return on its ability to invest and perform. The public sector wants to deliver services to the standard specified and make the best use of public resources.

PPP in itself is an innovation in public procurement, but the public sector must decide on the route which gives the best scope for the private sector to add value and in all cases adhere to key principles such as whole-life, VFM and optimum risk allocation. Through such an attitude and approach it will be possible to deliver public services in an efficient, effective and innovative way.

16.2   Innovation and Competitive Advantage in PPP

16.2.1   Innovation

'Innovation is an effort made by one or more individuals that produces an economic gain, either by reducing costs or through increased incomes' (Smith, 2003). An example of a direct and concise definition is provided by Cobbenhagen (2000) who presents 'renewal' with respect to products, markets and technological production processes, as one of the commonly used definitions of innovation.

Freeman, in *The Economics of Industrial Innovation* (1982) presents innovation as 'the actual use of a nontrivial change in a process, product or system that is novel to the institution developing the change'. According to Hobday (1998), innovation is of a heterogeneous nature, and several commentators point at the long inter-industry differences between the origins and processes of innovation. Accordingly, some innovation success factors are idiosyncratic to the specific environment of the construction organisation or procurement path. This chapter is not intended to study innovation in detail in the construction industry. It will concentrate on the innovations generated within a PFI/PPP project.

According to Rogers (1995) innovation is defined as 'an idea, practice, or object that is perceived as new by an individual or other unit of adoption'. As stated in Walker and Hampson (2003) innovation is 'part of a change strategy and is a decision-making process to enact change in technology process, services rendered or other management approaches' and 'a realisation that a current state must be changed in order to achieve competitive advantage'. King and Anderson (1995) see innovation from a wider perspective, defining it as 'a social process, involving interaction and communication within and between people in a whole range of social structures, from the immediate work group, through the department or division, to the organisation as a whole and the wider society'.

Organisational structure, which is defined by Child (1984) as 'the formal allocation of work roles and the administrative mechanisms to control and integrate work activities including those which cross organisational boundaries', has a considerable influence to facilitate innovation. Organic structures (Burns and Stalker, 1961; Lawrence and Lorsch, 1967) argued to be part of a normative prescription for facilitating innovation, in combination with participative leadership styles and cultural features, flattened hierarchies and maximised lateral communications.

Traditionally, it is accepted that construction is a cost-driven sector (Atkin, 1999). Many construction activities are carried out by local organisations that compete on the basis of lowest cost (Gann, 1997). Thus, work is won through finding ways of cutting costs. As a logical consequence of this working pattern the majority of the construction organisations tend to a greater extent to look at innovation as a means of reducing costs rather than enhancing value. This is the typical pattern in conventional procurement of construction projects. Barrett *et al.* (2001) found that the primary motivation for a small construction firm to innovate is to generate sufficient cash flow to survive in the short term.

The desire for innovation in the construction industry is well recognised (Atkin, 1999; Manseau and Seaden, 2001). In response to the findings of keynote reports by Latham (1994) and Egan (1998) a host of UK government-supported initiatives and programmes have been established to drive radical improvements in construction, including the Construction Research and Innovation Strategy Panel (CRISP), Partners in Innovation (PII) and Movement for Innovation (M4I).

In addition, Eaton (2000:1) declares, 'without innovation a business does not have a rational source of competitive advantage in construction'. Gann (2000:220) comments that construction firms need to improve their capabilities in managing innovation if they are to 'build reputations for technical excellence that set them apart from more traditional players'. Moreover, Barrett *et al.* (2001:1) remarked that successful innovation enables construction firms to better satisfy 'the aspirations and needs of society and clients, whilst improving their competitiveness in dynamic and abrasive markets'.

Part Three

Several studies have been undertaken in order to identify innovation success factors in construction. Tatum (1984) presented three conditions that occurred repeatedly within successful innovation:

1. Strong and unbiased management that were committed to selecting technologies best suited to serve project goals.
2. Early involvement of representatives with authority to commit resources to all parts influenced by the innovation.
3. The establishment of effective information flow within the project team to identify and resolve problems arising from the innovation.

Tatum (1989) found that innovative organisations tended to have a longer-term viewpoint and were prepared to accept development problems with an innovation as long as more enduring benefits remained apparent.

The authors believe that innovation is about creating value and increasing efficiency and is a key driver of competitive advantage.

The form of procurement is critical as it determines the overall framework embracing the structure of responsibilities and authorities for participants within the process. The traditional view of the construction industry is that demands of cost, time and quality necessary to meet client requirements on each individual project often limit opportunities for innovation. Innovation in the traditional procurement process (profession-led design procurement path) is less evident because the lead professional is usually conservative to see the eventual impact of innovative approaches upon production and long-term services of the constructed asset. This is due to the fact that traditional procurement is based on the rigid separation of the design and construction activities.

The structure of PPP projects often involves a complex web of contracts, linking a variety of different parties all with varying interests and involvement in the project. The structure of the contract will define the basis for the future long-term operational and managerial relationship between the authority and the concession company–special purpose vehicle (SPV). The public sector changes roles from service provider to service specifier and the private sector changes from asset provider to a service provider. Service provision for a 30–40-year concession period entails a change in both public and private organisational cultures. Both public and private have to adjust to the move to the service sector, and to the commitment to a long-term relationship. Within this organisational structure a partnering concept is created which provides a framework for the establishment of mutual objectives among the public and private parties which enthuses good relations, honesty, openness, trust, integrity and cooperation. This process of partnering in PPP attempts to establish working relationships amongst the stakeholders (public sector, construction contractors, maintenance and operation contractors, investors and finance providers, sub-contractors etc.) through a mutually developed, formal strategy of commitment and communication. The key to success is the effective communication of project objectives by the stakeholders and it requires a process of change, which must first be brought to the respective organisations and then incorporated into the team performance of the main stakeholders in PPP project organisation.

The private sector is no longer in a traditional construction project mode but moves into a new and diverse and pluralistic business culture in a consortium. The PPP concession company (SPV) is an autonomous legal unit. All contractual relationships of the concession company with other parties involved in the PPP have to provide for the extended life of the contract and establish measures to control it and establish dispute resolution procedures. The private sector must adjust their organisational cultures and structures to a long-term involvement instead of the traditional short-term and related temporary multi-organisations of the construction projects. The public sector too must change its role from service producer to that of the monitor of the performance and effectiveness of the service. Grant (1996) stated that PPPs are most successful when four pre-conditions occur, namely:

1. The partners are financially strong and organisationally stable.
2. The partners are willing to commit their best human resources to the project.
3. The project provides opportunities for all partners.
4. There is shared authority and responsibility.

The PPP approach offers the prospect of delivering the services required by public sector clients in a way that provides superior VFM than conventional procurement. This according to the House of Commons – Public Accounts – Twenty-Third Report (1999) is because the PFI/PPP approach can give scope for innovation in how services are delivered; because the client specifies what is required not how it is to be delivered, the supplier has scope to innovate. The public sector client must not unnecessarily restrict suppliers' scope of innovation, by prescribing in excessive detail how services are to be delivered. For the higher cost of private sector finance to be offset by bringing in private sector expertise, the public sector must be open to innovative ideas offered by the private sector. Private sector bidders need to be given as much freedom as possible to determine the best way to provide the services required. This issue will be detailed in the case studies below.

In its 2000 study, 'The role of cost saving and innovation in PFI/PPP projects', the Construction Industry Council (CIC) identified the role of innovation within construction-based projects. It stated that cost savings could be accrued from the use of innovative working procedures and new technologies. The results show an overall project saving in the region of 5–10% of which the highest average savings could be found from the construction phase. The savings on construction costs were also estimated to be 5–10% (CIC, 2000). This reduction in cost and/or improvement would have to come from either the transfer of risks or from improvements in the average unit of productivity. VFM accrues from the private sector being allowed the opportunity to be more innovative, in the sense of cost saving and product enhancement, than is likely to be found in traditional form of procurement.

Oluwoye and Lenard (1999) describe four key factors that affect the level of innovation that occurs in a project:

1. The client recognises the need for the innovation.
2. Contractual incentives that encourage innovation are put in place.

Part Three

3. A symbiotic learning environment is put in place.
4. Open communication is achieved at all levels.

The authors believe that construction project organisations operate in technological and market possibilities that arise in their own environment. The possibilities for innovation are directly related to the procurement path chosen to create a product or service and this is possible through the interaction with suppliers, clients (customers) and government agencies. Innovation must satisfy the criteria set by the regulatory framework, contract for the works, VFM and the quality of the output product/service set by the buyer.

PPP is a collaborative approach to solve project problems jointly through innovation.

In conjunction with the above, de Lemos *et al.* (2003) noted that in PFI projects the majority of innovations are derived from the following needs:

- To promote easy and cost-effective long-term maintenance of the constructed asset.
- To give the designer freedom to innovate in the aim of providing a service in the most effective way, thus increasing the project's profitability.
- On a whole-life cycle basis, an operator needs to consider the interaction between a more expensive design solution and lower operating and lifecycle costs or vice versa.

PFI/PPP is a contractor-led procurement system focused on design, build, finance and operate. The private sector offers a one-stop-shop service and therefore has the potential for increased integration within the project value chain. Figure 16.1 illustrates this approach.

The use of the PFI/PPP procurement system should, in theory, permit minimal disturbance to the project value chain, especially if the public sector (granting authority) has defined correctly the output specification (Male, 2002). In PFI/PPP procurement the SPV has to operate what they have financed, designed and built and their focus should be on ensuring continuity and integrity of delivery throughout the process. The SPV delivering the PFI/PPP project have to provide the correct balance of operational expenditure to increase the level of return on investment and it also provides the greatest opportunities to leverage the principles of demand and supply from the value chain management (ibid). Table 16.1 provides a typical value system and value chains together with primary activities and main stakeholders in a PFI project.

In order to understand the potential for innovation in PFI/PPP, it is necessary to distinguish between the PFI/PPP procurement process specific parameters and those variable resource inputs. Although certain parameters in PFI/PPP could be innovations in their own right, most of their value lies in the mechanism of their utilisation in the PFI/PPP procurement process that will deliver the quality service. Therefore the sustainability of any innovation in PFI/PPP depends upon the barriers that exist both internally and externally in the procurement system to prevent their implementation.

The PFI/PPP process is a process that delivers the service required throughout the lifetime of the project. It aligns the interests of the user, the service

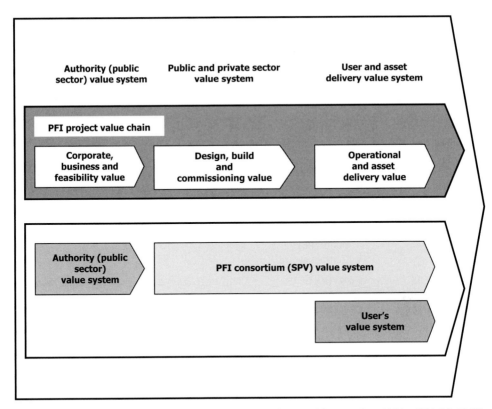

Figure 16.1   A generic PFI/PPP value system (adapted from Male, 2002 by Akbiyikli, 2005).

provider and the major financiers (it is in the financier's interests that the service is supplied to the agreed standard). The process establishes a relationship that is based on partnering, with the private sector determining the inputs required to achieve the output (services) specified by the public sector and the quality services are provided on a consistent basis. The private sector creates the asset and delivers the service in return for payment commensurate with the service levels delivered. Figure 16.2 shows a simplified PFI/PPP process.

The parameters that make PFI/PPP procurement achieve innovation when compared with the other procurement paths can be listed as:

- Delivery of quality services that provide VFM: PFI/PPP encourages a long-term approach to the creation and management of public services assets. Achieving VFM in the provision of a service requires that full account is taken of the risks and costs over a long term as opposed to short-term capital expenditure. Quality services can thus be sustained over many years at the lowest long-run cost.
- New options for public sector finances: demand from the public sector is growing in the UK both in quality and quantity for infrastructure projects. Therefore there are competing pressures for funds for new infrastructure and renewal of the existing ones. Competition for such funding is intense not just between infrastructure projects but also with the many other

Table 16.1 Value system and value chains for PFI (PPP) procurement phases (adapted from Male, 2002 and Davis Langdon & Everest, 2002 by Akbiyikli, 2005).

Public sector value system	Public–private value system	User value system	Delivery of asset value system
Primary PFI procurement phases			
1. Business need	6. Expressions of interests and publish *OJEU* notice	14. Construction post contract capital expenditure (CAPEX)	16. Delivery of asset
2. Appraisal options	7. Prequalification of bidders		
3. Business case and reference project	8. Selection of bidders	15. Operation and maintenance post contract operation expenditure (OPEX)	
4. Developing team	9. Refine appraisal		
5. Deciding tactics	10. Invitation to negotiation (ITN)		
	11. Evaluation of bids (BAFO)		
	12. Selection of preferred bidder (PB)		
	13. Contract award		
STAKEHOLDERS:	STAKEHOLDERS:	User value chain	Residual value risk
Authority's value system	Authority's value system	SPV's value chain	for both
Financiers/bankers' value system	SPV's value system	Regulatory authority's value chain	authority (public sector) and SPV (private sector)
Internal stakeholders' value system	Design and build contractor's value system	Financiers/bankers' value chain	
External stakeholders' value system	Operation and maintenance contractor's value system		
Regulatory authorities' value system	Suppliers' value system		
	Quality manager's value system		
	Project manager's value system		
	Regulatory authorities' value system		

demands on public sector finance. PFI/PPP has the additional benefit of relieving short-term pressure on the public finances, because PFI/PPP links public sector financial obligations to the delivery of the service.

■ Procurement efficiency: PFI/PPP projects should meet monetary and time budgets which are frequently overrun in conventional procurement. Any cost or time overruns have to be borne by the private sector.

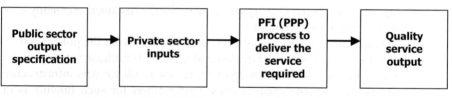

Figure 16.2 Simplified PFI/PPP process (Akbiyikli, 2005).

- Improved accountability: PFI/PPP encapsulates the proper consideration of the long-term ongoing liabilities that arise, avoiding the possibility of short-term policy decisions taken solely on a cash-accounting basis.
- Risk management: the awareness, identification and analysis of and response to risks allows PFI/PPP projects to proceed with a full range of risks being fully accounted for and priced into the service procurement contract.

The authors argue that in PFI/PPP projects the innovation for quality is focused essentially on performance-related features. This aim is 'to reflect the best understanding of what determines quality and to create a contractual framework that maximises cost effectiveness' (Chamberlain, 1995). Performance-related specifications aim to 'give better' levels of the long-term performance of the completed asset. Performance-related specifications also reflect the payment mechanism depending on the performance of the completed asset.

According to de Lemos *et al.* (2003) PFI/PPP improves quality essentially through two mechanisms:

1. Directly by the need to abide by the contracted service specifications.
2. Indirectly by allowing technical staff to focus on their core competences, rather than other management issues, improving the quality of the staff's technical activities.

As an example in PPP road construction projects there are few secrets from competitors. They see new road plans and projects and can watch construction details as roads are built. What possible innovations can a PPP road construction company have that can not be copied or even improved upon by a competitor? The answer is 'almost none!' But this does not necessarily mean that there is no innovation in PFI/PPP roads. The innovation will only remain an advantage for a period of time before it becomes replicated within co-operating organisations. Hence the principle of Sustainable Innovation is that the sources of Innovation are continually evolving (Akbiyikli, 2005).

Innovations of a road constructor will be adopted by others; sooner or later. Innovation exists in road construction only during the execution of a project and/or an activity in a project. In a PFI/PPP road project the innovation helps to achieve success in the award of the road concession. It is recognised that the innovation can be copied by competitors for subsequent PFI/PPP bids, but the SPV can still appropriate the innovation over the entire concession period rather than just the construction period (ibid).

The successful consortium (SPV) has demonstrated their innovation through the incorporation of innovations during the procurement phase. Selections include an appraisal of the 'innovativeness' of the SPV proposal. The contract awarded to the consortium, SPV, will always step up to the next level of innovation. The product (the constructed asset and the corresponding service) advantages are then sustainable over the PFI/PPP concession period. This is a major source of competitive advantage. The creation of a series of short-term competitive innovations during the concession period in order to deliver

Part Three

a superior service and reduce whole lifecycle costs enables the innovativeness to be appropriated over the entire concession period.

Innovation is not only in the product and process. It is also in the strategy, structure, system and behaviour. Innovation in the PFI/PPP deal is a holistic issue concerning all the stakeholders in the deal.

Creating an organisation with the capability to create a continuous series of innovations is not easy in the construction industry. A constructor's business success rests on its competency to create better innovations, make more improvements, and implement the changes faster than other constructors. Therefore it needs employees capable of team-working, to analyse improvement opportunities and put the changes into practice.

It is a management issue to create a continuous innovation process. The management must define innovation as a part of the company vision (business strategy) and communicate its importance to the organisation. The main idea is to create a culture that sustains the stakeholders' competitive advantage both in the process and in the service delivery (Akbiyikli, 2005).

The construction contractor in the case study PFI/PPP road projects, Morgan = EST, is a good example in the creation of a culture in the company continuously creating innovation and implementing it on successive road projects (ibid). (EST is an abbreviation of 'early solutions together'.) The successful organisations in the complex and turbulent construction industry are the ones with a more effective innovation process and an organisational culture that can sustain an evolving competitive advantage over the rivals and competitors on a continuing basis.

16.2.2   Competitive advantage

Sources of innovation that create potential competitive advantage derived from case studies for each stakeholder are presented in Figure 16.3 below. The categories of activity with potential for innovation are shown on the right-hand side of the figure.

The sources for competitive advantage in PFI/PPP road projects are: investment (financial model); innovation; VFM (value adding); partnering (honesty, openness and transparency: HOT); performance-related output; superior service product; project management; risk management; whole lifecycle costing; payment mechanism; and knowledge management (Figure 16.3). Another source is less disorder (clarity, communication, commitment) in the procurement structure that generates sources of potential innovation from the participating diverse stakeholders in the PFI/PPP concept. PFI/PPP enables increased innovation because it is a structured procurement which links strategy, structure, systems and behaviour. PFI/PPP enhances innovation because of the longevity of the concession that alters the way of thinking of the stakeholders in the process to recognise the longer-term pattern of uncertainty and change and the underlying structures producing those patterns. The individual and collective creativity and accumulated knowledge of all the stakeholders in the PFI/PPP process is an innovation itself. Another source of innovation is the understanding of 'win–win' mentality that all parties are satisfied with

STAKEHOLDER

COMPETITIVE ADVANTAGE (CA)

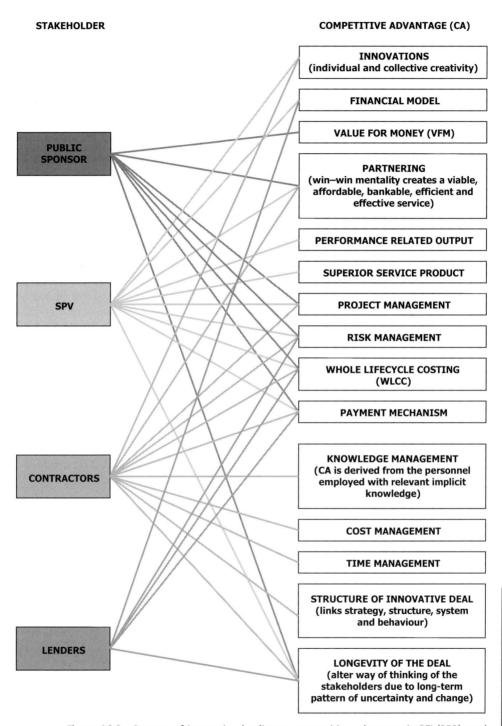

Figure 16.3 Sources of innovation leading to competitive advantage in PFI (PPP) road project (Akbiyikli, 2005).

Part Three

and which creates a viable, affordable, value-added, bankable, efficient and effective long-term service product.

The cumulative capabilities of all the stakeholders are considered as the basis of innovativeness in a PFI/PPP road project. It is this integrated and holistic capability that gives the ability of the PFI/PPP model to outperform the other less innovative procurement models.

16.3   Stimulants and Impediments to Innovation in PFI/PPP Projects

Based upon the established literature a hierarchical model of stimulants and impediments is presented in Table 16.2. The individual features and hierarchical structure were derived from the literature review. These stimulants and impediments were then utilised in the examination and evaluation of detailed case studies. The case studies were two UK prisons (combined as one study), a Portuguese bridge, a UK military development and a very small UK 'unbundled' primary school (Eaton, 2001; Eaton & Akbiyikli, 2005; De Lemos *et al.*, 2003, 2004). Each case study has been examined in detail and the innovation stimulants and impediments present within each project have been identified. The identification of the relevant features has been conducted independently of the case study compilation by a researcher having no affiliation with any party associated with the projects. A feature is only identified if it creates a 'significant' difference to the more 'traditional' approach to contracting the built facility. The detailed feature identification of stimulants is provided in Table 16.3 and the detailed feature identification of impediments is provided in Table 16.4. Thus a feature has been identified only if the two following conditions of the research protocol are met:

- The feature has materially affected the risk neutrality of the project.
- The feature has materially affected the substantive completion on time, to quality and to price.

Risk neutrality is defined as the basis of agreement for contract closure. Thus it is the aggregation of all of the terms and conditions negotiated. It therefore defines the contractual position of all the parties before delivery of the project. Thus a stimulant is a feature that has a positive effect on the positions of the parties. It may be that an innovation feature can deliver project operation before the target completion date, or it may create a cost saving on the original design that can be shared between the parties. In these circumstances when a stimulant was found to be present in a particular case study a (•) is shown in Table 16.3 against the identified feature. An obstacle has a negative effect on the project and when found to be present on a particular project it is indicated by an (x) in Table 16.4.

The four case studies have then been evaluated by a simple numeric count of the positive (+) stimulants to innovation and the negative (−) impediments to innovation that have been identified by the evaluation of the case study details. Table 16.5 presents the numeric count and Figure 16.4 presents the evaluation. The most effective innovation would occur when the stimulant (solid line) is as far from the axis as possible, and the impediment (dotted

Table 16.2   PFI theoretical stimulants and impediments.

PFI Stimulants	PFI Impediments
External level	
Clients	Client procurement route
Competition	Coalition nature of construction
Government	Lack of communication
Professional bodies	Legislation
Sharing of ideas in the industry	
Supply chain	
Organisation level	
Fair, constructive judgement of ideas	Destructive internal competition
Reward and recognition for creative work	Harsh criticism of new ideas
Mechanisms for developing new ideas	Conservatism and avoidance of risk
Clear shared vision	Rigid structures and strict processes
Encouragement of risk taking and risk management	Lack of mechanisms for developing new ideas
Attraction of creative people	Lack of rewards and recognition
Project level	
Supervisory encouragement	Format of project contract
Clear, appropriate goals	Rigid project demands
Motivation and commitment to the project work	Segmentation of project disciplines
Diverse and suitable background of individuals	Poor collaboration
Good communication	Poor communication
Openness to new ideas	Lack of openness and trust
Trust and help for others within the team	Poor project management
Constructive criticism of ideas	
Job role level	
Challenging and interesting tasks and projects	Extreme time pressures
Time control over work	Unrealistic expectations for productivity
High autonomy	Distractions from creativity
Freedom	Financial constraints
Access to appropriate materials and facilities	
Access to necessary information	
Adequate funds	
Training and development	
Creativity training	
Creativity element of job description and appraisal	
Conducive physical environment	

line) is as close to the central axis as possible. No work has yet been executed to quantify the proportional contribution of each feature. It treats all features in an identical manner.

Where the impediments exceed the stimulants (as in the prisons, bridge and military case studies) there is an indication that the 'incorporation of innovation' of the project has been impaired.

The simple numerical analysis by project is supplemented by case study analysis as presented in Table 16.6 and Figure 16.5. Table 16.6 presents the numeric aggregate count of the evaluation of stimulants and impediments by hierarchical level. The most effective innovation would again occur when the stimulant (solid line) is as far from the axis as possible, and the impediment

Table 16.3 Detailed stimulants of creativity in PFI case studies. (•, identified as significant in accordance with the research protocol as identified in the text).

Identified stimulants	Prisons	Bridge	Military	School
External Level				
Clients				•
Competition				
Government				
Professional bodies				
Sharing ideas in the industry				
Supply chain				•
Organisational level				
Encouragement of creative problem solving		•		
Fair, constructive judgement of ideas				
Reward and recognition for creative work				
Mechanisms for developing and implementing new ideas			•	
Clear shared vision				•
Encouragement of risk taking and risk management				
Attracting creative people				
Project level				
Supervisory role models				•
Clear, appropriate goals			•	•
Support for work group and individual contributions from supervisor				
Motivation and commitment to the project work		•		•
Diverse and suitable background of individuals				
Good communication				•
Openness to new ideas				
Trust and help for others within the team				
Constructive criticism of ideas				
Job role level				
Challenging and interesting tasks and projects	•	•	•	•
Time control over work				•
High autonomy				•
Freedom				
Access to appropriate materials and facilities		•		
Access to necessary information				•
Adequate funds	•	•		•
Training and development				
Creativity training				
Creativity in job description				
Conducive physical environment				
Total	+2	+5	+3	+12

Table 16.4 Detailed impediments to innovation in PFI case studies. (×, Identified as significant in accordance with the research protocol as identified in the text)

Identified impediments	Prisons	Bridge	Military	School
External level				
Client procurement route	×		×	
Coalition nature of the industry	×	×	×	
Lack of communication		×	×	
Legislation			×	
Organisational level				
Internal political problems	×			
Destructive internal competition			×	
Harsh criticism of new ideas	×			
Conservatism and avoidance of risk	×			
Rigid structures	×			
Strict processes and procedures	×	×	×	
Lack of mechanisms for developing and implementing new ideas	×		×	
Lack of rewards and recognition				
Project level				
Format of project contract	×	×	×	
Rigid project demands	×	×	×	
Segmentation of project disciplines	×	×	×	×
Poor project management			×	
Lack of communication and collaboration				
Lack of openness and trust		×	×	
Job role level				
Extreme time pressures	×		×	
Unrealistic expectations for productivity	×	×	×	
Distractions from creativity				
Financial constraints	×	×	×	×
Total	−14	−9	−15	−2

(dotted line) is as close to the central axis as possible. However, in this case there is an indication that the inter-relationship between the levels is also important. Intuitively the impediments are more significant in the descending order of external, organisational, project and role, whilst the stimulants appear to be more important at the organisation and project levels than at the external or job role levels. No work has yet been executed to quantify

Table 16.5 Summary of collated stimulants and impediments by project.

	Prisons	Bridge	Military	School
Identified stimulants	+2	+5	+3	+12
Identified impediments	−14	−9	−15	−2
Total	−12	−4	−12	+10

Part Three

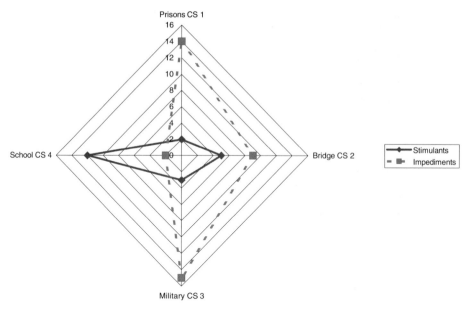

Figure 16.4   Comparison of stimulants and impediments to innovation by case study.

the proportional contribution of each level. It treats all levels in an identical manner.

The empirical study of the four cases suggests that to date the 'claimed' innovation associated with PFI/PPP is largely unrealised. There appears to be significant scope for innovation within the PFI/PPP projects.

In three of the case studies the numeric count of the impediments to innovation significantly outnumbers the stimulants and the aggregate impediments count exceeds the stimulants count at all but the job-role level. In three cases the construction contractor suffered significant cost overruns; however, the concessionaire with a guaranteed maximum price (GMP) obtained virtually complete protection against these cost overruns. The client was equally protected by the concession arrangement.

In the cases of the prisons and military projects the stimulant and impediment count is very similar (+2,−14: +3, −15), however, an evaluation of the 'success count' of each project, a crude measure of the perceived successful delivery of the projects, would yield a significantly different response, the prisons project being deemed overall, more successful, by all parties, than the military project. This is suggestive of an imbalance between the proportional contributions of individual features to the deemed success or otherwise of a project.

A further detail is that the higher-level stimulants, i.e. those at the external and organisational level, are noticeable largely by their absence, whilst the impediments to innovation occur at all levels in the hierarchy. One interpretation of this feature is that the senior management of PFI projects have not evolved sufficiently to recognise the difference between a major 'traditional' project and a major PFI project. Hence the senior management have not changed their patterns of behaviour despite the change in procurement process. This feature of organisational culture is currently being further examined.

Table 16.6 Summary of collated stimulants and impediments by hierarchical level.

	External level	Organisational level	Project level	Job role level
Identified stimulants	+2	+3	+6	+11
Identified impediments	−8	−10	−13	−9
Total	−6	−7	−7	+2

There appears to be some support within the analysis for the belief that innovation at the level of the job role is being achieved – it is surmised that this is the acclimatisation of individuals to the experience of the concept and operation associated with PFI.

The limitations associated with the findings are that a small sample of PFI projects has been utilised and no statistical analysis has been conducted. As stated previously each stimulant or impediment feature is treated equally. No ranking or relative weighting has been calculated. No ranking or weighting for the hierarchical levels has been calculated.

PFI/PPP is developing worldwide as a procurement mechanism. The elimination of unintentional constraints upon the potential innovation within the PFI/PPP project and the inclusion of stimulants by the use of this model can improve project quality, reduce costs and improve delivery times by minimising the risks associated with this form of procurement.

16.4   Innovation and Financial Issues in PFI/PPP Projects

For most people the most significant reason for achieving innovation is to improve the financial position of their organisation from the proposed project.

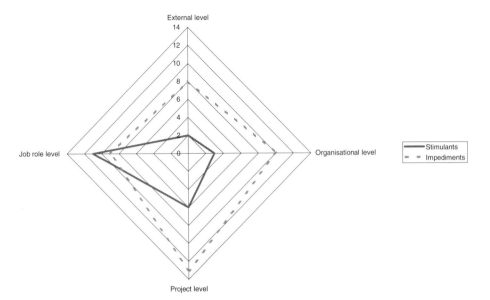

Figure 16.5 Hierarchical aggregated stimulants and impediments to innovation.

Part Three

With this intention uppermost in mind the authors present the findings from two further detailed PFI road projects.

Case Study 16.1   Newport Southern Distributor Road (NSDR)

Key innovation elements

Early contractor involvement

The Morgan–Vinci CJV (construction sub-contractor) were able to work with Newport City Council (NCC) and the supply chain throughout the bid process to identify the areas to add value. The innovative approach of the Morgan–Vinci CJV to whole-life cost financing; and the future maintenance of the road were key factors in the contract award. By early involvement in the NSDR scheme and close collaboration with NCC, the Morgan–Vinci CJV were able to add value in the areas detailed in Table 16.7.

Table 16.7   Total savings of early contractor involvement in NSDR (Akbiyikli, 2005).

Issue	Explanation	Saving/costs
Design issues	1. Modification of embankment design and re-programme approvals and construction.	£750k
	2. Refine pavement design based on additional site investigation works.	£250k
	3. Re-engineer Corporation Road Junction with council's team by avoiding need for full grade separation.	£1m
	4. Modify culvert designs	£350k
	5. Option selection review with council to select between signal controlled junctions and roundabouts.	—
	6. Refine council's outline to improve buildability and programme for the River Usk bridge.	£1m
	Sub-total Design issues	**£3.35m (6.09%)**
Use of recycled materials	1. Maximise use of project-derived and locally available recycled materials to produce *direct cost saving* (per tonne of aggregate) and *indirect cost saving* (from the avoidance of the waste disposal charges and landfill tax).	£2.00m

£2.00m (3.64%) |
	Sub-total Recycled materials	£1.00m
Commercial arrangements	1. Re-engineer statutory undertakers' works and negotiate improved commercial arrangements with undertakers and their contractors.	
	Sub-total Commercial arrangements	**£1.00m (1.82%)**
	TOTAL SAVINGS of early contractor involvement	**£6.35m (11.55%)**

Scheme objectives

The scheme is being delivered using informal partnering mechanisms and co-located council/designer/contractor team which has facilitated significant improvements by allowing all parties to work together early in the scheme's development and design.

- By working in partnership with the council from the outset, the Morgan–Vinci CJV have ensured that the council's requirements, and the needs of all stakeholders were defined.
- The Morgan–Vinci CJV have successfully achieved the *overall objectives* of the NSDR scheme, which are:
  - □ To enable traffic to avoid the town centre and inner residential areas
  - □ To improve the environment of the inner residential areas
  - □ To improve road safety
  - □ To improve economic development and regeneration
  - □ To improve access
  - □ To facilitate the provision of an improved public transport system
  - □ To impact particularly on lives of people living along and around the Corporate Road area
- The project team have optimised the opportunity to add value and enhance the scheme through value engineering of innovations such as:
  - □ Extended routes to schools
  - □ Adjustments to junctions and signs
  - □ Improved access to third party premises (e.g. local businesses)
- During pre-commencement the project partners combined to maximise the use of project-derived excavated materials. Opportunities were also sought to utilise locally available secondary aggregates from sustainable sources: by-products of the heavy industry historically located in the area. This has significantly reduced the environmental impact on the community by negating the demand for primary aggregates and reducing long distance haulage movements on the project. The NSDR scheme saved a considerable amount of cost (£2.0m) by using recycled material and secondary aggregates instead of purchasing primary materials (The Big Picture: WRAP, 2004). The specific cost savings by using recycled materials in highways maintenance and construction are:
  - □ The avoidance of waste disposal charges and Landfill Tax through the re-use of recycled and secondary materials
  - □ The avoidance of Aggregate Levy payments, from which recycled and secondary aggregates are exempt
  - □ Reduced cost of transporting aggregates when recovered materials are available locally
  - □ New recycling techniques have demonstrated cost and performance advantages (ibid).
- The scheme features two environmentally sensitive sites: the landfill area and the River Usk, which are both a Site of Special Scientific Interest and a candidate Special Area of Conservation. However, through active engagement with the local authorities and proactive management of the planning consent process, the Morgan–Vinci CJV has mitigated the impact of these sites on the construction programme. Due to the presence of the old tip (landfill) area on the planned highway route, the height of the foreseen embankment (up to 10 m), and the poor quality of the ground (4 m of made ground followed by 10–12 m of very soft alluvium) on both sides of the Usk river crossing, the direct construction of the road was not possible. The DGI-Menard, geotechnical sub-contractor of Morgan–Vinci CJV, proposed the use of the controlled modulus column (CMC) techniques. The concept of CMC is to install columns made of mortar in the existing soil so as to form a complex soil + inclusion behaving as a uniform soil having good geotechnical properties (DGI-Menard Inc, 2004).
- The Morgan–Vinci CJV and the supply chain partners have worked in collaboration with the Council and the Designer to ensure that the route is safe. Initiatives include:
  - □ The development of a junction and roundabout strategy
  - □ Accommodation of existing neighbourhood pedestrian routes

Part Three

□   Ground improvement systems used in the landfill area to minimise damage to the underlying strata and water table

Client collaboration

- The Morgan–Vinci CJV, through collaborative working with the council, designer and supply chain, have optimised processes and managed risk in order to deliver the NSDR scheme months ahead of the council's original programme.
- The co-location of the construction team, the council and the designer along with key members of the supply chain has fostered a non-adversarial team-based approach, enabling effective dialogue and integrated systems and processes to prevail throughout the project.
- Formal feedback mechanisms have also been employed to ensure that council requirements are met and that partnering and team working arrangements continue to deliver best value. The formal mechanisms include:
  □   Partnering workshops facilitated by an independent consultant
  □   Three-monthly partnering board meetings
  □   Monthly meetings of the core partnering team at the project level
  □   Monthly council board meetings
- Involvement of three out of five of the statutory undertakers in partnering workshops and the team approach has led to improved efficiencies on site and minimal abortive time.

Quality management

- Innovation has been encouraged throughout the scheme with both the designer and steel work fabricator being given incentives to generate innovative solutions.
- An agreement is also in place that details the share for council (client) and project partners of savings made through value engineering.
- The use of value engineering and innovation has resulted in approximately £1m savings on the council's (client) original illustrative design. Examples of innovative solutions developed are:
  □   Changed deck on the Usk crossing bridge (used composite deck instead of baffle)
  □   Shortened viaducts and incorporated reinforced earth
  □   Substantial reduction in the capital expenditure for statutory undertakers work due to value engineering and design opportunities
- As the format of the model contract is heavily incentivised to deliver a defect-free end product, i.e. permit of use will not be issued until the client is satisfied that no maintenance work will be required on the road within 12 months of opening, a rigorous self certification process has been implemented to ensure that all inspection and testing requirements are met.
- The self certification process:
  □   Began with the production of a defect-free design
  □   Each stage of the construction process is controlled by an exact inspection and testing regime to ensure that all elements meet the design specification
  □   Ensured ownership and commitment to the zero defect target at all levels of the supply chain
- Morgan–Vinci CJV met the zero defect target required at handover stage and hence the permit of use is issued on programme.

Predictability of cost control

- The NSDR DBFO project is a lump-sum contract with appropriate risk transfer to the concessionaire (SPV) and construction sub-contractor partner.
- The Morgan–Vinci CJV worked closely with the council (client) at all stages (tender, BAFO and financial close) to manage costs within their budget.

- The Morgan–Vinci CJV has actively sought savings throughout the duration of the scheme through creativity, innovation and continuous improvement. Significant savings have been made through environmental initiatives, value engineering and supply chain agreements giving cost certainty, as explained previously.
- All eligible changes (cost increases) are discussed with the council and value engineered before execution.
- The strategic partnering agreement between Morgan–Vinci CJV and the surfacing contractor allowed for a fixed price to be negotiated at tender stage, with a fixed, highly competitive level of inflation indexation over three years. This fixed price provided cost certainty.
- Capital expenditure and maintenance costs were considered throughout the various bidding stages in order to arrive at an effective whole life costing for the 40 years + 10 years residual life of the project.

Construction programme

- The co-location of the project team and several supply chain partners has facilitated regular interface and efficient decision making throughout.
- Partnering arrangements and existing relationships with the supply chain have engendered project focus ensuring delivery of programme.
- The relationship between the supply chain and the designer has also enabled the identification of opportunity to reduce cost and time.
- The use of recycled and secondary aggregates guaranteed security of supply and non-dependence on quarries. This minimised traffic movement through the route and minimised traffic disruption.
- The collaborative approach to problem resolution enabled the most cost-effective solution without detriment to the programme.
- Four revised programmes have been produced to incorporate the effect of changes to the project.

Results

The project team has proactively driven the above explained creativity, innovation and continuous improvement throughout the procurement and construction process resulting in:

- Bettering the NCC's completion programme by approximately 8 months
- £6.35m total savings and value adding in the scheme of early construction sub-contractor involvement.
- Award of the George Gibby Award for the Usk Crossing Bridge by the Institution of Civil Engineers in Wales.
- Winning of the Green Apple Award for sustainable construction and crowning as National Champions for Environmental Best Practice in the Building and Construction sector.

Case Study 16.2 Upgrade to Dual Carriageway Between Dundee and Arbroath

Innovations regarding sustainability, time, re-engineering issues

The construction sub-contractor's EST philosophy has been a guideline during the execution of the works to find the quickest, most effective way to make the A92 an efficient route. Affordability was a top priority for Angus Council and in order to satisfy this issue Morgan Est has identified a number of innovative ways to improve time and cost savings for Angus Council. These innovations were grouped under three headings (Morgan Sindall, 2004):

- Sustainable solutions
- Time savings
- Re-engineering the road

Sustainable solutions

The key requirement of the A92 construction was the large quantities of material needed to form the road structure. Redundant land in the form of a disused airfield provided a 'sustainable solution' for sourcing the necessary material which was recycled and used to form an improvement layer for the road. The unsuitable material from road excavation was used to fill the hole left in the airfield. The fill material was then covered with topsoil creating a new field that could be farmed.

Time savings

The early sourcing of suitable raw material meant that Morgan Est could work through the winter, coordinating the scheme four weeks ahead of schedule.

Re-engineering the road

Morgan Est has achieved further cost efficiencies by re-engineering the local authority roads around the scheme and therefore removing the need for an underpass which resulted both in money and time saving.

By working closely with the client and other stakeholders, Morgan Est's team has been able to add value by developing solutions that have saved time and money to the benefit of both the Angus Council, Morgan Est and the local economy.

16.5   Conclusion

It is the authors' contention that innovation and innovativeness are implicit characteristics of PPP procurement. The mechanism of PPP itself releases the stakeholders to concentrate on achieving VFM for the client and the constraints of narrow 'regulated' prescription are cast aside. However the evidence to date is that innovation within these projects is being impaired by reluctance from stakeholders to make best use of the potential opportunity and to go for safer 'tried and trusted' solutions. It is the authors' belief that with this reluctance to move away from 'tried and trusted' the potential benefits of the PPP procurement mechanism are not being achieved and hence 'best' VFM is not being attained.

References

Akbiyikli. R. (2005) *The Holistic Realisation of PFI Road Project Objectives in the UK*. Unpublished PhD Thesis, Research Centre for the Built and Human Environment, School of Construction & Property Management, University of Salford, Salford, UK.

Atkin, B.L. (1999) *Innovation in the Construction Sector*, ENBRI Report to Directorate-General Industry. Commission of the European Communities, Brussels.

Barrett, P., Sexton, M., Miozzo, M. *et al.* (2001) *Innovation in Small Construction Firms*. Base Report for EPSRC/DETR IMI Construction-LINK.

Birnie, J. (1998) Risk allocation to the construction firm within a private finance initiative (PFI) project. *Proceedings of 14th annual ARCOM conference*, September 9–11, University of Reading, Vol. 2: pp. 527–534.

Birnie, J. (1999) Private Finance Initiative (PFI) – UK Construction Industry Response. *Construction Procurement*, 5(1), 5–13.

Burns, T. and Stalker, G.M. (1961) *The Management of Innovations*. Tavistock Publications, London.

Chamberlain, W.P. (1995) *Performance-Related Specifications for HIGHWAY Construction and Rehabilitation*, NCHPR Synthesis of Highway Practice 212. Transportation Research Board, National Research Council, Washington DC.

Child, J. (1984) *Organization – A Guide to Problems and Practice*, 2nd edn. Harper & Row, New York.

CIC (2000) *The Role of Cost Saving and Innovation in PFI Projects*. Construction Industry Council, Thomas Telford Ltd., London.

Cobbenhagen, J. (2000) *Successful Innovation: Towards a New Theory for Management of Small and Medium-Sized Enterprises*. Edgar Elgar Publishing Limited, Cheltenham.

Davis Langdon & Everest (2002) *Generic Value Management Model for Construction*. Available at: http://www.davieslangdon-uk.com.

de Lemos, T., Almeida, L., Betts, M. and Eaton, D. (2003) An examination on the sustainable competitive advantage of Private Finance Initiative projects. *Construction Innovation*, 3, 249–259.

de Lemos, T., Betts, M., Eaton, D. and de Almeida, L.T. (2000) From concessions to project finance and the Private Finance Initiative. *The Journal of Private Finance*, Fall, 1–19.

de Lemos, T., Eaton, D., Betts, M. and de Almeida, L.T. (2004) Risk management in the Luseponte Concession – a case study of the two bridges in Lisbon, Portugal. *International Journal of Project Management*, 22, 63–73.

Debande, O. (1999) *Private Financing of Infrastructure. An Application to Public Transport Infrastructure*. Presented for the 6th International Conference on Competition and Ownership in Land Passenger Transport, Cape Town, South Africa, September.

DGI-Menard Inc (2004) http://www.dgi-menard.com/newport.html

Dupuit, J. (1844) On the measurement of the utility of public works. *Annales des Ponts et Chaussées*, 2nd series Vol.8, reprinted in Munby, D. (ed.) (1968) *Transport: Selected Readings*. Penguin, Harmondsworth.

Eaton, D. (2000) A detailed and dynamic competitive advantage hierarchy within the construction industry. *Proceedings of CIB W92-Procurement Systems Symposium-Information and Communication in Construction Procurement*, 24–27 April, Santiago, Chile.

Eaton, D. (2001) A temporal typology for innovation within the construction industry. *Construction Innovation*, 1, 165–179.

Eaton, D. and Akbiyikli, R. (2005) *Quantifying Quality*, A Report on PFI and the Delivery of Public Services. Royal Institution of Chartered Surveyors (RICS), London.

Egan, J. (1998) *Rethinking Construction*. HM Stationary Office, London.

Freeman, C. (1982) *The Economics of Industrial Innovation*, 2nd edn. Pinter, London.

Gann, D.M. (1997)*Technology and Industrial Performance in Construction*. Paper prepared for OECD Directorate for Science, Technology and Industry.

Gann, D.M. (2000) *Building Innovation: Complex Constructs in a Changing World*. Thomas Telford, London.

Part Three

Grant, T. (1996) Keys to successful public–private partnerships. *Canadian Business Review*, 23(3), 27–28.

Heald, D. and McLeod, A. (2002) Public expenditure. In: *Constitutional Law, The Laws of Scotland: Stair Memorial Encyclopaedia*. Edinburgh, Butterworths, para. 502.

Hobday, M. (1998) Product complexity, innovation and industrial organisation. *Research Policy*, 26, 689–710.

House of Commons (1999) *Public Accounts – Twenty-Third Report – Getting Better Value for Money from the Private Finance Initiative*. Public Accounts Committee Publications, Session 1998–99.

King, N. and Anderson, N. (1995) *Innovation and Change in Organisations*. Routledge, London.

Latham, M. (1994) *Constructing the Team*. Final Report of the Government/Industry Review of Procurement and Contractual Arrangements in the UK Construction Industry. HM Stationary Office, London.

Lawrence, P.R. and Lorsch, J.W. (1967) *Organisation and Environment*. Harvard University Press, Cambridge.

Levy, S.M. (1996) *Build, Operate, Transfer: Paving The Way For Tomorrow's Infrastructure*. John Wiley and Sons Inc, New York.

Male, S. (2002) Building the business value case. In: Kelly, J., Morledge, R. and Wilkinson. S. (eds.) *Best Value in Construction*. Blackwell Science Ltd., Oxford.

Manseau, A. and Seaden, G. (2001) *Innovation in Construction: An International Review of Public Policies*. Spon Press, London.

Morgan Sindall plc website, http://www.morgansindall.co.uk/access/general.asp?id = 914

OECD (2000) *Global Trends In Urban Water Supply and Waste Water Financing and Management: Changing Roles for the Public And Private Sectors*. CCNW/ENV (2000)36/FINAL. OECD/OCDE, Paris.

Oluwoye, J. and Lenard, D. (1999) Construction innovation: an overview of innovative construction methods. *Proceedings CIB W55&W65 Joint Triannial Symposium*, Cape Town, South Africa, September.

Quiggin, J. (2002) *Private Financing of Public Infrastructure*. Version 6 August, available at http://ecocomm.anu.au/quiggin.

Rogers, E.M. (1995) *Diffusion of Innovation*. Free Press, New York.

Smith, N.J. (2003) *Appraisal, Risk and Uncertainty*. Thomas Telford, London.

Tatum, C.B. (1984) What prompts construction innovation? *Journal of Construction Engineering and Management*, 110(3), 311–323.

Tatum, C.B. (1989) Organizing to increase innovation in construction firms. *Journal of Construction Engineering and Management*, 115(4), 602–617.

Vickerman, R. (2002a) *Private Financing of Transport Infrastructure: Some UK Experience*. Centre for European, Regional and Transport Economics, The University of Kent, Canterbury, UK.

Vickerman, R. (2002b) *Financing Schemes of Transport Infrastructure. Public and Private Initiatives in Infrastructure Provision*. Centre for European, Regional and Transport Economics, University of Kent at Canterbury, Paper for STELLA Workshop, Brussels, 26–27 April.

Walker, C. and Smith, A.J. (eds.) (1995) *Privatised Infrastructure: the Build Operate Transfer Approach*. Thomas Telford Services Ltd., London.

Walker, D. and Hampson, K. (2003) *Procurement Strategies – A Relationship-based Approach*. Blackwell Science, Oxford.

Winch, G.M. (2002) *Managing Construction Projects*. Blackwell Science, Oxford.

WRAP (2004) *The Waste and Resource Action Programme*. July. www.wrap.org.uk.

Part Three

17

Combining Finance and Design Innovation to Develop Winning Proposals

Colin F. Duffield and Chris J. Clifton

17.1   Introduction

This chapter focuses on how consortia seek innovative solutions to demonstrate that they offer the greatest value for money solution in response to an invitation to bid for a PFI/PPP project and considers synergies between financial structures and design innovation. Such innovation ranges from technical advancement, creative design that leads to whole of life efficiency and functionality, optimised risk allocation, (or for some governments, maximum risk transfer), corporate structures, operational improvements and efficiency and financial engineering to the most cost-effective outcome.

Discussion on design innovation draws from a workshop convened in conjunction with The Royal Australian Institute of Architects in 2006 (Clifton, 2006) and financial maturity in PFI/PPP projects is reflected upon through an analysis of recent Australian toll road projects.

The chapter commences with an outline of key drivers for PFI/PPP projects in a range of jurisdictions and then considers financing options adopted through the maturing of the PFI/PPP market, prior to discussing the relative merits of this and design innovation as it relates to the preparation of winning proposals. It concludes with a commentary on the importance of combining both design and finance to produce winning proposals.

17.2   The Drivers of PFI/PPP Projects in Different Jurisdictions

The motivation for major projects is both situational and project specific. Project drivers include: a functional requirement based on demand forecasts; safety or improved amenity; political motivation as influenced by perceived, or real, constituent pressure; strategic and commercial investment opportunity to deliver a financial return; and econometric outcomes that deliver productivity outcomes. The motivation for PFI/PPP projects not only includes the normal drivers but also overtly considers the economic and cultural climate of a country and such factors work themselves out through the form of finance, risk allocation profile and commercial viability criteria chosen for any given project. This section discusses the different drivers for PFI/PPPs in jurisdictions across Australia, UK, Canada, Hong Kong, United Nations and the World Bank. Interestingly, Australia, UK, Canada and Hong Kong have similar legal systems. They also have sophisticated financial markets and similar standings in respect to GDP per capita (being in the range of US$36 553 to US$39 213) (International Monetary Fund 2007) with world rankings from 13 to 17. In contrast, the GDPs of clients of the United Nations and World Bank commonly have GDPs per capita of below US$2000.

Recent PFI/PPP projects implemented in the developed world have focused heavily on achieving VFM outcomes for governments through the application of robust processes as articulated by clear service obligations through output style specifications. Commercial benefit, public interest and community acceptance are tested through the project procurement and bidding processes and are ratified contractually with terms and conditions that clearly detail service charge regimes, risk allocation and the expectations of all concerned at the time contracts are signed. Ongoing behaviour of the participants and incorporation of any necessary changes are managed through the administration of these contracts. This ongoing management is in its infancy and equitable and transparent techniques for this are still being developed.

From the perspective of a bidding consortium, the opportunity to participate in long-term PFI/PPP agreements is only possible if the consortium wins the bidding process. One of the keys to successful bidding is a clear understanding of what governments seek as part of their decision to pursue a PFI/PPP delivery model. A comparison of such drivers follows.

Current PFI/PPP processes for the evaluation and establishment of long-term outcomes are similar in jurisdictions such as the UK, Canada, the Netherlands, South Africa and Australia. The overall approach is similar to the process detailed in Figure 17.1 (Department of Treasury and Finance 2001; Industry Canada 2003; Department of Finance and Administration 2005; Sharp and Tinsley 2005), where a business case establishes the need for the project and a community's interest is quantified and tested via either an implicit or explicit public interest test. The financial benefits of the project are quantified through the establishment of a reference project and measured via a tool called the public sector comparator (PSC). The required service outcomes are specified in terms of an output specification that is released during the bidding process as part of the project brief and request for tender.

The expectations detailed in the project brief and request for tender are ultimately translated into terms and conditions of hard money contractual agreements involving both the performance standards expected and the financial structuring of the PFI/PPP.

The PSC has been used:

- To quantify that the decision to adopt a PFI/PPP procurement strategy is appropriate
- As a reference by which test and compare tender submissions in the ranking and evaluation of tenders.

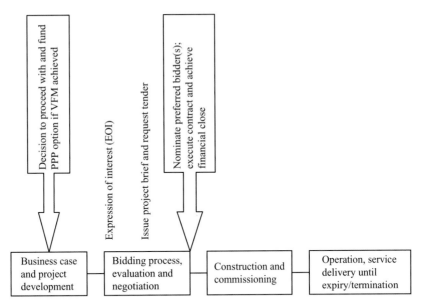

Figure 17.1   PFI/PPP relationship continuum.

The PSC is an important measure but the actual drivers behind a PFI/PPP process warrant equal reflection. The following sections discuss design innovation and finance initiative in detail.

17.3   Design Innovation: The Issues

Innovation is one of the key VFM drivers typically cited for a PFI/PPP, together with risk transfer, whole-of-life costing and asset utilisation. In this context innovation relates to design, specifically through focusing on output specifications, private sector bidders are given the opportunity to develop innovative design and other solutions so as to meet government's requirements at a lower cost (Department of Treasury and Finance, 2001).

17.3.1   Design inputs and value generation

The move to outsource the design function is a more general trend than just PFI/PPP projects and the result is that designers, such as architects, are often sub-contractors to others and may be treated simply as a provider of services rather than a generator of innovative solutions. In such arrangements the commercial pressure to deliver a project quickly places enormous pressure on designers to create innovative solutions and to do so within a specified budget. Done poorly, such pressure results in conservative designs and there are many examples where the design has not been fully thought through leading to quality issues as raised earlier. The flip side is that done properly the devolution of overall project control can facilitate innovation and a

Part Three

whole-life thinking into the process. There are many examples where properly scoped and specified PFI/PPP projects have led to very high-quality outcomes and done so with far greater cost certainty than traditional procurement techniques.

Some key areas for focus to achieve the positive outcomes include: specification; active involvement during the tender and procurement phases; and appropriate contractual mechanisms to ensure the correct motivators and abatements are in place to foster the standard of quality or service sought. Key to the overall success of an innovative design for PFI/PPP projects is the acceptance of such a design by the client. The issue of communication between client and bidder is of particular concern due to probity issues associated with the bidding process (see Figure 17.1).

The issues and possible solutions are discussed through the reporting of a number of case examples from a roundtable discussion held by The Royal Australian Institute of Architects in 2006.

17.3.2   PFI/PPP design roundtable discussions

A series of roundtable events have been held in Western Australia, in 2005, and Victoria in 2006 to identify ways in which the importance of good design can be brought to the forefront of the PFI/PPP procurement process. Both forums brought together participants in particular PFI/PPP projects from all sides including owner, government, private proponent, designers, constructors, financiers, operators and advisors. The group openly discussed issues and potential solutions to project difficulties in an endeavour to improve future projects. Two projects from the Victorian workshop (Clifton, 2006), are particularly relevant to this discussion and are reproduced as Case Study 17.1 and Case Study 17.2.

Case Study 17.1: Remand Centre, Ravenhall and Marngoneet Correctional Centre, Lara

The importance of good design and the design process

Design in prison facilities focuses on architectural and functional requirements through the provision of security and associated custodial services. A major issue in the design of the Remand Centre was the lack of exemplar prison projects in the world, and therefore limited reference schemes to assist bidders. Further dialogue to help convey the vision to the bidders would have helped.

In the prison project, output specifications were well developed including data, however in order to attain 'output specifications', no detailed drawings were included. As a result, despite several design workshops, none of the bid designs met expectation in the first round. This necessitated a further design-proving process, potentially adding to bid costs.

Social policy was a large driver in the process, whereby rehabilitation was seen as a key objective, as well as security and community safety. Architects had limited prior exposure to prison specifications of this nature, and greater interaction during early phase design between bidders and clients would have facilitated the design process. Notwithstanding this, the finished product exceeded expectations.

In social infrastructure, design should initially focus on operational management. Delivery systems were insufficiently addressed by bidders in the early design. Major consideration must be given to the operation of the facility for 20 years. A comment was made that PFI/PPP contracts give too much weight to legal and contractual issues rather than operational ones.

Probity also contributed to design limitations, as it limited interaction. The comment was made that, despite interactive sessions during the bid process, the probity regime can impair open discussion between the proponents and the state.

The complex design issues found in social infrastructure meant that the standard 9 weeks given to bidders to respond to a brief was seen as insufficient time to piece together all the complexities into an operational design. Despite this, innovative design was achieved, in accordance with government objectives; in future greater focus on interaction will provide even further scope for improved functionality.

Design risk

If the bid is more complex then it should be the responsibility of the government to define and make clear the issues. There were issues surrounding fitness for purpose clauses, and there would have been benefit in talking about outputs to identify specific fitness for purpose issues, such as delivery, movement and waste.

Bid costs

It was asked whether the teams pay the costs, success fees or some other formula to their consultants in the bids. Typically basic cost fees were paid in addition to a success fee. The legal costs in the bid are generally paid as a capped fee.

Innovation

All participants agreed it was very difficult to provide an innovative solution in an 8–10-week bid process. Innovation in the actual design was seen to be limited, though the design process and the financial structure proved to contain innovative solutions. The project was made a success through the hard work of all participants, and despite the shortfalls in parts of the overall process, the end result has exceeded expectations.

Case Study 17.2: Royal Women's Hospital Redevelopment

The importance of good design

There is a problem defining what good design is. The brief did not say the project was to be aesthetically striking, however the brief required that the project:

- Be a landmark building
- Focus on function
- Made a difference in design
- Recognise that some of the women attending hospital are 'well' women who are having babies
- Create a domestic setting
- Change the smell so it didn't smell like a hospital

The builder selected the architects, who are specialists in health design. There is only a small number of architects who have designed hospitals. Something different was desired for the project. The brief was good, it went into detail, and projected the vision that the project was not a hospital only for sick people. A major problem with the brief, however, was there were no pictures, only words. A film about childbirth was provided for bidders, and was seen as an excellent way to convey the vision.

Part Three

The project is a model case study, as it is on time, does not have variations and has maintained excellent relationships throughout, with a major emphasis on health planning.

The design process

There is still a need during the bid phase to facilitate greater interaction, and a better design outcome. This can be achieved as more is conveyed though a combination of a visual and text brief, containing videos and diagrams, not a specific schematic design. This is especially effective in complex areas, where pictures have proved very effective in conveying what is required. It was seen that if the state has a strong view about a specific design, it should inform the bidders of that, but it should not stifle the design process.

The Royal Women's Hospital tender processes yielded three bids with three distinctly different designs, which was seen as an excellent result and provided the state with choice and originality. It is important to see that the design reflects a whole-life approach, and whilst there has been a significant increase since early projects, there still needs to be more emphasis on operation at the design stage. The Hunter Technical College project (UK) was provided as an example of how good design can provide a hugely beneficial outcome to society. It proved the importance of focusing on the environments being created and not just the bricks and mortar. The project was about students and education, and was led by people with a vision. It created a huge amount of excitement from the design team through to the general public. The college initially had 600 students each year, but was designed for 800. Two thousand five hundred students are now enrolled due to the excellent results being achieved by students at the college.

The outcomes for health and prison projects are changing, though require inspirational people to bring about greater change. Hospitals are being designed for people who are sick, but want to get better. The consortiums being assembled have therefore become more specific in the architects they engage. Whilst it is viewed that Australian healthcare is not as advanced as in the UK, the state is learning from overseas but is not advising consortiums to supplement their teams with international expertise. Bidders may however see it is an advantage to have a specialist international architect to assist in preparing the bid.

Design risk

The designers went through the brief in detail, and found discrepancies which required clarification, however there were many barriers to gaining a true understanding of the client's needs. Again, it was seen that further flexibility to make changes at the design development stage was required, as currently the consultant risk remains unrealistically high.

Bid costs

Architects spend a lot of time in the bidding process, and the opportunity cost needs to be considered. High level resources are required, and it was thought that architects' exposure would be reduced if more design could be completed upfront.

17.3.3   Common themes for capturing innovative design

A number of common themes have emerged from the roundtable held in Western Australia, on 31 May 2005, the Victorian examples given above and the Fitzgerald review (2004) of Partnerships Victoria projects. Common findings are:

- A premium price does not have to be paid to achieve good design but a culture of investing for long-term outcomes is essential.

- Current PFI/PPP processes need improvement to allow more flexibility in the design process. Current timeframes and risk-allocation approaches constrain bidder's innovation which frequently results in the use of tried and true design solutions rather than developing creative, (and frequently riskier) solutions.

Specific areas for advancement are the advantages of using a design advocate, increasing interaction and refinement of typical PFI/PPP processes.

There appear to be benefits from the involvement of a design advocate, e.g. the Commission for Architecture and the Built Environment (CABE) in the UK. This is particularly the case in the early stages of a project where significant value can be added through involvement in the selection process of a proponent, and through input into the design brief. Likely positive outcomes include consistency in process and improved community transparency, and thus credibility.

Increased communication and interaction between clients and users is critical at each stage of the design process, and further development is required to understand the true needs from the end-users. The end-users can aid in the process, by outlining early their specific requirements, which assist architects in their design process.

Refining PFI/PPP processes may assist achieving better built outcomes, and increased dialogue may help eliminate stages such as 'best and final offers' (BAFOs). Incorporating an alliance style process was also proposed, though this was still seen as some way off (Clifton and Duffield, 2006). Whilst the client can transfer some risks, it still has a role in minimising that risk.

To achieve true innovation under the PFI/PPP model, designers must be aware of the issues and risks associated with the project prior to proceeding. This includes understanding the payment and abatement structures, pricing risks adequately and consideration of the risk-transfer mechanisms.

Industry suggestions on techniques to improve PFI/PPP policies in terms of design include:

- Engage the market before inviting expressions of interest.
- Standardise documents, particularly contracts.
- Ensure technical requirements are clearly defined.
- Communicate priorities, for example where there may be competing objectives of time, cost and quality.
- Defer the requirement for full bid documentation until later stages of bidding.
- Develop standard government processes.

17.4   Financing Arrangements for PFI/PPP Projects

Financing and innovation are synonymous with PPPs in Australia as evidenced by the majority of recent PPP projects being led by financiers. This section reflects on the value and innovation brought to projects through financing and dispels some myths that financing innovation is peculiar to specific countries based in their maturity in the PFI/PPP market or specific

Part Three

in-country circumstances such as taxation. A discussion on financing invariably raises the issues of balance between equity and debt, interest rate premiums and corporate structuring. The focus of this section is to provide an understanding of current approaches and products being adopted by financiers and to provide examples of the application of these products to specific projects.

The significance of finance within the overall PFI/PPP solution cannot be understated. Recently, innovative financial solutions that lower project risk ratings have allowed for cheaper finance to be achieved. This adds significant competitiveness to a PFI/PPP bid. This section investigates how finance drives a VFM outcome. It looks at a number of finance mechanisms adopted for PFI/PPP projects and considers the influence of project rating agencies prior to presenting specific examples and trends for project finance in a maturing market.

The first step in understanding project finance is to understand that finance is not free money. It is money and resources provided on account with the explicit expectation that the organisation providing the finance and resources will, on the balance of projects risks, receive a return on their investment commensurate with the risks undertaken. The first differential between PFI/PPPs and standard project financing is that there is the opportunity for organisations to assume business risks through the provision of equity to the proponent vehicle. Further PFI/PPP projects seek debt providers to assume a higher risk exposure via non-recourse finance (or realistically limited recourse finance) rather than a more secure form of finance where the lender has recourse to repayment via guarantees and or security.

Business investors and debt providers are always most interested in the likelihood of return on their investment and the security associated with their investment. Governments inherently structure PFI/PPP arrangements to transfer genuine business risk to the private proponents but these private proponents will only be associated with the successful tenderer if their structured financial arrangements overcome the inherent need to price risk yet to provide VFM (often measured in terms of cost effectiveness). This dichotomy stimulates innovation.

17.4.1   PFI/PPP financial products

The forms of finance are generally grouped by way of equity contribution and debt. Debt is further divided by way of the level of security provided. The blend of debt and equity relates directly to the risk profile of a PFI/PPP project and this risk profile changes significantly over the life of a project. Figure 17.2 schematically presents the risk profile and associated cash flow for a project over its life.

Significant stages throughout the life of a project being: pre-financial close is when a consortia does not know if it will be successful in winning the project; detailed design–construction–service commissioning is the period where the hard assets are detailed and procured, these hard assets provide the functionality and operational efficiencies to delivery a particular service;

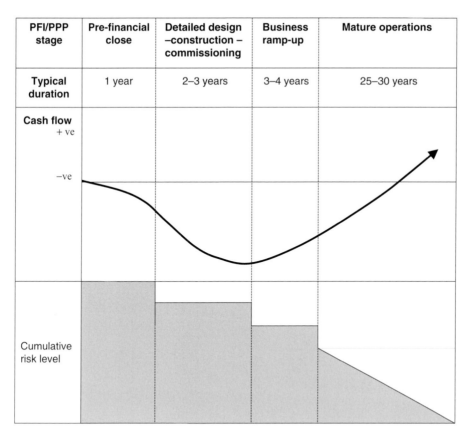

PFI/PPP stage	Pre-financial close	Detailed design –construction – commissioning	Business ramp-up	Mature operations
Typical duration	1 year	2–3 years	3–4 years	25–30 years

Figure 17.2 Schematic of PFI/PPP cash flow and associated risk profile over its life.

business ramp up is the period where the service business is developing its output capacity; and mature operation is where the PFI/PPP project would be better termed an operating business with a sunset clause on the term of the operation.

Various sectors of the finance market have differing levels of appetite for risk (and associated level of return on their investment) and therefore the financial structure and associated participants involved in the various PFI/PPP stages change over the life of a project. Formalised assessment of a project's financial risk profile is typically undertaken through the use of rating agencies such as Standard & Poors or Moody's. Project ratings provided by these agencies help to establish the investment worthiness of particular projects and likely interest rate premiums associated with particular risk profiles. A range of finance options is summarised in Table 17.1.

Specific products always involve the inter-relationship between security, duration and terms of the agreement. For example, bank debt may be provided as a style of credit secured against assets, it may be a bullet structure where a term loan, with periodic instalments of interest, has the entire principal due at the end of the term as a final payment, or it may be in the form of tenors of durations of 15 years plus. Bonds represent a promise to pay money back on a

Part Three

Table 17.1 A range of finance options.

Investment	Typical appetite for risk	Description	Examples of providers
Equity	Fully exposed to risk – high return expected	The proprietor's capital investment in an enterprise or undertaking. It is generally unsecured, high risk and illiquid if securities in the enterprise are unlisted	Investors, major companies, passive investment via superannuation funds, trade Investors, e.g. Laing and specialist PFI/PPP firms such as Bilfinger Berger BOT and Pleanary
Initial public offering (IPO)	Subject to the scrutiny of market researchers and the vagaries of the stock market	Public raising of funds through the sharing of ownership	The investment markets
Senior debt	The highest ranking for repayment, security, or action, i.e. lowest risk exposure	Bonds market, loan accounts	Banks, superannuation funds and capital markets
Mezzanine finance, sometimes referred to as non (or limited)-recourse finance	High risk and totally dependent on the success of a project	Lenders rely on the project's cash flows and security over the project vehicle's assets as the only means to repay debt service	Capital markets, specialist banking sectors, e.g. Macquarie, Deutsche, Monoline insurers, e.g. MBIA, AMBAC, FSA, XL, short-term or discount securities
Bonds	Risk profile ranges depending on the terms and styles of the agreement	Bonds are typically long term (greater than 1 year) short term, and potentially junk bonds (really a form of speculative investment)	Banks, governments and capital markets
Hybrid instruments: debt/equity	These instruments are high risk and they are designed to maximise taxation efficiency and thus may influence a PPP's financial arrangement	A capital raising device which has both features of debt and equity (Review of Business Taxation, 1999)	These include classes of preference shares, convertible notes, capital protected equity loans, profit participating loans, perpetual debt, endowment warrants and equity swaps

specified future date – the maturity date and the bondholder is also normally entitled to regular 'coupon' interest payments. Capital market products include: medium term (5–10-year) having floating rates, long term (15–35-year) fixed-rate bonds that are adjusted and linked to Consumer Price Index, and wrapped or unwrapped structures. Wrapped structures are those where balance sheet strength or guarantees from one organisation are provided to support a particular investment, thus lowering its overall risk profile and thereby provides access to lower risk premium loans. Wrapped structures have become one of the more common innovations in PFI/PPP projects and therefore warrant further discussion.

17.4.2   Monoline wrap

A monoline wrap has been commonly used as the underwriting arrangement for social infrastructure projects such as schools and hospitals in Europe, but had been applied only to major public infrastructure ventures such as roads in Australia to date (AFR, 2006). Monolines exhibit strong financial fundamentals in terms of earnings, asset quality and their capital role (Knepper, 2006).

Monolines essentially guarantee transactions by lending their balance sheets, with the guarantee usually irrevocable and unconditional, resulting in the guarantor stepping into the place of the issuer where they guarantee payment in accordance with the original transaction schedule. Subsequently, in the event of the default of the underlying issuer, or where the issuer fails to pay the coupon and/or principal on a timely basis, the investor has recourse to the financial guarantor (also known as the wrapper) in that they will pay the coupon and/or principal in accordance with the terms of the affected bond issue.

Whilst there are significant benefits to wrapping transactions, the one major negative is the price of the wrap. The biggest benefit in wrapping a transaction is the rating, in that a transaction that is wrapped carries the ratings of the guarantor rather than the underlying issuer. In addition to enhancing the liquidity and marketability of the issue, this has obvious pricing implications.

17.4.3   Financial engineering

Financial engineering involves the balancing act between provision of finance, security and return on investment. The balance between equity and debt (gearing) is determined by:

- Agreeing base case assumptions for all relevant items which affect a project's cash flow.
- Obtaining agreement regarding the base cash flow required for sensible operation of the business before servicing debt.
- Selection of appropriate debt cover ratios (risk adjustment).
- Detailed scenario analyses to stress test cash flows to ensure that coverage ratios are reasonable.

Part Three

Table 17.2   Key reasons for wrapping from issuer and investor perspective (Knepper 2006).

Issuer	Investor
Pricing benefits often outweigh cost of guarantee with transaction carrying the ratings of the guarantor	In the event of the issuer failing to pay interest and/or principal, investors have recourse to the guarantor to make full and timely payment of interest and/or principal
Helps to maintain issuer confidentiality in that the issuer may not want to disclose proprietary information to investors and by wrapping, the credit focus shifts to the guarantor rather than the issuer	Investors benefit from the surveillance expertise of the guarantor and the comfort that the guarantor is sharing the risk by lending their credit quality to the issue
Issuer benefits from the expertise and experience of the guarantor in that they wrap a vast array of transactions	Benefits from the added scrutiny brought to the transaction by the wrapper both in the development process and the continued surveillance throughout the life of the transaction, as the wrapper is on risk for the duration
Helps to broaden market acceptance of new or complex transactions. Helps also in the secondary market by promoting liquidity	The investor also benefits from the rating agency scrutiny in that they analyse both the transaction and the wrapper. Investors in unwrapped tranches also benefit indirectly in that the wrapper monitors and assesses the whole transaction

- The process of 'engineering' a financial arrangement is iterative and somewhat rule of thumb.
- If equity return is too low, government contributions are sought.

Considering the cash flow and risk details provided in Figure 17.2 along with the finance options detailed in Table 17.1 results in a variety of financial products being adopted during the various phases of a project. The initial pre-financial close period has been historically dominated through the provision of hard to obtain and expensive equity, this is then supplemented by mezzanine style finance, and when some form of security (e.g. physical assets or confidence that real business return is likely) starts to emerge then longer-term (and cheaper) forms of debt also emerge in the market.

Financial innovation is therefore based around market confidence and products that bring relatively cheaper finance to a project early. A useful way to understand financial innovation is to reflect on past projects.

Evolution of financing arrangements of PPPs in Australia

A selection of Australian projects is used as the basis to demonstrate the innovation (and changes) brought to PFI/PPPs. Early projects relied heavily on equity and debt arrangements, though as the market has matured the key financiers have looked to offset their risks through broadening of the investor base. This has specifically included private investors, superannuation and trusts. This is leading to the development of a new asset class in its own right. Case Study 17.3 (The Melbourne City Link project) is a good example of an early PFI/PPP project where there was little or no pre-existing market, nor appetite from the financial market to invest early. Equity for the project was primarily provided by the constructors for the project with early mezzanine finance provided by stapled securities. Case Study 17.3 also demonstrates how the banks sought to mitigate their risk through the sharing of the financing with partners and co-arrangers.

Case Study 17.3: Melbourne City Link: Toll Road

Contractual arrangement

The Transurban City Link entity commenced as a joint venture initiative between Transfield and Obayashi of Japan in 1995. Transurban was a single project company with the sole purpose to finance, design, construct, operate and maintain the $1.9bn Melbourne City Link project that was completed in 2000. Transurban City Link was floated on the Australian Stock Exchange in March 1996 after it secured a BOOT contractual arrangement for the project.

Financial arrangements

Prior to the contract being let, Transurban arranged Stapled Securities (Infrastructure Bonds) to supplement the equity provided by the joint venture partners. Holders of the Stapled Securities had the opportunity to convert to shareholders when the company was floated, approximately 1 year after winning the contract. After floating, the original joint venture partners each retained a 10% shareholding in Transurban City Link.

Lead lenders were three of Australia's major banks, namely ANZ Banking Group, Commonwealth Bank and Westpac. Co-arrangers included Banque Nationale de Paris, Credit Lyonnais Australia and IBJ Australia Bank. A second tranche of money was arranged by the National Australia Bank.

Mandiartha (2007) undertook an interesting analysis of equity interests in recent transport PPP projects in Australia, such projects account for some 25% of the Australian PPP market (Eggers and Startup, 2007). He analysed five major tollroad projects: Southern Cross Station (Victoria) – contract date July 2002; Westlink M7/ Western Sydney Orbital (New South Wales) – contract date August 2003; Cross City Tunnel Sydney (New South Wales) – contract date July 2004; Lane Cove Tunnel (New South Wales) – contract date July 2004; and Eastlink (Victoria) – contract date November 2004. The equity investors in these projects have been categorised as financiers, constructors, operators, superannuation funds, direct capital market investment and other, the results are presented as Figure 17.3.

The differences between the financing arrangements for Case Study 17.3, commenced in 1995, and the projects presented in Figure 17.3, 2002–2004, is stark. The innovation in 1995 was provided through the initiative of construction companies whilst the primary driver for each of the later projects was the financier. Also of interest is the innovation that the financiers have delivered over the window of projects detailed in Figure 17.3. The direct equity provision by the financiers in the early phase of the project has dropped from as high as 100% to zero for the Eastlink project where the equity was fully provided by the release of an IPO on financial close. It is also interesting to note that the constructors' direct equity involvement has remained

Part Three

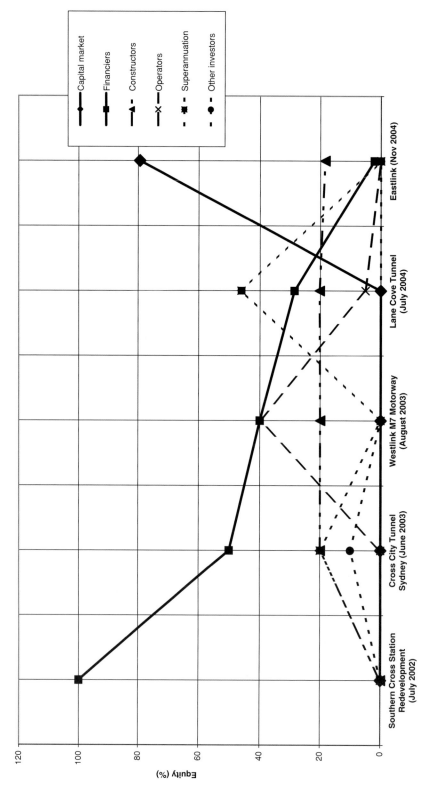

Figure 17.3   Equity investors in recent Australian toll road projects.

constant at about 20% and that the historically conservative superannuation funds are becoming more confident to invest early in PFI/PPP projects. The lack of equity involvement in the projects by the operators is surprising given the strong link between downstream long-term business outcomes and the performance of the operator.

The innovative contribution of financiers to the PFI/PPP market is certainly not confined to Australia as evidenced by the following section.

Innovative finance – a US perspective

The term 'innovative finance' can be broadly defined as a combination of special funding initiatives, though it has become synonymous in the transportation industry with techniques that are specifically designed to supplement the traditional methods used to finance highways (US Department of Transportation, 2004). US Department of Transportation (US DOT) innovative finance initiatives are intended to augment rather than replace traditional financing techniques.

Policy makers recognised they could increase development and expand the base of available resources for transportation projects by:

- Removing barriers to private investment
- Bringing the time value of money into federal programme decision making
- Encouraging the use of new revenue streams, particularly to retire debt obligations
- Reducing financing and related costs, thus freeing up savings for transportation system investment

Over the past decade, innovative finance has undergone several transformations including innovative debt financing, credit assistance and PPPs. With the advent of dedicated public funding for highways, the private sector involvement in highway financing and construction slowed somewhat; however, there has been renewed interest in private sector involvement in recent years as highway budgets have been stretched. Institutional models currently used include:

- Concessions for the long-term operation and maintenance of individual facilities or entire highway systems.
- Purely private sector highway design, construction, financing and operation.
- PPPs in designing, constructing, and operating major new highway systems. While a few states currently account for the majority of private sector financing, many more states have expressed interest in the potential for greater private sector involvement.

It is also interesting that a North American PPP deal won accolades for the 2005 project finance deal of the year for the joint venture between the Army and Actus Lend Lease to provide 7894 Army family houses in Hawaii (Editor, 2006). This particular deal is based around a 50-year lease. The pre-financial close period was financed by bridging loans from Goldman Sachs and Bank of America. Post-financial close floating rate bonds were floated with the aid of monoline insurers. The use of credit-linked certificates enabling reinvestment of the proceeds from the bonds was also considered innovative.

Common themes for innovative PFI/PPP financing

Common themes in achieving innovation from the financing arrangements for PFI/PPP projects revolve around the confidence the market has in terms of risk, security and return. There is a consistent message that breaking into new markets requires direct early equity investment. Areas of innovation include:

- Encouragement of long-term debt providers to participate as early as possible in projects. This reduces the extent of mezzanine style finance.
- Encouragement of equity providers to participate in new and emerging markets.
- Utilisation of wrapped finance as a form of structuring.
- Working closely with rating agencies to ensure project risks are kept to a minimum and that subsequent project credit ratings are improved.
- Early and direct engagement of capital markets not only provides important project equity but it stimulates market interest in infrastructure as an asset class.

17.5   The Theory and Practice of Winning Proposals

Winning proposals are those that are adjudged to provide the 'best' VFM based on the scope of a project as detailed in the brief and to do so within the constraints of the project and the market place. PFI/PPP tenders are generally evaluated in terms of technical functionality, finance and commercial outcomes. Frequently the detailed evaluation for each of these elements is undertaken independently by experts and thus there is an opportunity to improve the process as the synergies between the various elements are better understood.

The links between the elements are clear and strong. The technical design not only complies with the specification but also provides the mechanism for the proponent to structure the efficiencies and outputs for their service outcome. These efficiencies and outputs govern the commercial deliverables for the PFI/PPP business and thus they have a large bearing on the cash flow and ultimate return on the commercial viability of any investment. Similarly, the finance costs and arrangements are a function of the return on investment. Thus, improved and innovative design solutions that improve business performance should give rise to innovative finance that is more cost effective than would otherwise have been the case.

This chapter has detailed a range of areas where design innovation can be improved through better communication with operators and clients, greater time to deliver creative solutions, ensuring technical requirements are clearly defined prior to finalising documentation, through the use of standardised processes wherever possible and through a real understanding of the commercial drivers for the delivery of the service over the whole of its life.

Financial innovation is strongly linked to market acceptance and also to knowledge of the latest 'smart' approaches. It seems from the analysis of Australian transport projects that innovation and market acceptance builds quickly and therefore an appropriate balance between debt and equity for any particular market is an essential element for success.

To illustrate the salient points presented in this chapter, Case Study 17.4 is provided on Melbourne Convention Centre in Victoria.

Case Study 17.4: Melbourne Convention Centre

Project description

The new 5000-seat Melbourne Convention Centre is a Victorian State Government PFI/PPP including the design and construction of the new centre, adjacent to the existing exhibition centre, and ongoing maintenance for 25 years of both the new convention centre and the existing exhibition centre. The project commenced in June 2006 and it will be ready for operations in 2009. The Convention Centre will be delivered as part of a $1bn integrated mixed-use precinct in a single stage of development which includes:

- A 319-room hotel
- An 18 000m^2 office and residential tower
- A 10 000m^2 riverfront promenade of lifestyle retail, incorporating cafes, bookstores and tourism retail
- A 50 000m^2 premium brand homemaker retail complex that will be one of the largest single-stage retail developments ever completed in the Melbourne CBD
- An investment in public spaces including a partnership with the National Trust for a revitalised Maritime Museum

The project was awarded by the Victorian Government to the Multiplex/Plenary consortium consisting of:

- Plenary Group – consortium lead, equity investor and project management
- Deutsche Bank – financial underwriter
- Austexx – commercial development partner
- Multiplex Constructions – builder
- Multiplex Facilities Management – service delivery over the 25-year concession period
- Hilton International – hotel operator
- NH Architecture/Woods Bagot/Larry Oltmanns – architecture and urban design

Financial structure

The Melbourne Convention Centre contained a significant degree of financial innovation, through the use of an Australian-first financial guarantee mechanism for a social infrastructure project, which has led to the $192m bond issue being awarded a AAA risk rating, the highest possible. The AAA rating comes after several major projects in Victoria, including the PFI/PPP redevelopment of Spencer Street Station, have run into cost and timing issues (AFR, 2006). Deutsche Bank, which is also responsible for the $480m senior debt on the project, is managing the bond issue with National Australia Bank. The inflation-linked bonds will be unconditionally underwritten by US company Financial Security Assurance.

The project's credit risk rating without the arrangement with FSA would be BBB, reflecting the exposure to an unrated builder, some risks with third parties at the precinct and a good relationship with the state government. This exposure to an unrated builder is mitigated by the non-complex nature of the works and an appropriate contractual structure.

Part Three

Design and commercial innovations within Case Study 17.4 included the creative use of the government's proposed siting of the Convention Centre to meet all the stipulated requirements and then add value to the commercial developments of hotel and tower to the site. This design initiative leveraged the land into a part development project. The innovation added significant value to the site and overall development and the resultant savings provided opportunity for sharing with government, thus enhancing the bid. Further, adjacent to the Convention Centre there is an existing exhibition centre. Through detailed discussions with the owner and operator of the Convention Centre an overall strategy for the management of the combined complexes was developed, thus further saving operating costs over the whole of life.

Finance innovation was achieved via the financial guarantee mechanism allowing debt for the project to be raised on the basis of a AAA credit rating, this is the same credit rating as the state of Victoria.

In conclusion, winning proposals should always be innovative and in some way an enhancement on what has gone before. The consistent assertion presented has been that the integration of finance and design innovations will provide greater opportunity for successful bidding on PFI/PPP projects.

References

Australian Financial Review (AFR) (2006) Standard and Poor's gives bond issue top rating. 20 June 2006.

Clifton, C. (2006) *PFI/PPP Roundtable: Summary of Proceedings*. The Royal Australian Institute of Architects.

Clifton, C. and Duffield, C.F. (2006) Improved PFI/PPP service outcomes through the integration of alliance principles. *International Journal of Project Management: Governance Issues in Public Private Partnerships*, 24(7), 573–586.

Department of Finance and Administration (2005) *Public Private Partnerships Guideline: Commonwealth Policy Principles for the Use of Private Financing – Business Case Development*, Financial Management Guidance No. 17. Australian Government, May 2005.

Department of Treasury and Finance (2001) *Partnership Victoria – Practitioners' Guide*. Melbourne, DTF, State of Victoria.

Editor (2006) North American PPP Deal of the Year 2005 – Army Hawaii: Building Base. *Project Finance*, February.

Eggers, W.D. and Startup, T. (2007) *Closing an Infrastructure Gap: The Role of Public Private Partnerships*. A Delloite Research Study.

Fitzgerald, P. (2004) *Review of Partnerships Victoria Provided Infrastructure*. GSG Strategy & Marketing, Melbourne.

Knepper, L. (2006) *Unwrapping the Wrappers*. Barclays Capital: Securitisation Research. http://www.mbia.com/investor/publications/wrappers.pdf.

Industry Canada (2003) *The Public Sector Comparator: A Canadian Best Practice Guide*. Ottawa.

International Monetary Fund (2007) *World Economic Outlook Database*, http://www.imf.org/

Mandiartha, I.P. (2007) *Project Financing and Risks in Transportation Projects*, Research Report 421642. Department of Civil and Environmental Engineering, The University of Melbourne.

Review of Business Taxation (1999) *A Platform for Consultation*, Discussion Paper 2 – Building on a Strong Foundation, http://www.rbt.treasury.gov.au.

Sharp, L. and Tinsley, F. (2005) PPP Policies throughout Australia: A Comparative analysis of Public Private Partnerships. *Public Infrastructure Bulletin*, Issue 5.

US Department of Transportation (2004) *Status of the Nation's Highways, Bridges, and Transit, 2004 Conditions and Performance*. US Department of Transportation – Federal Highway Administration, Washington, DC.

Part Three

18

The Application of a Whole-Life Value Methodology to PPP/PFI Projects

John Kelly

18.1   Introduction

Whole-life value of projects has two primary stages; first, the definition of the project and the generation of options to satisfy the project; and second, the evaluation of the options to derive the best VFM solution. These primary stages have four distinct attributes: the identification of a project and its place within the strategies and programmes of a client organisation; the definition of the project in explicit functional terms; the value criteria by which the project will be judged a success; and finally the method of calculation for determining which of the competing options best satisfy the functional values defined. The functional definition of the project is a necessary precursor to the generation of options to satisfy the functional requirements. The options having been generated are judged in terms of their value to the client and their whole-life cost. These two evaluations combine to form a whole-life value evaluation.

This chapter discusses the four attributes and sets out methods determined by research by which each might be satisfied. The practical application of the four attributes is illustrated by reference to the stages in the development of a PPP/PFI project.

18.2   Projects

18.2.1   The definition of a project

A project is defined as 'an investment by an organisation on a temporary activity to achieve a core business objective within a programmed time that returns added value to the business activity of the organisation' (Kelly *et al.*, 2004). This is a useful definition as it recognises the temporary nature of projects which often use resources different to those used by the client in the core business but with the aim of changing and enhancing the core business.

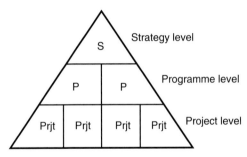

Figure 18.1 Projects are nested within programmes. Programmes are developed from strategies.

The cost of undertaking the project should be less than the total value gained to the core business by the undertaking of the project.

18.2.2   Strategies, programmes and projects

Before launching a project it is important to consider whether the exercise under review is indeed a unitary project rather than a strategy or a programme of projects. A unitary project is typified by its definition in terms of a single clear mission statement, the ability to estimate a start and completion date and the resources necessary to undertake the project. If these characteristics are not present then consideration should be given as to whether the envisaged task is a strategy or a programme of projects. Projects are often nested within programmes which themselves make up a strategy, as illustrated in Figure 18.1.

A strategy can be defined in terms of a means–ends relationship where the means are the set of rules to guiding organisational decision makers and ends are the measurable objectives against which organisational performance can be quantified. Strategy is also seen as a cultural web, a set of collective beliefs shared by an organisation about the direction in which it is going (Johnson and Scholes, 1999).

A programme is a set of nested projects which have a common objective and are fixed within the means–ends relationship of the strategy. Value is associated with the strategy at an organisational level and reflects corporate values in the undertaking of an overall policy. Projects have specific values as a subset of strategic values which relate to an activity which will ultimately achieve strategic fit with the core business of the organisation.

18.2.3   The attributes of the project

Morris and Hough (1987) undertook a study of major projects internationally to determine those factors which, if not considered fully, could lead to project failure. The factors identified have been adapted for use in a whole-life value methodology as information necessary prior to the definition of project function and are as follows:

- Organisation: the identification of the client's business, the place of the project within the business, and the users of the project (who may not

Part Three

necessarily be a part of the client organisation). Under this heading there would be an investigation of the client's:

- □ Hierarchical organisational structure, and the client's key activities and processes that would impact the project
- □ Decision-making structures and how this will interface with the project teams and the communication networks anticipated for controlling the project
- □ Delegation of executive power to the project sponsor or project manager
- □ Future dynamics in the context of organisational change

- Stakeholder analysis: identification of all those who have a stake in the project. Stakeholders should be listed and their relative input/influence assessed.
- Context/culture: the context of the project should recognise such factors as culture, tradition, social aspects relating to the local population, the local environment and/or the relationship between the client and local organisations.
- Location: the location factors relate to the current site, proposed sites or the characteristics of a preferred site where the site has not to date been acquired. All projects, whether construction or service projects, will have a location.
- Community: it is important to identify the community groups who may require to be consulted with respect to the proposed project. Some market research may need to be undertaken to ascertain local perceptions. The positioning of the project within the local community should also be completely understood.
- Politics: the political situation in which the project is to be conceived should be fully investigated through the analysis of local government and central government policies and client organisational politics.
- Finance: at the formative stages of the project all options should be considered with regard to the financial structuring of the project. This will include traditional capital purchase, prudential borrowing and private finance and will impact the source of funding, the allocation of funding, and the effects of the project cashflow on the cashflow of the client organisation.
- Time: under this heading are the general considerations regarding the timing of the project including a list of the chronological procedures which must be observed in order to correctly launch the project. In situations where the project is to be phased, time constraints for each stage of the project should be recorded. This data becomes the basis of the construction of a time line diagram for the project.
- Legal and contractual issues: all factors which have a legal bearing on the project are listed under this heading including data relating to the client's ongoing partnership agreements with suppliers and contractors.
- Project parameters and constraints: it is important to understand the boundaries of the project and the constraints that will impact its development.
- Change management: all projects by definition will change the working practices of the client organisation. The activities involved in change

management are evaluating, planning and implementing, usually through education, training, communication, team and leadership development. At the inception of the project a change management plan should be developed which makes explicit the potential changes to the client organisation in terms of working practices and employment structures. There should also be a methodology for implementing change to the project as it progresses including the approval and organisational structures.

Once all of the above data has been gathered and appraised then the primary and secondary functions of the project should be plain to see. The project, having been defined in functional terms, is then conducive to the generation of options to satisfy the functional requirements. The options are then judged in terms of their value to the client and their whole-life cost. These two evaluations combine to form a whole-life value evaluation.

18.3   Client Value System

The client value system is a description and ordering of those facets which will subsequently be used to judge the success of the project and therefore need to be made explicit at its inception. It is suggested that the procedure involves the construction of the diagram shown in Figure 18.2. The stages in the construction of the diagram are:

1. Identify the client: the diagram to be constructed represents the views of the client corporate, therefore those constructing the diagram should be the client stakeholders involved with the project's outcomes once it becomes a part of the client's mainstream business activity. This does not include those consultant advisers to the client employed to enable the construction of the project.
2. Decide the value criteria: the diagram will typically be composed of up to nine variables proved to be the key criteria against which client value relationships can be made explicit (Kelly, 2007). Additional value criteria suggested by stakeholders should be examined to determine that they are in no way correlated with one of the nine variables. Ease of maintenance, for example, may be suggested as an additional value criterion but this would be highly correlated with OPEX and correlated to a lesser degree with environmental impact and comfort. Value criteria within the list should be explored for relevance; for example, exchange may not be relevant to a hospital that is to remain in public ownership. The nine variables are:
   - Capital costs (CAPEX) are all costs associated with the capital costs of the project. In some situations, particularly PFI, the capital investment is subsumed within the operating cost and therefore the capital cost variable is omitted. In other projects, particularly in the public sector, the budget is fixed. In this situation the question has to relate to the space provided, for example, are you (the client) willing to reduce functional space to achieve a more expensive environmentally friendly solution. If the answer is yes then 'E' would be inserted in

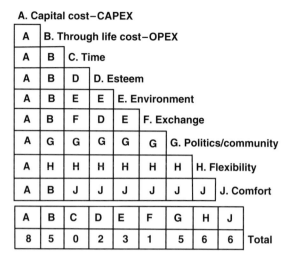

Figure 18.2 Paired comparison matrix for a special school project reflecting the client's value system.

the matrix illustrated in Figure 18.2. In the illustrated situation the answer was no and therefore 'A' is inserted in the matrix.

- Operating cost (OPEX) refers to all costs associated with the operations and maintenance implications of the completed project as it moves to an operational product within the client's core business. In the context of a building this includes facilities management which may be limited to maintenance, repairs, utilities, cleaning, insurance, caretaker and security, but may be expanded to include the full operational backup such as catering, IT provision, photocopying, mail handling and other office services.
- Time in this context is specifically the length of time between the present, i.e. contemporary with the client value system exercise, and the point in time when the project is complete and is absorbed into the core business of the client. Most commonly this latter date is referred to as the date of practical completion.
- Environment refers to the extent to which the project results in a sympathetic approach to the environment, measured by its local and global impact, its embodied energy, the energy consumed through use and other 'green' issues.
- Exchange or resale is the monetary value of the project. This may be viewed as assets on the balance sheet, the increase in share value, capitalised rental or how much the project would realise were it to be sold.
- Flexibility represents the extent to which project objectives require that the design has to reflect a continually changing environment. Flexibility is generally associated with changing technology or organisational processes or both.
- Esteem is the extent to which the client wishes to commit resources for an aesthetic statement or portray the esteem of the organisation, internally and externally.

Part Three

- Comfort is the physical and psychological comfort of the building as a place for working and living.
- Politics is an external dimension that refers to the extent to which community, popularity and good neighbour issues are important to the client.

3. Organisational or project values: in many situations it is necessary to undertake the value exercise with the project characteristics set to one side, i.e. to obtain the value criteria for the organisation. Following this the question can be asked, 'How does this vary for the project?' The reason for doing this is that the particular project may have value criteria that may be different to the core business organisation.

4. Undertake the paired comparison by asking the questions in turn, for example, which is more important to you CAPEX or OPEX. If CAPEX is more important 'A' is inserted in the appropriate box as shown in Figure 18.2.

5. To calculate the client's value system of the project the number of A's, B's, etc. are summed and the total entered in the total box.

6. The rank order of the variables represents the client value criteria. For example in Figure 18.2 the client is stating that the building must be provided with the required functional space, within the budget, it must be flexible and comfortable. Of less importance are factors such as the earning capacity of the building and the time taken to realise the project. It is important to note that the numbers in the totals box are ordinal and not interval values. This means that care has to be observed in using the numbers as numbers in later exercises.

The example shown in Figure 18.2 was derived at a value management workshop charged with examining the brief for a special school to replace an existing school. The outcome is the value criteria of the client which are the success criteria against which the project will be judged on completion. The method is explicit and auditable. At the workshop the procurement method was confirmed as a two-stage tender with a design–build contractor brought on board at an early stage to confirm the economic viability of the functional requirements. The contractor would be required to agree to a contract based upon a guaranteed maximum price.

The explicit statement of the client's value system is used as one part of the evaluation to determine whole-life value of competing options generated in answer to the functional requirements of the project. The second element to cost the options is undertaken by reference to whole-life costing.

18.4   Whole-Life Costing

The objective of this section is not to undertake a traditional or holistic literature review of whole-life costing but to examine the literature specifically to identify common and primary issues which have relevance to the option appraisal of solutions generated in answer to the functional requirements of the project. Common issues are those debated in all of the reviewed literature

Part Three

and primary issues are those which are fundamental to whole-life costing to the extent that the process would fail without their incorporation.

Ruegg *et al.* (1980) state that from the perspective of the investor or decision maker all costs arising from the investment decision are potentially important to that decision and that those costs are the total whole-life costs and not exclusively the capital costs. Another important premise is that whole-life costing is concerned with evaluating alternatives. Ruegg *et al.* outline five basic steps to making decisions about options:

1. Identify project objectives, options and constraints.
2. Establish basic assumptions.
3. Compile data.
4. Discount cash flows to a comparable time base.
5. Compute total lifecycle costs, compare options and make decisions.

The basic assumptions referred to are related to the period of study, the discount rate, the level of comprehensiveness, data requirements, cash flows and inflation.

18.4.1   Costs

Marshall and Ruegg (1981) give recommended practice for measuring benefit-to-cost ratios and savings-to-investment ratios based on the five-step process above. In 1986 the Quantity Surveyors Division of the RICS produced a guide which listed the costs to be included within a whole-life cost calculation. All expenditure incurred by a building and during its life were described as:

- Acquisition costs: total cost to the owner of acquiring an item and bringing it to the condition where it is capable of performing its intended function.
- Disposal costs: total cost to the owner of disposing of an item when it has failed or is no longer required for any reason.
- Financing costs: cost of raising the capital to finance a project.
- Maintenance costs: cost of maintaining the building, to keep it in good repair and working condition.
- Occupation costs: costs to perform the functions for which the building is intended.
- Operating costs: costs of, for example, building tax, cleaning, energy, etc. which are necessary for the building to be used.

18.4.2   Life

In the RICS guide, life is defined as the length of time during which the building satisfies specific requirements described as:

- Economic life: a period of occupation which is considered to be the least-cost option to satisfy a required functional objective.
- Functional life: the period until a building ceases to function for the same purpose as that for which it was built.

- Legal life: the life of a building, or an element of a building until the time when it no longer satisfies legal or statutory requirements.
- Physical life: life of a building or an element of a building to the time when physical collapse is possible.
- Social life: life of a building until the time when human desire dictates replacement for reasons other than economic considerations.
- Technological life: life of a building or an element until it is no longer technically superior to alternatives.

Flanagan *et al.* (1989) state that two different timescales are involved in whole-life costing: firstly the expected life of the building, the system or the component; and secondly the period of analysis. Flanagan states 'it is important when carrying out any form of life cycle costing to differentiate between these two timescales, since there is no reason to believe that they will be equal: for example the recommended period of analysis for federal buildings in the US is 25 years, considerably less than any reasonable building life'. This introduces a further element to the above list, namely the period of study.

Ruegg and Marshall (1990) confirm seven study periods:

1. The investor's holding period: the time before selling or demolishing.
2. The physical life of the project: specifically relating to equipment.
3. The multiple lives of options: recognising that options having exactly the same total costs over one period of time will have different total costs if the cash flows are taken over different periods due to replacement and maintenance occurring at differing points in time.
4. Uneven lives of options: recognising that where alternatives have different lives and cash flows then residual values have to fully compensate particularly over short study timeframes. A note is also made of the dangers of using annual equivalent discount models where alternatives have uneven lives.
5. Equal to the investor's time horizon: the period of interest the investor has in the building.
6. Equal to the longest life of alternatives.
7. The quoted building life.

Kelly and Hunter (2005) recommend that a whole-life cost calculation should not extend beyond 30 years. This reflects the view of the authors that buildings change significantly both functionally and economically within a 30-year period to the extent that the costs and functions known at year zero (Figure 18.3) cannot reflect those costs and functions in 30 years hence. Examples are given for retailing, which has changed significantly within 30 years, and healthcare, which is practised entirely differently today from how it was practised in 1977. The exception may be housing.

18.4.3   Data

Kelly and Hunter (2005) and Flanagan and Jewel (2005) cite the same basic data sources as: data from specialist manufacturers suppliers and contractors; predictive calculations from model building; and historic data. All

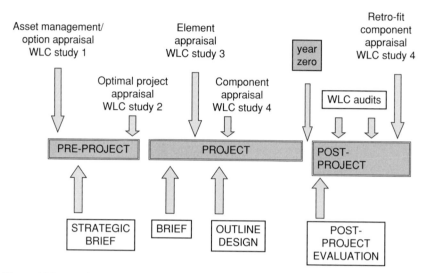

Figure 18.3 Application of whole-life costing at differing stages in the evolution of a project.

authors highlight the danger associated with data used for whole-life costing. Flanagan and Jewel state:

- Data are often missing.
- Data can often be inaccurate.
- People often believe they have more data than actually exists.
- It can be difficult to download data for subsequent analyses and for data sharing by a third party.
- There will be huge variation in the data, sometimes for the same item.
- Data are often not up to date.
- Data input is unreliable: the input should be undertaken by those with a vested interest in getting it right.

Both Kelly and Hunter and Flanagan and Jewel quote the UK Office of Government Commerce (OGC, 2003), which states that it is important to focus on future trends rather than compare costs of the past. Where historic data is available it may provide misleading information, such as the past mistakes in the industry and focusing on lowest price. Irrespective of whether or not historic cost information is available, it is always preferable to estimate the cost from first principles and only to use historical cost information as a check.

18.4.4   Discount rates

Ruegg and Marshall (1990) consider in detail the discount rates to use, citing business discount rates for commercial decisions and public discount rates for public decisions. They also introduce the theory of risk adjusted discount rates. Boussabaine and Kirkham (2004) take this further and introduce methods of assessing and blending the risk methodology with whole-life cost calculations.

The evidence from the literature in the context of this chapter gives support to the development of whole-life costing taking account of all relevant costs, over a given time period for all options being considered, using contemporary data, with appropriate discount rates and taking into account risk. In the context of time a 30-year limit is considered appropriate for all calculations with a formula to take account of residual values. The periods of up to 30 years will commence at year zero defined as the date of occupation by the user, see Figure 18.3.

The standard, ISO/FDIS 15686-5: 2006(E) Buildings and Constructed Assets – Service Life Planning Part 5 – Life Cycle Costing, in its draft form, has the objective of helping to improve decision making and evaluation processes, at relevant stages of any project. Other key objectives are to 'make the life-cycle costing assessments and the underlying assumptions more transparent and robust' and 'provide the framework for consistent life-cycle costing predictions and performance assessment which will facilitate more robust levels of comparative analysis and cost benchmarking'. These three objectives, out of 14 listed, are considered the most important in the context of this chapter. The standard describes lifecycle costing as 'a valuable technique which is used for predicting and assessing the cost performance of constructed assets'.

The standard describes three levels of application:

1. Strategic level, relating to the structure, envelope, services and finishes
2. System level (elemental level), relating to floor wall and ceiling finishes, energy, ventilation, water capacity, communications, cladding, roofing, windows and doors, foundations, solid or framed walls and floors
3. Detail level (component level) for example ceiling tiles, floor coverings, electrical and mechanical plant, etc.

This is a useful categorisation but it ignores the level of asset management which is described elsewhere in the standard as 'life-cycle costing is relevant at portfolio/estate management, constructed asset and facility management levels, primarily to inform decision-making and comparing alternatives. Life-cycle costing allows consistent comparisons to be performed between alternatives with different cash flows and different time frames. The analysis takes into account relevant factors throughout the service life, with regard to the clients' specified brief and project specific service life performance requirements.'

18.4.5   Summary of the review of whole-life value

Whole-life value of projects has two primary stages: the definition of the project and the generation of options to satisfy the project; and the evaluation of the options to derive the best VFM solution. The section above describes the theoretical understanding of the project in functional terms, the generation of technical options to satisfy the functional requirements and methods of assessing the value and cost of each option. The next section applies this theory to the PPP/PFI process.

Part Three

18.5   The Application of Whole-Life Value to the PPP/PFI Process

The inception of public sector projects occurs from a demand for enhanced or rationalised services arising from either a population change, a strategic change delivered by the vote, central government legislation, local legislation, rising expectations, a response to a quality audit and/or cultural change. Whatever triggers the project, it will go through the following four-stage processes characteristic of all projects:

1. Recognition, strategic planning and business definition
2. Project planning and the establishment of systems
3. Tactical design of the component parts of the project
4. Acceptance into core business

The identification of the four-stage process is useful both from the perspective of the logic of the PPP/PFI process and from the application of appropriate whole-life value techniques. The following section describes the whole-life value processes applicable to the appropriate stage and actions involved in PFI.

18.5.1   Phase one – asset management

CLAW (2003) define asset management as 'optimising the utilisation of assets in terms of service benefits and financial return'. In the context of planning, asset management makes explicit the resources invested in property assets and promotes strategies and programmes which make best use of the assets in terms of the efficiency and effectiveness of the services supported. A principal aim is to minimise the opportunity cost of resources embedded in land and buildings. The UK has a public sector asset base of approximately £658bn (2004) to support public services (HM Treasury, 2004a).

The assets held can be described typically as:

- Those supporting direct service, e.g. housing, schools, residential care homes, waste management sites, etc.
- Those supporting the administration of services, mainly offices and maintenance/vehicle depots
- Non-operational property awaiting sale or utilisation
- Infrastructure, mainly roads and public open space

Existing assets should be the minimum necessary to support the delivery of services in the most efficient and effective manner. The asset management plan should be detailed for each asset and show the projected spend profile over a given period of time. The asset management plan is an important constituent of the option appraisal exercise necessary in the development of the outline business case to support a PPP/PFI project.

The Audit Commission (2000) states 'property tends to be expensive to acquire, inflexible in use, time-consuming to manage and costly to run. As such, it *should* receive significant corporate attention'. Criticism is made of the public sector for its lack of good asset management compromising financial and managerial decisions. Specifically cited is that property is not regarded as a strategic resource, and its attributes are neither well defined nor recorded.

The lack of an asset management plan compromises but does not invalidate the option appraisal exercise.

18.5.2   Phase two – need for a project identified

The fact that a project may be necessary for one of the reasons outlined above triggers the first stage in the whole-life value exercise. This first stage requires an exercise to determine:

- Confirmation that what is being perceived as a project is indeed a unitary project and not a strategy or programme of projects
- The definition of the project in explicit functional terms
- The value criteria against which the project will be judged

HM Treasury (2004b) refers to programmes and projects collectively; however the view taken in this chapter is that the focus should be on the evaluation of the project. Where a number of projects comprise a programme then the evaluation of the programme should be based upon the summation of the values for each project. A programme should not be evaluated in the absence of knowledge of the constituent projects.

At the end of phase two the need for a project has been identified, the relevant data has been uncovered and the project has been made explicit in functional and value terms. The next stage is to consider the options available to answer the functional requirements of the project. The options will be evaluated using the value criteria and whole-life costing techniques.

18.5.3   Phase three – options appraisal and the outline business case

The outline business case is defined as that early stage in the project's development when: the strategic context can be described; the information impacting the project has been made explicit; the mission of the project has been established and agreed; and therefore options for the proposed facility or service can be generated. The outline business case provides the justification for the choice of a specific option having evaluated all competing options in value and cost terms. All options will be considered against performance criteria which, when met at the lowest cost, provide optimal VFM (Kelly and Hunter, 2005). The methodology proposed here is conducive with the guidance on VFM issued by HM Treasury (2004b).

Generation of options

With the project's strategic context and functional mission made clear, radical options are developed. Many options are generated, most effectively through a formal brainstorming session, and recorded for evaluation. In order to test the parameters of feasible solutions, restrictions are not applied within the brainstorming session and any idea is considered valid.

Evaluation

The options generated are evaluated with reference to the value system. A structured evaluation may use the value system as a part of a structured weighting and scoring system in order to select the most promising options and thereby keep the appraisal process manageable. This logical process is designed to prevent the elimination of the optimal solution before it is given full consideration (HM Treasury Green Book).

Development

The most promising options are evaluated further using whole-life costing applied at an elemental level. HM Treasury Green Book, Chapter 5, states that costing for option appraisal, including that used to determine the balance between investment and benefits, should specifically take account of:

1. Relevant opportunity costs: it is important to explore those opportunities that may exist within existing assets, for example an opportunity within an option being considered may be to use land in a different more valuable way than its current use.
2. Costs already incurred, which are irrevocable or sunk should be ignored even for the purposes of an appraisal.
3. Depreciation and capital charges should not be included in the appraisal of whether or not to purchase the asset which gives rise to them.
4. Residual and terminal values should be included and tested for sensitivity.
5. Options that expose the client to contingent liabilities, e.g. options which may involve redundancy payments to staff, are to be researched to determine the extent of such liability and the probability of it occurring.

The option appraisal evaluations should distinguish between:

- Fixed costs, i.e. those that remain constant over wide ranges of activity for a specified time period e.g. an office building.
- Variable costs, i.e. costs that vary according to the volume of activity e.g. manpower costs.
- Semi variable costs, i.e. those that include both a fixed and variable component but are characterised by linear progression.
- Step costs, i.e. those costs that are fixed for a given level of activity but eventually increase by a given amount at some critical point, e.g. a cohort of 33 children requires one teacher and one classroom but a cohort of 34 may require two teachers and two classrooms.

Discounting is used to convert all future costs and benefits to 'present values', so that they can be compared. The current (July 2007) recommended discount rate is 3.5%. Calculating the present value of the differences between the streams of costs and benefits provides the net present value of an option. HM Treasury Green Book states that for projects with very long-term impacts of over 30 years, a declining schedule of discount rate should be used. However, as rehearsed above, the view put forward here is that value in an

Table 18.1   Recommended adjustment ranges for optimism bias (HM Treasury, 2003).

| | Optimism bias (%) | | | |
| | Works duration | | Capital expenditure | |
Project type	Upper	Lower	Upper	Lower
Standard buildings	4	1	24	2
Non-standard buildings	39	2	51	4
Standard civil engineering	20	1	44	3
Non-standard civil engineering	25	3	66	6
Equipment/development	54	10	200	10
Outsourcing	N/A	N/A	41	0

element remaining at year 30 should be transmuted to a residual value and included in the evaluation as such.

Public sector comparator

In the context of PPP/PFI the outcome of the option appraisal exercise is a recommendation of a preferred option based upon the outcome of two evaluations, one based upon the value criteria determined to measure success and the other based upon the net whole-life cost. This preferred option is the public sector comparator. The costs included in the whole-life cost study at the outline business case stage will be subject to an addition for optimism bias. This adjustment, based upon historic data of the accuracy of initial estimates, is added to the final estimated whole-life costs (Table 18.1).

Exemplar design

An exemplar design is a design undertaken by a design team commissioned by the client to accompany the design brief to be sent to prospective bidders. An exemplar design brings a level of certainty to the process and permits whole-life value to be illustrated and whole-life cost to be more accurately calculated. There is an argument for a reduction in optimism bias where whole-life cost figures can be related back to an exemplar design. The exemplar design informs bidders of what is acceptable to the client as an interpretation of the design brief but does not preclude innovation, particularly technical innovation on the part of individual bidders.

Approval of the outline business case

The approval of the outline business case confirms that:

1. The project is best procured through a PPP/PFI procurement route.
2. That a market for the PPP/PFI project exists and is competitive.
3. That the whole-life value criteria and evaluations are robust.

Part Three

4. The public sector comparator has done its job and should not be referred back to once the outline business case has been approved.

18.5.4   Phase four – design development

In the context of a PPP/PFI project the design development stage is undertaken by the bidders (after BAFO the preferred bidder). It is at this stage that detailed whole-life costing is carried out on the competing options to meet the technical criteria specified by the project brief. Whether or not the elemental whole-life cost calculations undertaken at outline business case stage are made available to bidder is a moot point. Its usefulness would be as a reference and control document to enable the bidder to develop appropriate solutions which would meet the value criteria at an appropriate price.

The preferred bidder moving to contract close will use the whole-life cost calculation to:

1. Enable smoothing of cashflow during the concession period.
2. Ensure sufficient unitary charge is received before incurring significant maintenance expenditure.
3. Enable sensitivity analysis to be carried out to understand and minimise risk.
4. Ensure that an optimal position is taken on the selection of components to meet the specification.

18.6   Discussion

Whole-life value as a concept and as a methodology has assumed a much greater significance since the publication in 2004 of documents by HM Treasury which have elevated the outline business case to the stage at which decisions on value, budget and procurement are taken. The fact that the public sector comparator ceases to be a reference document after the outline business case is significant. This brings into focus the importance of whole-life value and its two primary activities:

1. The definition of the project and the generation of options to satisfy the project.
2. The evaluation of the options to derive the best value for money solution.

These two primary activities have four distinct attributes:

1. The definition of the project and the place of the project within the strategies and programmes of a client organisation. Although HM Treasury treat programmes and projects collectively it is the view here that all evaluations must be carried out at a project level. Individual projects are summed to achieve an evaluation of a programme of projects.
2. The definition of the project in explicit functional terms. This is a requirement which allows options to be derived by brainstorming. HM Treasury make the valid point that brainstorming should not be constrained such

that the options become a variation on a theme but be widely sought to define the boundary of possibilities.

3. The value criteria by which the project will be judged a success. This is fundamental to the value part of the whole-life value solution. A pragmatic method by paired comparison is discussed.

4. The method of calculation for determining which of the competing options best satisfy the functional values defined. This represents the quantitative whole-life cost part of the whole-life value equation. The methods used at the outline business case stage will be focused on the evaluation of options at construction elements level. Later during design development the focus will be on components of construction representing the technical solution.

The whole-life costing exercise requires decisions to be taken regarding costs, the life of the study, the discount rates to be used, the data to be derived, and an understanding of what is appropriate at each stage of the project. All these issues are discussed in the main body of this chapter.

The final issue to be addressed is the reason for doing whole-life value and whole-life costing exercises. The exercises are undertaken to ensure best VFM. They are a means of judging between options and an auditable method of justifying decisions reached. The methodology has no other function.

References

Audit Commission (2000) *Hot Property: Getting the Best From Local Authority Assets.* http://www.audit-commission.gov.uk/Products/NATIONAL-REPORT/910C1FD0-F8D4-486C-8EBD-0C680929E154/archive_nrhotprp.pdf. (Accessed June 2007)

Boussabaine, A. and Kirkham, R. (2004) *Whole Life-Cycle Costing: Risk and Risk Responses.* Blackwell, Oxford.

CLAW (2003) *Supplementary Guidance to Asset Management Planning in Wales,* http://www.claw.gov.uk/fileadmin/claw/Asset_Management_Guides/CLAW_AMP_Supp_Guidance_English_.doc. (Accessed June 2007)

Flanagan, R. and Jewel, C. (2005) *Whole Life Appraisal for Construction.* Blackwell, Oxford.

Flanagan, R., Norman, G., Meadows, J. and Robinson, G. (1989) *Whole Life Costing – Theory And Practice.* BSP Professional Books, Oxford.

HM Treasury (2003) *Supplementary Guidance on the Treatment of Optimism Bias.* http://www.hm-treasury.gov.uk/media/D/B/GreenBook_optimism_bias.pdf. (Accessed June 2007)

HM Treasury (2004a) *Towards Better Management of Public Sector Assets.* http://www.ogc.gov.uk/documents/Towards_better_management_of_public_sector_assets_-_Sir_Michael_Lyons.pdf. (Accessed June 2007)

HM Treasury (2004b) *Value for Money Assessment Guidance.* http://www.hm-treasury.gov.uk/media/95C/76/95C76F05-BCDC-D4B3-15DFDC2502B56ADC.pdf. (Accessed June 2007)

HM Treasury (undated) *Green Book.* http://www.hm-treasury.gov.uk/economic_data_and_tools/greenbook/data_greenbook_ index.cfm. (Accessed June 2007)

Johnson, G. and Scholes, C. (1999) *Exploring Corporate Strategy: Text and Cases*, 5th edn. Prentice Hall Europe, Hemel Hempstead.

Kelly, J. (2007) Making client values explicit in value management workshops. *Construction Management and Economics Volume*, 25(4), 435–442.

Kelly, J. and Hunter, K. (2005) *A Framework for Whole Life Costing*. SCQS, Huddersfield.

Kelly, J., Male, S. and Graham, G. (2004) *Value Management of Construction Projects*. Blackwell Publishing, Oxford.

Marshall, H.E. and Ruegg, R.T. (1981) Recommended Practice for Measuring Benefit/ *Cost and Savings-to-Investment Ratios for Buildings and Building Systems*. US Dept of Commerce, National Bureau of Standards.

Morris, P.W.G. and Hough, G.H. (1987) *The Anatomy of Major Projects: A Study of the Reality of Project Management*. Wiley, Chichester.

OGC (2003) *Whole-life Costing and Cost Management*, Procurement Guide Number 7, Achieving Excellence in Construction, London.

Royal Institution of Chartered Surveyors (1986) *A Guide to Life Cycle Costing for Construction*. Surveyors Publications, London.

Ruegg, R. and Marshall, H. (1990) *Building Economics: Theory and Practice*. Van Nostrand Reinhold, New York.

Ruegg, R.T., Petersen, S.R. and Marshall, H.E. (1980) *Recommended Practice for Measuring Life-Cycle Costs of Buildings and Building Systems*. US Dept of Commerce, National Bureau of Standards.

Part Three

19

Best Value Procurement in Build Operate Transfer Projects: The Turkish Experience

Irem Dikmen, M. Talat Birgonul and Guzide Atasoy

19.1   Introduction

The BOT approach was developed at the end of the 1970s as a way to acquire necessary infrastructure investments for the developing countries with limited borrowing capacity and budgetary constraints. The concept was first initiated by Turgut Ozal, Prime Minister of Turkey, in 1984 as a part of the Turkish Privatisation Programme. The BOT model provides extensive benefits for the government in the implementation of their infrastructural development programme with the minimum possible financial burden and risks, and in the reduction of unit cost of services through new technology and the private sector's innovative management techniques. The advantages of BOT for the public sector are summarised by Li and Akintoye (2003) as: enhancing the government's capacity to develop integrated solutions; facilitating creative and innovative approaches; reducing the cost to implement the project; reducing the time to implement the project; transferring certain risks to the private project partner; attracting larger potentially more sophisticated bidders to the project; and accessing skills, experience and technology. However, unsuccessful implementations of the model have demonstrated that benefits are attainable only if some country factors are positive (economical and political stability, mature legal frameworks) or the government provides guarantees for the investors to reduce risks associated with negative country factors.

Many urgent energy and transportation projects using the BOT model could not be realised for several reasons: lack of adequate legislation; poor organisation of governmental agencies in packaging projects; ineffective tendering and evaluation procedures employed by client organisations; lack of coordination between private and public sectors; and the unwillingness of the Turkish government to provide guarantees against risks originating from the unstable economical and political environment experienced in Turkey (Ozdoganm and Birgonul, 2000). Canakci (2006) states that insufficient legal framework, administrative difficulties and lack of a systematic approach about risk allocation between the public and private sectors are the major factors that bar the way of BOT projects in Turkey.

In 1984, Turkey established Law No: 3096, one of the first legal arrangements in the world to organise the participation of private sector in infrastructure investments. Based on this law, BOT arrangements were put into practice within the energy sector, a high-priority issue at that time. Subsequently, in 1988, Law No: 3465 was brought into force. Under this legislation 21 highway service stations were constructed using the BOT approach. With Law No: 3996 (1994), the legislative system concerning the implementation of all kinds of investments and services within the context of BOT model has been facilitated. Moreover, with the introduction of this new legal arrangement, governmental guarantees have been organised in a detailed manner for the first time.

BOT would seem to account for purely privately financed projects but in reality this is not the case. No bidder volunteers to become involved with a BOT project without financial commitment from the host government. This fact enabled the successful utilisation of several BOT agreements in energy and airport passenger terminal projects. In the energy sector, the government has guaranteed that if a certain level of demand is not reached, they will buy the excess energy. So far 18 hydroelectric power plants have been constructed and seven projects are still in progress with the BOT model. Whilst four airport passenger terminals have been constructed and put into service by the private sector, the number of successfully realised BOT projects is rather low in the transportation sector. Currently, the General Directorate of Highways (GDH) has only implemented the BOT model in the Gocek Tunnel Project, one of the case studies described later in this chapter. According to the Ministry of Public Works and Settlement, a third Bosphorus Bridge and Istanbul-Bursa-Izmir Highway are in the agenda of the Turkish government for tendering based on a BOT approach in the near future.

The poor performance of the BOT model in the transportation sector can be attributed to the uncertainty of traffic demand in the future and lack of government guarantees as well as poor tender evaluation mechanisms. To increase the success of BOT projects, it is felt that a framework that systematically allocates risks between the parties and an effective tender evaluation mechanism that takes into account all project success criteria will be required. In this chapter, a methodology will be proposed to facilitate the evaluation of tenders in the transportation sector. The two case studies presented will demonstrate that:

- The success of a BOT project significantly depends on the right tender evaluation method.
- There is no single formula that will work in all BOT projects. The appropriate methodology depends on project features like size and technical complexity as well as the expectations of the government from the project (technical innovation, minimum cost etc.).
- 'Best value' procurement that takes into account costs, risks, required level of government guarantees and capability of concessionaires should be used to select the best offer.

19.2   Bid Evaluation in BOT Projects

To provide the highest benefits for the public it is of vital importance to select a project suitable for PPP. Ashley *et al.* (1998) developed a project scoring table based on nine high level evaluation criteria to assess the suitability of a project for PPP. These criteria are grouped into nine clusters:

1. Political clearance
2. Partnership structure
3. Project scope
4. Environmental clearance
5. Construction risk allocation
6. Operational risk allocation
7. Financing package
8. Economic viability
9. Developer financial involvement

After the viability of the project is proved, the next step is to award the bid to the most capable contractor. Selection of the best contractor requires a sophisticated bid evaluation system in which the criteria that describe the 'best value' are clearly identified.

In the literature, different methods have been suggested to evaluate the competitive tenders. Zhang (2004) summarised these methods as follows:

- *Simple scoring method*: evaluation criteria and maximum possible scores are determined with each criterion assumed to have equal importance. Each bidder is rated according to these criteria and the bidder with the highest total score is awarded the project.
- *Net present value (NPV) method*: the bidder offering the lowest NPV for the concession period (i.e. the lowest cost to the public) is selected. Using this method only the financial and economic aspects of each tender are considered.
- *Multi-attribute analysis*: criteria are decided in the same way as for the simple scoring method, but each of these factors is divided into sub-categories with relative importance weights assigned. After multiplying the weights and the assigned scores of each bidder, the bidder with the highest maximum score is selected.
- *Kepner-Tregoe decision analysis technique*: this technique evaluates proposals based on criteria identified as 'musts' and 'wants'. The 'musts' are the mandatory needs for the project and are expressed in the form of 'yes/no' questions. Bidders satisfying the 'musts' are then evaluated based on the 'wants' using a simple scoring or multi-attribute scoring method.
- *Two envelope method*: bidders are expected to submit two different envelopes; the first providing technical information with the second providing cost information. Initially the technical offers are evaluated and then, for those approved, the financial envelope is opened. If the cost is within the acceptable range as defined by the client, that bidder is chosen.

Part Three

- *NPV and scoring method*: with this approach two different evaluations are undertaken. NPV is used for financial evaluation and the scoring method is then used for evaluation of any unquantifiable information.
- *Binary and NPV method*: bidders are first evaluated with 'musts' criteria and those passing this step are then evaluated according to their NPVs.

Every country selects the method most suitable for their needs and expectations. For example, the simple scoring method has been used in PPP transportation projects in California, multi-attribute analysis has been used in PFI projects in the UK, and Kepner-Tregoe decision analysis has been used in BOT tunnel projects in Hong Kong (Zhang, 2006). The success of the evaluation system largely depends on the selection and utilisation of the right criteria in the evaluation process.

Zhang and Kumaraswamy (2001) identified the main tender evaluation criteria of the Hong Kong government as:

- The level and stability of the proposed toll regime
- The proposed methodology for toll adjustments
- The robustness of the proposed works programme
- The financial strength of the tenderer and its shareholders, their ability to arrange and support an appropriate financing package, and the resources they are able to devote to the project
- The structure of the proposed financing package including the levels of debt and equity, hedging arrangements for any interest rate and/or currency risks, and the level of shareholders' support
- The proposed corporate and financing structure of the franchisee
- The quality of the engineering design, environmental considerations, construction methods, including traffic control, surveillance, electrical and mechanical installation, ventilation and lighting systems
- The ability to manage, maintain and operate effectively and efficiently
- The benefits for the government and community

Referring to the California toll roads, Levy (1996) suggested that the best option should be selected by considering: the degree to which the proposal encourages economic prosperity; the degree of local support for the project; the relative ease of proposal implementation; the experience/expertise of sponsors and support team; the support for environmental quality and energy conservation; the degree to which non-toll revenues support proposal costs; the degree of technical innovation displayed in the proposal; and the degree of support for achieving civil rights objectives. Although all of these criteria reflect government expectations and may act as a strong foundation for tender evaluation, some difficulties may be faced in practice due to the unavailability of objective information while assigning ratings to the determined criteria. Whilst the evaluation will be based on many factors which can be expressed in monetary terms (such as toll rate), others can only be evaluated subjectively (such as level of innovation). Apart from the difficulty of assessing the importance of a factor and rating of an alternative with respect to a given factor, 'interrelations between the factors' pose a further challenge. For example, 'level of experience' affects most of the risk parameters, however, its

Figure 19.1 Map of Turkey.

impact should be reflected in the ratings rather than treating it as a separate parameter. Moreover, assumptions that underlie bid proposals, particularly those regarding government guarantees and risk allocation principles, should be checked before arriving at a decision on the best option.

Two extreme case studies are now presented to demonstrate the complexity of the evaluation process and how the success of a BOT project may be affected by the choice of evaluation criteria. If a single parameter is used to determine the best option, typically the maximisation of NPV or minimisation of the operation period, then the evaluation of proposals will be a relatively easier task. However, if the investors are given free rein concerning the construction method, risk-allocation/-sharing schemes and duration of the operation period in their proposals, their evaluation can turn out to be a highly complex process.

19.3   Case Studies

The first attempted implementation of a BOT approach by GDH was the Izmit Bay Crossing project in 1994. Unfortunately this project was cancelled due to lack of preliminary design and adequate information, legislative problems and an inadequate tendering process. After this failure, the second trial in the transportation sector was the successfully completed Gocek Tunnel project. The geographical locations of these two projects are shown on the map of Turkey in Figure 19.1.

Case Study 19.1: Izmit Bay Crossing Project

In 1994, the Turkish Government announced the Izmit Bay Crossing project and in June 1994 the GDH issued a pre-qualification document that outlined the scope of the project and qualifications necessary to be considered for the GDH's shortlist. These prequalification criteria for participating companies were:

Table 19.1 Requirements/'musts' set by GDH.

	Requirements	Explanations
Financial	Participation in expropriation	The selected company should cover $30m of the expropriation cost
	Non-recourse to government	The bidder does not have the right to recourse the financing to government
Technical	Requirements	The technical requirements specified in the tender document should be satisfied such as the bridge width or design speed
Legal	Equity/debt ratio	20/80 ratio should be satisfied
	Arbitration	Disputes between the parties should be settled by the Turkish courts, international arbitration is not allowed

- Proven experience in the design and construction of major infrastructure projects, particularly on long-span bridges constructed in seismic zones similar to Izmit Bay
- Proven technical experience and administrative capability in managing major transportation projects
- Necessary financial strength and ability to secure a sound financial package
- Experience and managerial capability in traffic management, operation and maintenance of tolled highways and bridges

In April 1995, six companies complying with these requirements were shortlisted. In February 1996, these companies were invited to submit bids. Being uncertain which legal regulation should be followed the Japanese, Italian, American and French consortiums did not submit their bids. Whilst one bid was accepted, this was later retracted as there were too few bidders for a fair comparison. In December 1996, the process was repeated, however this time only three companies provided bids.

When selecting the best offer, GDH considered the overall viability of the bid, the financial liability on the government, and the 'musts' given in Table 19.1. Although they were not announced before the submission of tenders and no explanation was given about their relative importance weights, the GDH considered the following criteria when evaluating the bids:

- Technical viability: construction, operation and maintenance, seismic, environmental, technological and accident risks
- Financial viability: financial package structure, cost estimation method, toll rate and its structure, construction and operation periods
- Required guarantees: subordinated loan, no second facility, political risk, revenue/demand, senior loan, international arbitration guarantees and tax exemption
- Company-related factors: expertise, reputation and soundness of the bid

The Izmit Bay Crossing project consisted of the bay crossing and motorway that would be constructed on both sides of the bay. The government did not specify the construction method for the crossing except for the technical requirements such as width and design speed. Three alternative methods offered by the bidders were divided into multiple schemes according to the construction method to be used and inclusion of the motorway. For example, one company proposed 14 alternatives based on two different construction methods, whether a motorway would be constructed or not, and varying lengths of the motorway. The designs offered by the bidders for this project included a tube tunnel, suspension bridge and cable-stayed bridge. The bids fell within the range of $937m to $1.41bn, depending on the proposed construction technique and alternative routes

Table 19.2 Evaluation scheme and comparison of two alternatives.

Criteria	Alternative A	Alternative B
Technical viability	High	Low
Financial structure	Satisfactory	Not satisfactory
Toll rate:		
for automobile	$11	$9
for truck	$48.85	$45
Toll rate structure	Not clear	Clear
Construction + operation period	27 years	22 years
Guarantees asked	Low liability to the government	High liability to the government
Expertise	Satisfactory	Not satisfactory
Reputation	Good	Good
Soundness of the bid	Realistic/consistent	Unrealistic/inconsistent

employed. It was a difficult task to compare the alternatives due to the diversity of technical solutions, and different operation periods and toll structures. A sophisticated design could result in higher construction costs that would need to be supported by higher toll rates whereas a low bid might result in lower toll rates but produce a lower-quality structure (Levy, 1996).

To demonstrate the complexity of the evaluation process, data from two offers is shown in Table 19.2. For commercial confidentiality reasons, the names of the companies and the chosen alternatives are withheld.

After examining the bids, Company A (Alternative A in Table 19.2) was invited to participate in further negotiations. Company B (Alternative B in the Table 19.2) subsequently took the case to court. With the vagueness of the selection criteria and subjectivity of the bid evaluation procedure, the GDH could not defend its decision and the project was cancelled.

Case Study 19.2: Gocek Tunnel Project

The aim of the Gocek Tunnel project was to connect Dalaman Airport, through the mountainous terrain, to the tourist resorts located along the Mediterranean coast in the south-west of Turkey. The tunnel is composed of 830 m of tube and 130 m of cut and cover tunnel construction and has significantly decreased transfer time to the tourist resorts. Gocek Tunnel is expected to increase the number of tourists coming to the region as well as improve travel for the local people.

The Gocek Tunnel is the first BOT transportation project. In 2002, the government invited bidders for the project and the contract was finally awarded to a Turkish consortium. This project was relatively small scale with $10m expenditure. In the invitation to bidders, the project was introduced as shown in Table 19.3.

In the tender documents the construction method and the toll rate were fixed and they included details of pre-qualification criteria about technical and financial viability. The bids were evaluated on the basis of a single criterion, the duration of the operation period. The tender evaluation was relatively easy and the project was awarded to a consortium that offered 2 years of construction and 26 years of operation period.

Being the first BOT transportation project, the success of the Gocek Tunnel may be attributed to the ease of its tender evaluation process. Fixing all other criteria except the operation period left no space for claims. However, single criterion evaluation may not be the best strategy for the procurement of all BOT projects. It worked well for the Gocek Tunnel as the project's size was relatively small and it did not embrace any technological complexity.

Part Three

Table 19.3   Invitation to bidders for the Gocek Tunnel Project.

Method of tender	Build operate transfer (within the framework of Law No: 3996 and governmental decree No:94/507)
Details of the work	The preparation of the application projects of the road, including 960 m of Gocek Tunnel
	The construction of the toll collection area, tollbooth and systems related with the safe operation and maintenance of buildings (signalisation, illumination, ventilation etc.) in compliance with the project that will be approved by the client
	The preparation of the projects for the connection roads which will be used by the vehicles which do not prefer to use the toll road
	The procurement of all equipments, machinery and device needed for the construction and operation of the tollbooth and all other structures
	The operation and at the end of concession period, transfer of the facility free-of-charge to GDH, in good operating condition, usable and without any debt or liability
	Existing route is 7200 m and the tunnelled route will be 4450 m with the roads at the entrance and exit of the tunnel. Total tunnel length is 960 m

There is always a trade-off between the ease of comparison of alternatives and level of innovation required. If all other criteria (such as construction method, technology etc.) are fixed except the duration, this will limit the ability of the private sector to propose innovative solutions. In the following section, a tender evaluation methodology will be proposed for cases where the government's value system can be expressed by a number of attributes that cannot be expressed in monetary terms.

19.4   Best Value Procurement in BOT Projects

Definition of a client's value system at the early phase of a project and implementation of a value management system are stated among the critical success factors for PFI projects (Kelly, 2003). The idea of the 'best value' procurement is that the client should select the proposal that has the highest overall value rather than the one with the lowest cost. The 'value' is defined according to the needs and preferences of the client. Usually, a number of attributes that define the client's needs are set and a multi-criteria evaluation method is used to assess the overall value of each proposal. Zhang (2006) proposed a 'best value' procurement strategy for BOT projects and discussed that the client's objectives should be expressed in terms of 'best value' contributing factors against which alternative proposals are evaluated; consequently, a sound and defensible contract award decision can be made. Akintoye *et al.* (2003) pointed out that clients must secure Value for Money (VFM) and identified the procurement process (high cost of the PFI procurement process, lengthy and complex negotiations, potential conflicts of interests among those involved in the procurement) as one of the major impediments to the success of a project.

In this chapter a 'best value' procurement approach combined with a multi-criteria evaluation methodology will be proposed for the evaluation of BOT projects. The methodology was developed in the light of the interviews conducted with experts experienced in the procurement of BOT projects between December 2006 and February 2007. The basics of best value contracting and multi-attribute rating technique are explained and each step of the methodology is described based on suggestions made by the experts who considered the experiences of GDH regarding the procurement of previous BOT projects.

It was concluded that a methodology which considers both quantifiable and unquantifiable factors is needed. The cost to the public can be calculated by carrying out an NPV analysis and a subjective rating approach may be developed for the factors which can not be incorporated into the NPV analysis:

- To ensure that all bidders make the same assumptions about the risk allocation between the parties and the government and guarantees that will be given to the concessionaires, standard risk-allocation tables and lists should be prepared and included in the invitation to bidders. This is an extremely important issue as in earlier projects it was found that the lowest cost/duration bidder may not be the most economic choice since the bidder may have assumed that some guarantees would be given by the government and did not include a risk premium in their offer. Strategically, some bidders may propose unrealistically low prices/durations to be the first company to start negotiations with the client organisation and assume that they may adjust their price according to different risk-allocation schemes that will be agreed upon between the parties during negotiations.
- Company factors (such as experience, reputation, resources etc.) should be considered in such a way that their impact is reflected in the subjective rating. If the company factors are not favourable, the risk rating should be escalated by a certain percentage. A company factor should not be considered as an independent attribute but as a factor that is related with manageability of risk, thus it should be reflected in the risk rating.
- As it is very hard to assign subjective ratings to each proposal along a number of dimensions, if possible, they should be categorised and every proposal in the same category must be given the same rating.
- The criteria used for evaluation should be explicitly defined and the bidders should be informed about standard tables to be used for subjective rating of the proposals.

The basic steps of the proposed methodology are explained below:

1. *Selection of the short-listed companies*: the 'must conditions' such as the technical, legal and financial requirements should be set. Companies that do not meet the requirements should not be invited to submit a proposal.
2. *Declaration of government's risk-sharing principles*: bidders should be informed about the risk-allocation scheme between the concessionaire and GDH as well as a list of guarantees that will be given by the government.

Part Three

3. *Ranking the companies for further negotiations*: companies should be ranked in the descending order of their 'best value' offers. 'Best value' is defined in terms of monetary and non-monetary factors. Both NPV and multi-attribute analysis should be used for comparison of alternatives. Negotiations should be carried out with the parties according to the best value ranking.
4. *Selection of the best offer*: depending on the outcome of the negotiations, an evaluation framework should be repeated and the bidder providing the 'best value' should be chosen.

19.5   An Application of the Proposed Methodology

Experts were consulted to understand how the above procedure could be put into operation on a real project. As the first step, it was agreed that the GDH should prepare and announce a pre-qualification checklist before the tender phase to facilitate the shortlisting of qualified bidders meeting the requirements of the government. This can significantly reduce government's time and costs associated with evaluating a large number of bids.

The second step is the provision of the same risk-allocation scheme and list of government guarantees to all bidders. This facilitates the comparison of alternatives as each bid proposal will be based on the same assumptions of risk-sharing principles and government guarantees. In preparation of the risk-allocation table, the government should investigate the probable risks. Although the risk sources and impacts may change according to different project features, governments may use a standard risk checklist and decide on the content of a risk-allocation scheme that may be revised according to the specific requirements of the project. In Tables 19.4 and 19.5, examples of a risk-allocation scheme and a list of guarantees are presented. However, these tables should not be considered generic tables that could be used in every BOT project. They reflect the opinion of experts and the current practice according to BOT law in Turkey.

The third step in the selection process is the rating of the short-listed companies according to a multi-attribute evaluation framework and deciding on the ranking of the companies that will be invited to the negotiations. There must be a limit on the number of companies that will be invited to the negotiations to reduce the government's efforts in terms of time and cost. According to the experts interviewed, in concessionaire selection, financial aspects are the most important issues to be considered and can be analysed using NPV. In NPV analysis, the financial outcome of the bids showing the cost to the public can be calculated as a function of the construction period, operation period, the level of toll rate at the start of the operation period, the proposed methodology and escalation indices for toll rate adjustment during the operation period and the interest rate. The demand projection of the traffic should be provided by the government to the bidders so that their toll rate calculations are based on the same amount of traffic for the ease of financial comparison. The interest rate to be applied will be determined by the government as a fixed value for all bidders. The experts also mentioned that price

Table 19.4   An example of a risk-allocation table.

Risks	Risk allocation		
	Government	Contractor	Shared
Technology		✓	
Financial		✓	
Legislative			✓
Design error		✓	
Delay in approvals	✓		
Construction		✓	
Operation and maintenance		✓	
Force majeure			✓
Quality		✓	
Delay in land acquisition			✓
Health and safety		✓	
Environmental		✓	
Inflation		✓	
Exchange rate		✓	
Ground conditions		✓	

elasticity of demand should also be considered. If the toll rates are too high, demand can be expected to be lower than initially anticipated. To ensure that the bidders stay in an acceptable range, the upper limits of the toll rates may also be provided to them.

As previously stated, the multi-attribute rating method converts the non-quantifiable data into numbers using a set of attributes that define the value system of the client. The initial task is to determine the evaluation criteria together with their sub-categories and assign relative importance weights for each criterion. Then, the scale for scoring the bids is decided (such as 1–5 Likert scale). The weights and the assigned scores are then multiplied to find an overall score for each option.

In the current application there were two factors considered by the experts: technical and financial viability. The 'best value' option was defined as the one that had the minimum technical and financial risks. The technical risk factor was assumed to have the following sub-factors:

Table 19.5   An example of a list of guarantees.

Guarantees	Provision
Subordinated loan guarantee	Yes
Tax exemption	No
No second facility guarantee	No
Political risk guarantee	Yes
Revenue/demand guarantee	No
Senior loan guarantee	No
International arbitration guarantee	Yes

Part Three

- Construction risk: the probability of having technical problems during project implementation due to the proposed construction technology and methods
- Design risk: the probability of having design changes due to design errors, vagueness of design or poor constructability
- Operation and maintenance risk: the probability of having problems due to the proposed operation, maintenance and inspection methods or poor maintainability
- Seismic risk: the probability of having seismic damage due to potential earthquakes with respect to the selected construction technology
- Environmental risk: the probability of having adverse environmental impacts due to the selected technology, lack of environmental policy and management plan
- Safety risk: the probability of having accidents during the construction and operation periods due to the selected methods, unqualified personnel or lack of a safety plan
- Transfer risk: the probability of having problems after the transfer of the facility to the government due to a poor-quality transfer package (no training etc.)

All of the above-mentioned risks are dependent on the proposed construction method as much as the expertise of the company. The experience should not be considered as a different evaluation criterion, instead it is a factor influencing the magnitude of all risks. Thus, to consider its effects it should be mathematically reflected in the ratings. It is suggested that the bidders should be informed about the weights before they submit their proposals.

Table 19.6 demonstrates how the ratings change according to 'experience' such that if experience in similar projects is low, the risk rating will be higher (for example, rating value may be increased by 1, on a scale of 1–5) whereas it will be the same as the initially defined value if the experience is high. The weights in Table 19.6 reflect the subjective judgments of the experts interviewed.

With regard to financial viability, the factors affecting the magnitude of financial risk were determined to be:

- Financial commitment by the project company (the financial strength of the project company, the amount of resources committed to the project)
- The soundness of the financial analysis (realistic revenue, cost and time plans)
- The structure of the financial package (hedging arrangements for currency risks, sources and currencies of loans, standby loan agreement, fixed and low interest rate financing, insurance and financiers' abilities)

Depending on the status of the bidders, each of the above mentioned criteria can be scored as high–low, yes–no, and good–poor respectively. Bidders are then grouped into eight categories where all bidders within the same category are assigned the same rating. Table 19.7 shows a sample table that may be used during evaluations. It must be stressed that the groups that appear in

Table 19.6 Sample of a technical risk rating table.

Factors (i)	Weight (w_i)	Rating (R_i) (1–5 scale)	Experience	Revised R_i (1–5 scale)*
Construction risk	0.2	R_1	High	R_1
			Low	$R_1 + 1$
Design risk	0.2	R_2	High	R_2
			Low	$R_2 + 1$
Operation and maintenance risk	0.2	R_3	High	R_3
			Low	$R_3 + 1$
Seismic risk	0.1	R_4	High	R_4
			Low	$R_4 + 1$
Environmental risk	0.1	R_5	High	R_5
			Low	$R_5 + 1$
Safety risk	0.1	R_6	High	R_6
			Low	$R_6 + 1$
Transfer risk	0.1	R_7	High	R_7
			Low	$R_7 + 1$
Technical risk rating (TRR) $= \sum (w_i * \text{revised } R_i)$				

*The maximum value of revised R_i is 5 (e.g. if R_i is 5, revised R_i will also be 5, although the experience is low)

this table are subjectively defined by the experts interviewed and cannot be generalised.

After the calculation of the technical and the financial risks a combined risk value can be calculated by assigning relative weights to financial and technical risks and multiplying these weights by the pre-calculated risk scores. The assigned weights may change with respect to project factors such as size and technical complexity. If the weights of the technical risk and financial risk are denoted by w_1 and w_2, respectively, the final risk rating (RR_{Final}) is calculated by the following formula:

$$RR_{Final} = w_1 * TRR + w_2 * FRR$$

The 'best value' offer is the one providing the minimum cost to the public and minimum risk, that is the minimum NPV expressed in monetary terms

Table 19.7 Sample of a financial risk rating table.

Groups according to financial risk	Financial commitment by the project company	The soundness of the financial analysis	The structure of of the financial package	Financial risk rating (FRR) (1–5 scale)
Group 1	High	Yes	Good	FRR_{Group1}
Group 2	Low	Yes	Good	FRR_{Group2}
Group 3	High	No	Good	FRR_{Group3}
Group 4	Low	No	Good	FRR_{Group4}
Group 5	High	Yes	Poor	FRR_{Group5}
Group 6	Low	Yes	Poor	FRR_{Group6}
Group 7	High	No	Poor	FRR_{Group7}
Group 8	Low	No	Poor	FRR_{Group8}

Part Three

and RR_{Final} expressed on a scale of 1–5. A formula that combines these two indicators is needed to select the best option. It is clear that a generic formula that applies to all situations cannot be suggested as it depends on organisational and country-specific needs. Different alternatives can be proposed as follows:

- To eliminate the high-risk tenders, the ones having RR_{Final} higher than 4 (on a scale of 1–5) may be discarded.
- After eliminating the high-risk options, different methods can be applied to combine the NPV and RR_{Final} of the remaining ones. RR_{Final} can be multiplied with NPV and the one that has the minimum value can be selected. Alternatively, the offer that has the minimum 'risk-adjusted NPV' can be chosen. To find the risk-adjusted NPV, offers can be categorised such as low risk (RR_{Final} between 1 and 2), average risk (RR_{Final} between 2 and 3) and high risk (RR_{Final} between 3 and 4) offers and NPVs of the bids may be increased by the same percentage (such as 10% for the low-risk group, 15% for the average-risk group and 20% for the high-risk group) in each category.

This procedure can be used to rank the companies according to the value offered by the bidders before the negotiations. The consortium that offers the 'best value' will be invited to the negotiations first, which creates a substantial competitive advantage for that consortium. However, the final ranking of the companies may change significantly after the negotiations.

Although an example procedure for 'best value' procurement is explained in this chapter by referring to the experiences of GDH, it can not be claimed to be the best procedure. It is clear that the evaluation method may be improved in time as a result of lessons learnt during the procurement phase of different BOT projects. The government may review successful projects and try to find a correlation between the methods used to select bidders and the actual performance of the concessionaire in the project.

19.6   Concluding Remarks

The major idea of this research is that governments should try to implement effective tender-evaluation strategies in order to increase the success of the BOT model. In developing countries like Turkey, because of legal and bureaucratic problems, the procurement phase may extend over several years, leading to a considerable delay in the realisation of urgently needed infrastructure projects. In Turkey, the number of successfully realised projects in the transportation sector is rather low when compared to energy projects. This may be attributed to the vagueness of the risk-allocation principles between the private and public sectors, high level of demand risk and lack of systematic procurement procedures. The two cases, the Izmit Bay Crossing and Gocek Tunnel project, represent unsuccessful and successful cases in terms of procurement strategy. However, the case studies imply that there is no single recipe for a successful procurement strategy. In relatively small projects (like the Gocek Tunnel) where there are no alternative design and technology options, only one criterion (e.g. operation period) may be used for the selection of the best proposal by fixing all other variables. However, in bigger projects where alternative technologies are possible, a multi-attribute assessment, based on best value evaluation, may be preferred.

A 'best value' procurement approach has been introduced and, in light of the lessons learnt from previous experiences, its application to transportation projects has been examined. The proposed system is based on consideration of both tangible costs to the public and intangible attributes, mainly risk factors. Bidders should be made aware of the risk-allocation scheme and government guarantees so all bid proposals are prepared based on the same assumptions. To minimise the subjectivity and maximise the transparency of the process bidders should be informed about certain rules before they submit their bids. However, subjectivity cannot be totally eliminated as the determination of rules, rating procedures and assigned ratings involves some level of subjectivity. Implementation of standard procedures is not always successful as special cases can hardly be evaluated. For example, fixing the risk-allocation schemes at the start may prevent effective risk-mitigation strategies that may be proposed by the private sector. It should be remembered that the proposed procedure is only for the identification of shortlisted companies that will be invited to negotiations and, during the course of the negotiations, special circumstances might appear that could substantially change the final ranking.

It should also be emphasised that the proposed procedure may not be applicable for other organisations because of the potential differences between the value systems of client organisations. The rating process and the identified criteria are expected to change from organisation to organisation and even within the same organisation it may change over time. Some criteria like benefits to the government may also be considered (for example, one company may propose profit sharing with government) as well as opportunities (technological innovation, reputation etc.) which are not mentioned during technical and financial viability assessment. Thus, the analysis may be based on the assessment of opportunities and benefits (to the government) as well as costs and risks, a potential topic for further research studies.

References

Akintoye, A., Hardcastle, C., Beck, M. *et al.* (2003) Achieving best value in private finance initiative project procurement. *Construction Management and Economics*, 21, 461–470.

Ashley, D., Bauman, R., Carroll, J. *et al.* (1998) Evaluating viability of privatized transportation projects. *Journal of Infrastructural Systems*, 4(3), 102–110.

Canakci, I.H. (2006) The speech of the Treasury Undersecretary. In: *The International Public-Private-Partnerships Conference*, September 2006, Ankara, Turkey.

Levy, S.M. (1996) *Build Operate Transfer*. Wiley, New York.

Li, B. and Akintoye, A. (2003) An overview of public–private partnership. In: Akintoye, A., Beck, M. and Hardcastle, C. (eds.) *Public–Private Partnership*. Blackwell, Oxford.

Part Three

Kelly, J. (2003) Value management in public–private partnership procurement. In: Akintoye, A., Beck, M. and Hardcastle, C. (eds.) *Public–Private Partnership*. Blackwell, Oxford.

Ozdoganm, I.D. and Birgonul, M.T. (2000) A decision support framework for projects sponsors in the planning stage of BOT projects. *Construction Management and Economics*, 18, 343–353.

Zhang, X.Q. and Kumaraswamy, M.M. (2001) Hong Kong experience in managing BOT projects. *Construction Engineering and Management*, 127(2), 154–162.

Zhang, X.Q. (2004) Concessionaire selection: methods and criteria. *Journal of Construction Engineering and Management*, 130(2), 235.

Zhang, X.Q. (2006) Public clients' best value perspectives of public private partnerships in infrastructure development. *Journal of Construction Engineering and Management*, 132(2), 107.

Part Three

20

Application of Risk Analysis in Privately Financed Projects: The Value For Money Assessment through the Public Sector Comparator and Private Finance Alternative

Tony Merna and Douglas Lamb

20.1   Introduction

Several countries utilise Private Finance/Public-Private Partnerships (PF/PPP) to encourage investment in public services; however, many governments have formed stringent economic assessments to appraise the validity of private investment in public services. Central to the assessment is the VFM and the associated transference of risk. Current practices associated with the key inputs to VFM differ according to country and sector. This chapter outlines a quantitative approach to analysis of risk, and discusses how this approach can be applied to the formation of the public sector comparator (PSC) and the private finance alternative (PFA) to form robust appraisals of VFM.

The analysis of risk in traditional construction contracts often operates purely over the construction and commissioning timeframes in comparison to those of a project finance nature which offer several contract structures, PF being one. In PF projects the emphasis of project analysis for principal (client) organisations focuses upon the holistic delivery of a service to the public sector via a project agreement, typically 25–60 years in duration. This long timeframe introduces a greater challenge with regard to modelling the perceived risks facing a project delivered by a promoter (contracted) organisation. In PF contracts lenders often support the promoters through the future revenue stream of the project or non-recourse financing, with the added flexibility to incorporate collateral to form limited recourse financing. This places additional pressure on the lenders to monitor and promote successful design, construction, operation and maintenance of service, with their involvement being a vital ingredient to the efficient application of funds, resources and risk management. This further supports the public sector in the monitoring and implementation of the project agreement, considered by Lane (2000) as a critical value generative area to PF/PPP contracts. It has not been acceptable practice to expect principal organisations to undertake the majority of the risk-management exercise, or expect tendering periods of a matter of weeks as outlined in Smith (1999). Instead tendering periods have been extended, to between 1 and 3 years (Lamb and Merna, 2004a) with greater timeframes being made available for both the principal and promoter to undertake risk management.

Central to the introduction and justification of private sector involvement in public services, is the private sector's ability to either exceed, or meet the same standard, cost and timing in delivering a service compared to that of the public sector delivery model. The ability to compare alternative procurement routes has been contested by some analyses (NAO, 2003a). However, such assessments form an essential element to the PF/PPP project appraisal process (HM Treasury, 2003).

As a means of assessing PF/PPP procurement options, the concept VFM represents the optimum combination of whole-life costs and quality to meet the user requirement (OGC, 2003). However, a number of sources (HM Treasury, 1998; Akintola et al., 2003; Broadbent et al., 2003; Heald, 2003) view the allocation of risk as a critical determinant to the VFM. Furthermore, Froud and Shaoul (2001) identified limitations to the systems implemented to assess the VFM on account of a lack of generally accepted methodologies for appraising VFM. Reports into the performance and assessment of VFM have addressed risk in both qualitative and quantitative terms, with the philosophy and development of VFM assessments still ongoing. Combining this with specific business sectors which conduct their own internal assessments (NAO, 2003a, 2003b; Audit Commission, 2003) creates further difficulties in justifying PF (Broadbent and Laughlin, 1999). The approach adopted for the appraisal of VFM is further complicated by uncertainty in the economic variables used, the value management techniques adopted (Merna and Lamb, 2004) and the various qualitative and quantitative risk-assessment techniques available (Raz and Michael, 2001). This is also reflected in the guidance produced internationally, with countries such as Australia, South Africa, Netherlands and Canada adopting different approaches to a number of issues (NSWG, 2000; National Treasury, 2001; Partnership Victoria, 2001; PPPU, 2001). These include:

- The procurement process supporting the production of the PSC
- Options appraised or considered
- The level of disclosure of the PSC
- Risk-assessment techniques applied
- Decision-making methodologies incorporated
- Economic parameters used in the assessment

In the UK, guidance has been released in the form of the 'Green Book' (HM Treasury, 2003) outlining practices for the creation of a PSC, which acts as a benchmark to assess the VFM of a PFA. However, it does not bridge the gap between the VFM assessments made during the procurement process and the reappraisal of VFM during operational performance. Additional guidance in the form of *Value for Money Assessment Guidance and Quantitative*

Assessment, User Guide (HM Treasury, 2004) has been produced to offer further assistance. However, these documents merely focus on the application of the techniques identified by HM Treasury, without identifying critical weaknesses associated with such aspects as optimism bias; Mott MacDonald (2002) and Department of Finance and Personnel (2004) outline the qualitative interaction associated with the determination of the optimism bias. Additional efforts to improve the accuracy of the optimism bias may be tenable through additional project information (Al-Momani, 2000). When forming a PSC it is essential to appreciate the weakness of specific elements of the analysis that may form bias within the model.

20.1.1   Definition of a comparator

The authors suggest that for the purpose of privately financed projects a comparator is any parameter, variable, system, tool or technique used to compare tangible or intangible assets and liabilities present in any project or entity to that of another at a specific point in time, making every effort to make the comparison relative, quantifiably measurable and as equitable as possible.

In terms of the procurement process, a comparator is the PSC that compares a publicly procured project to that of a private finance nature. However, comparators may be used throughout the lifecycle of a project, from project appraisal to review. In fact comparators may be used to compute an earned value analysis, advising principals of appropriate periods to renegotiate the concession agreement. They may also be used in the market testing, refinancing and restructuring appraisal of a project, or anywhere that may require significant contractual renegotiation of the original contractual agreement.

20.2   The PSC and the PFA

A PSC is defined as a hypothetical risk-adjusted costing, by the public sector as a supplier, to an output specification prepared as part of a procurement exercise. According to the Treasury Task Force (TTF, 1998):

- It is expressed in net present value (NPV) and/or net present cost (NPC) terms.
- It is based on recent actual public sector methods of providing defined output (including any reasonably foreseeable efficiencies the public sector could make).
- It takes full account of the risks which would be encountered by that method of procurement.

There are several variances with regards to the scope and content of the PSC, including the economic parameters which describe whether an output or input specification is adopted for the production of the PSC. In fact, the output specification cannot physically be translated into NPV or NPC terms until an input specification is generated. Some of these problems have

Part Three

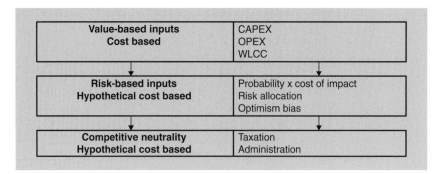

Figure 20.1   Forming the public sector comparator.

been addressed through the derivation of simultaneous procurement strategies promoted within the UK National Health Service in the form of Local Improvement Finance Trust (PUK, 2004) and Department for Education and Skills through the Building Schools for the Future programme (4Ps, 2004). Where bundled schemes are simultaneously procured using PF/PPP, contracts providing real-time responses and data are used for comparisons. Whilst this has its benefits when principals are seeking to redress stock, which consists of multiple distinguishable assets, the same strategies may not be applicable to individual asset procurements. Furthermore, they can fail to address the fundamental issue of attaining competitively priced work unless the timing and programming of the work is reappraised, especially if the strategy still promotes large-scale investment influxes promoting boom and bust cycles. In the following section an approach to the PSC is utilised where the NPC, NPV and internal rate of return (IRR) form the economic metrics from which VFM is assessed.

A generic model for the PSC and PFA is illustrated in Figure 20.1. The capital, operational and whole lifecycle costing expenditure incurred on the project forms the primary cost basis, which is referred to in Figure 20.1 as the value-based inputs. These typically form between 70 and 80% of the final PSC NPC depending upon the nature of the PPP concession. The process of developing the value-based inputs of the PSC has been addressed by earlier works of the authors (Lamb and Merna, 2004a,b; Merna and Lamb, 2004). The risk-based inputs have several structures in which to approach quantitative modelling on a project basis (Williams 1994; Simon et al., 1997). An approach proposed by MoF (2002), separates risk into two categories, namely pure and spread risk, that impact the project. In terms of modelling such structures have the capacity to operate on both a spreadsheet- and network-based appraisal and are readily incorporated into a risk register, which may also incorporate the optimism bias. A model is produced using both the pure and spread risks to form a deterministic and stochastic appreciation of the expected NPC. The type of information available for modelling PSC and PFA typically conforms to the following conceptual model in Figure 20.2.

As depicted in Figure 20.2, the project moves through the lifecycle whereby specific qualitative or quantitative information sources appreciate

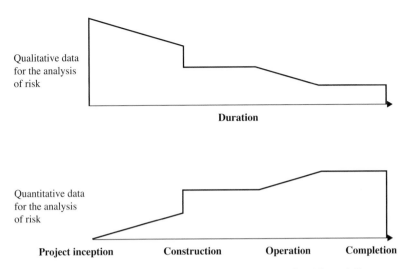

Figure 20.2 Qualitative and quantitative information sources for risk modelling.

and depreciate according to project activities, changes to the project team, advances in analysis techniques and knowledge retention systems.

Risk-based inputs typically form 10–20% of the final PSC NPC. Finally the competitive neutrality elements are added to address differences associated with taxation or administration structures between the public and private delivery models, which is typically between 0 and 10% of the PSC NPC. Forming the PSC is an iterative process, adjusted throughout the procurement process and submitted during the outline business case (OBC) and the final business case (FBC) (HM Treasury, 2003).

To form the VFM assessment, the PSC must be compared to the PFA. This may not be produced until bids are received from promoting parties, which may be acceptable where private companies are used to identify potential projects in the country (National Treasury, 2001). However, in the UK it is common practice for the PFA to be produced during the OBC by the public participant.

The structure of the PFA as illustrated in Figure 20.3 is similar to that of the PSC, except that the model addresses the financing charges and potential revenue sources and quantities. This may take the form of either the unitary payment or market demand required by the promoter to service the costs, risks and profits. From the risk matrix submitted in the concession agreement, a risk allocation structure and risk register may be formed to identify the risk management plan. The promoter aims to minimise the hypothetical element to the analysis of risk.

20.2.1   Assessing value for money

Figure 20.4 conceptually illustrates how a single point estimate of VFM can be calculated and the cost elements considered during the development of the PSC and the PFA.

Part Three

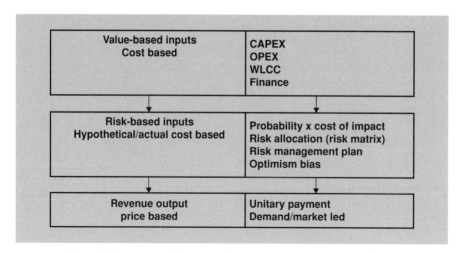

Figure 20.3   Forming the private finance alternative.

20.2.2   Traditional public model (PSC)

Each element of the conceptual model depicted in Figure 20.4 contributes to the VFM assessment. It is critical to appreciate bias or degrees of uncertainty associated with each element to gauge the accuracy and robustness of the final assessment regarding VFM. This applies in particular to the factors discussed below.

Retained risk

Risk retained by the public sector, refers to the risks that are managed more efficiently within the public sector, within a traditional public procurement contract.

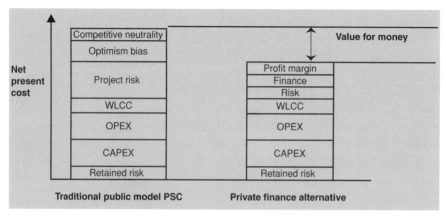

Figure 20.4   Conceptual model of value for money.

Capital, operational expenditure and whole-life cycle costing

The capital and operational expenditures are generated based upon the base case model. The expenditures do not make any allowances for risk or uncertainty to prevent double counting.

Optimism bias

There is evidence that there is a widespread tendency for appraisers to be over-optimistic when preparing proposals (HM Treasury, 2003). The optimism bias is associated with the estimates submitted for public works during the business cases. The optimism bias is added to the capital element of the PSC. Note the adjustment only caters for the risk associated with inaccuracy of the assumptions and estimates held by the traditional business case to that of the final outturn cost.

Project risk

Risk within the project can be categorised as follows (MoF, 2002):

- Pure risk (chance of occurrence multiplied by the financial consequence should it occur)
- Spread risk (uncertainty associated with the market and technical estimates made within the project)

Pure risks can be adequately managed and totalled within a risk register. A typical calculation is shown in Table 20.1. All the pure risk values are totalled and added to the traditional procurement option. The risks transferred and probabilities of occurrence are normally based upon the historical performance of contracts previously used to deliver projects and services. Pure risks may be formed from historical data or valuations placed by experienced professionals. Such quantitative and qualitative assessments need to be identified and segregated to support the probity of the models proposed.

Spread risks are those risks concerned with the uncertainty surrounding the estimated amounts. Probability, sensitivity and scenario analysis may be used to interrogate the effects of spread risk upon the project. For example, in PFA finance interest may fluctuate according to the inter-bank borrowing rates. This may be modelled using a random walk function through a Weiner and Poisson process (Vasudevan and Higgins, 2004). This may be an

Table 20.1 Calculating a pure risk.

Risk	Description	Allocation	Probability	Financial impact	Value
Technical	Delay incurred by poor ground conditions	Promoter	5%	£6 090 000	£304 500

Part Three

acceptable solution, especially where inter-bank borrowing rates have low rates of volatility. The degree of uncertainty may depend on the volatility and how far one looks into the future (Ho and Liu, 2001), contrasted to how far one looks back and the specific external parameters that influenced the volatility then compared to how they will now and in the future.

The degree of accuracy associated with pure and spread risks may be brought into question, with elements of the analysis potentially relying upon both qualitative and quantitative data sources. Modellers should therefore distinguish between acceptable degrees of confidence intervals, to provide several layers of analysis, allowing decision makers to appreciate both quantitative and qualitative outputs.

Whilst a number of contributions (see Simon *et al.*, 1997) have focused on the modelling of risk, limited assessment has gone into the contractual interpretation and influence over the risk output profile. This is a reflection of the traditional application of risk management, where the models are used to form contractual strategies or tactics in relation to the risk that are then used to structure or negotiate the contract. Instead the PSC and PFA are assessing the hypothetical performance of an already constructed contractual structure in relation to the risk present. The assessment of risk must accommodate the polarisation of risk via the inclusion of such hypothetical contractual structures. For example, analysis that may be conducted on a continuous probability basis may find the contractual triggers operate purely on a discrete output basis. Risk registers and the like need to take into account either the risk matrix attached to the concession agreement, or specific output profiles contained within the payment mechanism, performance standards and output specification elements of the agreement. Thus, when modelling not only a project but a specific contractual strategy, modellers should be aware of such factors and make the appropriate allowances where possible.

Competitive neutrality

The purpose of adjusting the PSC to allow for taxation and administration costs is to allow for differential tax receipts and any bias that may stem from them. Adjustments are made to the PSC to allow for taxation, which impedes the efficiency of private finance solutions. In the UK, steps have been taken to estimate the expected cost of taxation on a PFP by investigating: the degrees of soft services; capital value of the PFP; tax treatment of the project expenditure; and riskiness of the project (KPMG, 2002).

Based upon the variables and accounting characteristics of the project, a percentage increase to the overall NPC of the scheme can often be identified. In situations where the tax difference between the public and private option may be material to the appraisal such costs need to be stripped out of the models.

20.2.3   Private finance alternative (PFA)

For the PFA there are two periods in which an option appraisal can be developed: before contract negotiation or during contract negotiation. The UK prefers to derive the PFA appraisal post contract negotiation, allowing an assessment of VFM to be formed before the private sector has been contacted. This is submitted usually in an outline business case to support the appliance of private finance as a procurement solution. However, countries such as South Africa prefer to derive the PFA through the private sector bids. This harbours a risk that VFM may not be achieved, which may be an acceptable strategy in developing countries, where public funding is simply not available for such projects and the key priority is attaining affordable investment in traditional public services.

The capital, operational and lifecycle expenditures are usually based on previous projects, with a further risk or contingent sum being inserted. This contingent sum may form a whole or elements of the future profits.

Conducting a VFM assessment through the derivation of a PSC and PFA

Before a VFM assessment can be conducted the following steps must be undertaken in order to protect the probity of the model proposed:

- Identify the variable, parameter and methodologies to be used to assess VFM.
- Select appropriate modelling software to outline key weakness.
- Identify the output or input specification, performance standards and payment mechanism for the project.
- Identify the contracts base costs, programme and network of activities (value-based inputs).
- Identify and insert risks linking them to specific activities (risk-based inputs).
- Adjust for competitive neutrality (hypothetical cost-based inputs).
- Carry out tests and simulations of the model.
- Analyse simulated outputs normally in terms of economic parameters.

Case Study 20.1: Street Lighting of a Major Municipality

A large municipality is considering the application of PFI in the design, installation, operation, maintenance and finance of its current street lighting, signage and street furnishings. The area services approximately 4 million people and its current stock is a mixture of new and old. The current policy attempts to renew priority areas identified by principal (a government department) followed by a continued refurbishment and rehabilitation programme. The successful bidder is expected to take over the current staff, sites, lighting columns, signs and street furnishings which will be transferred back to the principal in an 'as new state' at the end of the 32-year concession period.

The principal proposes to assess VFM using the following economic parameters: IRR, NPV and NPC. The PSC and PFA will be assessed, based on these parameters, in order to determine the VFM of either the PSC or the PFA.

Whilst there are several methodologies adopted for the appraisal of projects, such as optioneering, cost-benefit analysis, the preferred methodology to be adopted for this street lighting scheme is optioneering, as it focuses upon the technical solutions.

A network-based modelling software system is used to model the activities undertaken in the project, allowing risks to be linked to specific activities. However, the programme utilised does

Part Three

Table 20.2   Value-based inputs for the PSC.

Activities	Description	Cost £(M)	Revenue £(M)	Duration (Months)
1	Start	0	0	1
2	Site Planning and Procurement	15.3	0	24
3	Design and Bidding	57.4	0	18
4	Site Mobilisation	25	0	4
5	Installation of Signs and Street Lights	382.5	0	60
6	Initial Operation and Maintenance	106.2	0	60
7	Initial Unitary Payment	0	0	60
8	Commissioning	15	0	5
9	Unitary Payment (Secondary)	0	0	300
10	Operation and Maintenance (Secondary)	512	0	300
11	Finance Debt	0	0	240
12	Finance Equity	0	0	240
13	Third Party Revenue	0	0	360
14	Closure	0	0	1

not allow the model to cater for sculpted repayment profiles of CAPEX, which forms the majority of the principal in debt instruments, resulting in lower returns on equity and higher debt yields. The availability of such instruments is also limited. The model can also assign pure and spread risks through probability distributions with such models being inherently incapable of dealing with anything that lies beyond a probability distribution. Aspects of project modelling that do not conform to probability theory according to Pender (2001) include:

- The reliance on randomness, whereas many of the interactions are planned
- Projects are unique reducing the reliability of statistical aggregates
- Uncertainty and ignorance in relation to the risk
- Communication of results difficulties

The PSC

The base case costs, network and programme are formed for the street lighting project. Many of the activities are ghost activities only used for specific elements of the VFM calculations. For example the NPC does not require any of the revenue-based activities to be used. However, for a hypothetical PSC IRR the revenue activities may be used.

The total base cost of the project pre-risk adjustment as depicted in Table 20.2 is £1113.4m, with an expected duration of 388 months. There are no expected revenues from the PSC and there is no allocated cost to finance, as the project is funded centrally through the government organisation.

The network of activities in Figure 20.5, establishes the precedence and programme of works. Both cost and time models may be computed and risk allocated to specific activities. Due to the fact that the project is to be undertaken as an ongoing concern, with the principal maintaining ownership of the assets, whilst also completing renewal, rehabilitation and refurbishment of the current stock, a complex precedence network is formed. The value-based inputs may be enhanced, through the identification and allocation of risk.

The PSC risk adjust inputs

Due to the format of the model, pure and spread risks are identified in accordance to the activities they impact upon. They are allocated to the network as illustrated in Figure 20.5. Table 20.3 is a

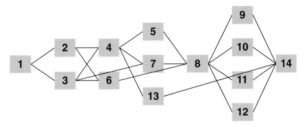

Figure 20.5   Network of activities outlining precedence.

representation of the pure risks that are considered to affect the project. These are calculated as outlined in Table 20.1. The pure risks are then assigned to the activities, making sure the spread risks that are to be assigned in the future do not affect the pure risk, thus resulting in double counting. The expected value of risks is attached to specific activities within the network, to form the pure-risk adjusted model. The expected value of risk for activity 6 totals £6.2m and for activity 10 totals £24.8m.

Spread risk with distinct probability distributions are then assigned to the project activities to allow a Monte-Carlo simulation to be conducted. Table 20.4 illustrates the spread risk considered in the PSC, outlining the probability distributions assigned and the ranges by which the cost of a

Table 20.3   Pure risk attached to the network.

Activity impacted	Expected value £(M)	Description of risks
2	0.8	Optimism bias, accuracy of inventory, change in the input and output specification, planning consent delay, land purchasing orders, public inquiry time and cost
3	1	Site access waivers, preparation and mobilisation costs, force majeure, site purchasing and sale, hire and leasing fluctuations, optimism bias,
4	5.2	Poor design in terms of, installation, performance, changes to design codes, design costs, skills availability, research and development requirements, innovation and performance enhancement, optimism bias
5	33	Latent defects, funding availability, waste and environmental management, force majeure, contractor default, long lead items, guarantees, industrial action, access and charges, supply network failure, accidental damage or loss, theft, optimism bias, health and safety failures
6, 10	31	Technical obsolescence, fire and vandalism, public liability and claims, insurance and uninsurable events, availability and performance of stock, repair, refurb, renewal costs, health and safety failures, legislative alterations, supply and offtake fluctuations, change in law, industrial action, latent defects, theft, force majeure, operator default, optimism bias.
8	0.5	Health and safety, standards and performance of current stock, change in law, labour or expertise shortage, authoritative consent, down times, handover procedures, optimism bias.

Part Three

Table 20.4   Spread risk.

Activity impacted	Distribution	Description of risks	Upper limit	Lower limit
3	Uniform	Design cost and delay overruns	15%	0%
	Uniform	Labour and material charges	10%	0%
4	Skewed triangular	Vandalism	9%	0%
5	Skewed triangular	Construction cost and delay overruns	15%	0%
	Skewed triangular	Material and equipment escalation	20%	0%
	Skewed triangular	Output specification failures	15%	0%
	Skewed triangular	Vandalism	9%	0%
6	Uniform	Performance standards	25%	0%
	Uniform	Operation and maintenance increases	24%	−2%
	Skewed triangular	Electricity fluctuations	22%	−1%
	Skewed triangular	Vandalism	9%	0%
8	Skewed triangular	Commissioning cost and delay overruns	10%	−3%
10	Skewed triangular	Performance standards	30%	0%
	Skewed triangular	Operation and maintenance increases	18%	0%
	Skewed triangular	Electricity fluctuations	25%	−5%
	Skewed triangular	Vandalism	15%	0%

specific activity is affected. Combining the pure and spread risks, a risk-adjusted PSC is constructed, which is illustrated as a cumulative frequency distribution of the NPC.

The cumulative frequency data illustrated in Figure 20.6 outlines the variance of likely NPC outcomes for the PSC. Finally the costs neutrality element is combined to the risk adjustment. This adds additional cost throughout the project lifecycle, resulting in a total uplift of £90m.

Using a 75% percentile analysis to remove outlying data from the quantitative assessment, the worst and best case scenarios for the cost of publicly delivering the service are illustrated in Table 20.5 alongside the minimum and maximum NPC data. These figures may then be compared to that of the PFA to form a quantitative assessment of the VFM offered by the public or private solution.

The private finance alternative

In this case study the PFA is developed from bidders' responses to the invitation to negotiate. The bidders had to respond using a *pro forma*, which allowed greater uniformity in comparing the cost. The value-based inputs were derived for the project, again using a network-based model to link specific activities and cost in the project.

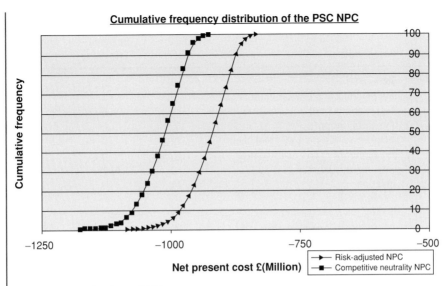

Figure 20.6   Risk-adjusted PSC.

The value-based costs depicted in Table 20.6 are considerably lower than those of the PSC, which initially warranted further investigation. However, the model figures could be confirmed. In this base case the total NPC is estimated to be £947.4m. The cost of debt finance achieved is extremely low due to the ongoing revenue generated through the current asset stock, which the promoter would take over post concession signature. This allows revenues to minimise the capital requirement. The promoter also foresees the development of tertiary revenue sources such as advertisement and the creation of a street furniture and sign business. No allowances have been made with regards to the cost of equity, as the yield from the equity is determined based on the IRR of the project once the senior debt has been serviced. Therefore, negative NPV and IRR would indicate a zero payment towards equity holders and default on either debt or operational performance depending upon the timing and size of the risk impacting the cash flow.

The financing structuring proposed by the promoting party holds an approximate debt:equity ratio of 80:20, which is dependent upon the actual cash lock-up attained at installation completion and commissioning. This generates a senior debt facility of £196m and £173.5m in interest repayments. The debt holds a term of 20 years with a margin 250bps over LIBOR (London Inter Bank Offer Rate). The promoter provides a £40m investment in the form of subordinate debt and equity with excesses in available cash forming the dividend to the shareholders.

The network for the PFA includes an additional activity to that of the PSC that accounts for taxation. The PFA network as depicted in Figure 20.7 has an additional activity to cater for, namely

Table 20.5   PSC NPC results.

Scenario	Risk adjusted NPC £(M)	Competitive neutrality adjusted NPC £(M)
Maximum	−1085.59	−1175.59
Worst case	−975.59	−1065.59
Base case	−808	−898
Best case	−875.59	−965.59
Minimum	−835.59	−925.59

Part Three

Table 20.6   Value-based inputs for the PFA.

Activities	Description	Cost £(M)	Revenue £(M)	Duration (Months)
1	Start	0	0	1
2	Site planning and procurement	5.3	0	26
3	Design and bidding	27	0	18
4	Site mobilisation	5	0	4
5	Installation of signs and street lights	282.5	0	60
6	Initial operation and maintenance	75.3	0	60
7	Initial unitary payment	0	265	60
8	Taxation	41	0	206
9	Commissioning	5	0	5
10	Unitary payment (secondary)	0	1301	300
11	Operation and maintenance (secondary)	290.3	0	300
12	Finance debt	173.5	0	240
13	Finance equity	0	0	240
14	Third party revenue	0	158.9	360
15	Closure	0	0	1

taxation. The programme is set and the pure and spread risk may be identified and allocated to the network.

The PFA risk inputs

Identification of the pure risks must disassociate them from the risks that are being dealt with through the spread analysis. In this example a greater number of pure risks are identified for the PFA as illustrated in Table 20.7, compared to that of the PSC. However, the valuation of the risks within the model are substantially lower than those seen in the PSC.

The risks are attached to the network as before. The spread risks are now identified and allocated to the network to allow a Monte Carlo simulation to check the validity and bankability of the value-based inputs. Specific risk tools such as SWAPS and hedges are used to limit the upper and lower limits of the financial package as depicted in Table 20.8. Further risk mitigation steps are taken to address electricity price fluctuations, with the promoter forming a consortium, which included an electricity supplier.

After the initial appraisal of risks, and a study of the project's economic outputs, as illustrated in Table 20.9, the unitary payment structures provided an acceptable structure for the senior

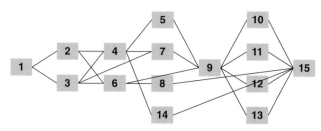

Figure 20.7   Network diagram for the PFA.

Table 20.7   Pure risk for the PFA.

Activity impacted	Expected value £(M)	Description of risks
2	0.8	Accuracy of inventory, change in the input and output specification, planning consent delay, land purchasing orders, public inquiry time and cost, vires
3	0.75	Site access waivers, preparation and mobilisation costs, force majeure, site purchasing and sale, hire and leasing fluctuations, health and safety
4	3.2	Poor design in terms of, installation, performance, changes to design codes, design costs, skills availability, research and development requirements, innovation and performance enhancement
5	14	Latent defects, funding availability, waste and environmental management, force majeure, contractor default, long lead items, guarantees, industrial action, access and charges, supply network failure, accidental damage or loss, theft, health and safety failures
6, 11	12	Technical obsolescence, fire and vandalism, public liability and claims, insurance and uninsurable events, availability and performance of stock, repair, refurb, renewal costs, health and safety failures, legislative alterations, supply and offtake fluctuations, change in law, industrial action, latent defects, theft, force majeure, operator default
8	4	Change in law, variation to charges, exemptions and enhancements, grace periods, notifications, audit and irregularities, payment dates and timing
9	0.5	Health and safety, standards and performance of current stock, change in law, labour or expertise shortage, authoritative consent, down times, handover procedures
7,10	15	Performance standard failures, late payments, principal default, electricity supply default, latent defects
12,13	2	Draw down and availability, marriaging of funding, default of supply, default of repayment and charges, administrative charges
14	1	Industrial actions, labour increases and retention, technical advanvances, competition, overheads and rates, material price fluctuations

and subordinate financiers. Based on the economic performance of the PFA, the private sector promoter is prepared to commit to a gearing of 80:20. This forms the basis of the bankability test. If financiers and investors did not accept the degree of risk associated with their investment, then an additional iteration would have to be carried out, which would see an adjustment to either the risk allocation structure or the value-based inputs such as the unitary payment.

The payment plan is derived from the initial unitary payment and secondary unitary payment values illustrated in Table 20.6. This is then discounted by the principal organisation and compared to the NPC of the PSC, to form an initial appraisal of the VFM.

Part Three

Table 20.8   Spread risks.

Activity impacted	Distribution	Description of risks	Upper limit	Lower limit
3	Uniform	Design cost and delay overruns	0%	0%
	Uniform	Labour and material charges	0%	0%
4	Skewed triangular	Vandalism	0%	0%
5	Skewed triangular	Construction cost and delay overruns	8%	0%
	Skewed triangular	Material and equipment escalation	5%	0%
	Skewed triangular	Output specification failures	7%	0%
	Skewed triangular	Vandalism	0%	0%
6	Uniform	Performance standards	10%	0%
	Uniform	Operation and maintenance increases	8%	0%
	Skewed triangular	Electricty fluctuations	5%	0%
	Skewed triangular	Vandalism	6%	0%
7	Triangular	Exchange rate	10%	−10%
	Uniform	Performance standard deductions	0%	−25%
8	Skewed triangular	Revenue fluctuations	20%	0%
9	Skewed triangular	Commissioning cost and delay overruns	5%	0%
10	Triangular	Exchange rate	10%	−10%
	Skewed triangular	Performance standard deductions	20%	0%
11	Skewed triangular	Operation and maintenance increases	8%	0%
	Skewed triangular	Electricty fluctuations	5%	0%
	Skewed triangular	Vandalism	4%	0%
12	Skewed triangular	Finance floatation	12%	−8%
14	Skewed triangular	Business performance	8%	−8%
	Triangular	Advertisement demand	10%	−10%

VFM assessment

From the payment plan which is discounted at a rate of 3.5% as stipulated in the Green Book (HM Treasury, 2003) a NPC of £−752.18m is attained over the life of the concession. Combining this with the risk adjusted NPC of the PSC a VFM assessment table is formed (Table 20.10).

The VFM expected for assigning the PFA is depicted in Table 20.10, and suggests that under all the proposed risk scenarios the PFA would offer VFM compared to that of the PSC, limiting the future need for further analysis or sensitivity testing. The payment plan that contains the programme for the unitary payment made by the principal to the promoter must now be checked set against affordability. This will test the payment plan based on current sources of revenue available to the

Table 20.9   Economic parameters of the PFA.

Economic parameter	Base case	Mean	Standard deviation	Minimum	Maximum
IRR	18.4	14.4	1.9	8.8	19.6
NPV	363.2	287	47.8	138.4	410.6
Cash lock up	136.4	174.8	16	133.2	227
Pay back period (years)	10.2	11.4	0.7	10	14.2

Table 20.10   VFM assessment table.

Scenario	Risk adjusted NPC£(M)	Competitive neutrality adjusted NPC£(M)	VFM assessment £ (M)	Percentage
Maximum	−1085.59	−1175.59	−423.46	36.02%
Worst case	−975.59	−1065.59	−313.46	29.42%
Base case	−808	−898	−145.87	16.24%
Best case	−875.59	−965.59	−213.46	22.11%
Minimum	−835.59	−925.59	−173.46	18.74%

principal, to ensure they are able to service the future liabilities. Often, sinking funds may be used to address marriage problems between the payment plan and the revenue

When forming the model, the cost and revenue considerations of both the public and private sector form biases when discounting the model (see Grout, 1997). However, to say that the PFA experiences true market risk is inappropriate, especially where projects in the past have actively reduced the performance deductions that may be incurred upon the revenue stream (NAO, 2003c). Therefore, this lends itself to unique pricing structures in relation to risk, which may be further enhanced through the application of PPPs. Discount rates should never be a decisive factor when determining the VFM of a specific procurement route (see, e.g., the concerns about the credibility of the PSC and PFA aired by Spackman, 2003).

20.3   Conclusion

PSC and PFA are formed from value- and risk-based inputs. There are several systems available to produce and test the inputs that are assigned for such models. However, practitioners must be aware of the deficiencies that may reside in such techniques and determine appropriate output displays.

The model produced provides an insight into the quantitative techniques used to construct the PSC, but the PSC is just one instrument used to test the validity of PF projects. Further analysis in the form of bankability and affordability must also support the PSC to establish private sector interest and public sector ability to service the liabilities that are granted by the concession. Grace periods, tax holidays, guarantees, counter-trade, risk acceptance and tax breaks may all be granted by principal organisations to promote the application of PF in their host country. However, the principal's ability to service such support must be established.

Whilst the model proposed utilises the NPC, NPV and IRR to construct the PSC and PFA to form the VFM assessment, further development in the PSC may be used to allow a theoretical IRR to be produced. Assigning the PSC revenue, which would be of a comparable nature to that of the PFA, according to the performance standards and payment structures of the concession would allow the PSC to produce a theoretical IRR. However, such models may only become useful if the operational management systems of the public and private sector are standardised, operating upon measurable and monitorable performance standards.

Part Three

References

4Ps Public-Private Partnership Programme (2004) *Building Schools for the Future – an Overview*. 4Ps, London. www.4ps.co.uk/home.aspx?pageid:3.1.3#bsf&Lgov.

Akintola, A., Hardcastle, C., Beck, M. *et al.* (2003) Achieving best value in private finance initiative project procurement. *Construction Management and Economics*, 21, 461–470.

Al-Momani, H.A. (2000) Construction delay; A quantitative analysis. *International Journal of Project Management*, 18, 51–59.

Audit Commission (2003) *PFI In Schools: The Quality and Cost of Buildings and Services Provided by Early Private Finance Initiative Schemes*. London.

Broadbent, J. and Laughlin, R. (1999) The PFI; clarification of a future research agenda. *Financial Accountability and Management*, 15(2), 95–114.

Broadbent, J., Gill, J. and Laughlin, R. (2003) Evaluating the Private Finance Initiative in the National Health Service in the UK. *Accounting, Auditing and Accountability Journal*, 16(3), 442–445.

Department of Finance and Personnel (2004) *Economic Appraisal Guidance: Optimism Bias Calculator.* Http://www2.Dfpni.Gov.Uk/Economic_Appraisal_Guidance/Ob-Calculator-Civil-Eng.Xls.

Froud, J. and Shaoul, J. (2001) Appraising and evaluating PFI for NHS hospitals. *Financial Accountability and Management*, 17(3), 247–270.

Government Of South Australia (2004) *Partnership SA Guidelines*. Department Of Treasury And Finance Project Analysis Branch, Melbourne.

Grout, A.P. (1997) The economics of the Private Finance Initiative. *Oxford Review Of Economic Policy*, 13(4), 53–66.

Heald, D. (2003) Value for money tests and accounting treatment in PFI schemes. *Accounting, Auditing and Accountability Journal*, 16(3), 342–371.

HM Treasury (1998) *How to construct a Public Sector Comparator*. Treasury Task-force, Technical Note No 5. Office of Government Commerce, London.

HM Treasury (2003) *The Green Book: Appraisal and Evaluation in Central Government*. HM Stationary Office, London.

HM Treasury (2004) *Value for Money Assessment Guidance and Quantitative Assessment, User Guide*. HM Stationary Office, Norwich.

Ho, P.S. and Liu, L.Y. (2001) An option pricing-based model for evaluating the financial viability of privatised infrastructure projects. *Construction Management and Economics*, 20, 143–156.

KPMG (2002) *Supplementary Green Book Guidance: Adjustment for Taxation in PFI vs PSC Comparisons*. HM Stationary Office, London.

Lamb, D.J. and Merna, A. (2004a) *A Guide to the Procurement of Privately Financed Projects: an Indicative Assessment of the Procurement Process*. Thomas Telford, London.

Lamb, D.J. and Merna, A. (2004b) Development and maintenance of a robust public sector comparator. *The Journal Of Structured And Project Finance*, 10(1), 86–95.

Lane, J.E. (2000) *New Public Management*. Routledge, London.

Merna, A. and Lamb, D. (2004) *Project Finance: The Guide to Value and Risk Management in PPP Projects*. Euromoney Books, London.

Ministry of Finance (2002) *Public Private Comparator*. The Hague Netherlands, Ministry of Finance PPP Knowledge Centre. www.Minfin.Nl/Pps.

Mott Macdonald (2002) *Review of Large Public Procurement in the UK*. Published Under HM Treasury, London.

Part Three

National Audit Office (2003a) *The Operational Performance of PFI Prisons*. HM Stationary Office, London.

National Audit Office (2003b) *PFI: Construction Performance*. HM Stationary Office, London.

National Audit Office (2003c) *Northern Ireland Court Service PFI: The Laganside Courts*, HM Stationary Office, London.

National Treasury (2001) *PPP Manual – Guidelines on Dealing with Unsolicited Proposals for National and Provincial Government PPPs*. National Treasury, South Africa. http://Www.Treasury.Gov.Za/Organisation/Ppp/Manual/K.Pdf.

New South Wales Government (2000) *Working with Government: Guidelines for Privately Financed Projects*. NSW Government, Sydney.

Office Of Government and Commerce (2003) *Procurement Guide 06: Procurement And Contract Strategies*. London.

Partnership Victoria (2001) *Public Sector Comparator: Technical Note, Department Of Treasury And Finance*. Melbourne Victoria.

Partnerships UK (2004) *PFI In Primary Care (NHS LIFT)*. Partnerships UK, London. www.Partnershipsuk.Org.Uk/Refer/Index.Htm#Lift.

Pender, S. (2001) Managing incomplete knowledge: why risk management is not sufficient. *International Journal Of Project Management*, 19, 79–87.

Public-Private Partnership Unit (2001) *Private Participation in the Provision of Public Services: Guidelines for the Private Sector*. Department Of Treasury And Finance, Melbourne.

Raz. T. and Michael, E. (2001) Use and benefits of tools for project risk management. *International Journal Of Project Management*, 19, 9–17.

Simon, P., Hillson, D. and Newland, K. (1997) *Project Risk Analysis and Management Guide*. The Association For Project Management Group Ltd, Norwich.

Smith, N.J. (1999) *Managing Risk in Construction Projects*. Blackwell Science Ltd, Oxford.

Spackman, M. (2003) Public-private partnerships: lessons from the British approach. *Economic Systems*, 26, 283–301.

Treasury Taskforce (1998) *Technical Note No. 5: How to Construct a Public Sector Comparator*. Office of Government Commerce, HMSO, London.

Vasudevan, S. and Higgins, B. (2004) Strategic energy risk management for end users. *Journal of Structured and Project Finance*, 10(1), 74–78.

Williams, T.M. (1994) Using a risk register to integrate risk management in project definition. *International Journal of Project Management*, 12(1), 17–22.

Part Three

Developing a Framework for Procurement Options Analysis

Darrin Grimsey and Mervyn K. Lewis

21.1 Introduction

PPPs are a valuable procurement option but never have been, nor will they ever be, the dominant form of infrastructure provision. There has been some discussion of what project types might be most suitable for PPPs, but the question invariably comes down to a case-by-case analysis. This chapter develops a framework for making this decision in a systematic way and illustrates the proposed approach using a case study of a procurement analysis for a hospital project.

21.2 What do PPPs Bring to Procurement?

PPPs involve the provision of public assets and services through the participation of the public sector, the private sector and members of the community. Generally speaking, PPPs fill a space between traditionally procured government projects and full privatisation. PPPs are not privatisation, because with privatisation the government no longer has a direct role in ongoing operations, whereas with a PPP the government retains ultimate responsibility. Nor do PPPs involve simply the one-off engagement of a private contractor to provide goods or services under a normal commercial arrangement. Instead, the emphasis of PPPs is on long-term contracts and the term PPP covers a variety of transactions where the private sector is given the right to operate, for an extended period, a service traditionally the responsibility of the public sector alone.

However, the defining characteristic of a PPP is not private sector involvement in itself, but 'bundling'. Under traditional methods for procuring infrastructure, the public sector obtains new assets – for example, roads, bridges, schools, hospitals, buildings etc. – separately from services. The associated services have then been delivered by public sector organisations either by using their own workforce or by outsourcing or contracting out, fully or in part, the service provision to other specialist operators. External contracting out and outsourcing has grown over the last two decades as public (and private sector) organisations have searched for ways to enhance efficiency and make better use of resources. A partnership agenda takes this idea further and offers a different approach to traditional procurement because the acquisition of infrastructure assets and associated services is accomplished with one long-term contract, under which the initial capital outlay and ongoing services are financed by the private sector.

One of the major objectives of a PPP is to harness private sector management expertise, and the market disciplines associated with private ownership and finance, for the provision of public services. Of course, private sector skills are also employed under traditional procurement when the public sector engages private sector design skills and private constructors. What the PPP adds to such arrangements is a different type of inducement for those involved. The private sector entity is encouraged to plan beyond the bounds of the construction phase and incorporate features that will facilitate operations and maintenance, within a cooperative framework. Under the terms of the contract, the private sector partner is paid for the delivery of the services to specified levels and must itself organise all the managerial, financial and technical resources needed to achieve the required standards. Importantly, the private sector bears the risks of meeting the service specification.

There is a long history of publicly procured contracts being delayed and turning out to be more expensive than budgeted. Transferring these risks to the private sector under a PPP structure, and having the private sector bear the cost of design and construction overruns, is one way in which a PPP can potentially add Value for Money (VFM) in a public project. There are also risks attached to site use, building standards, operations, revenue, financial conditions, service performance, obsolescence and residual asset value, amongst others, to be taken into account when evaluating whether the PPP route to public procurement constitutes good VFM.

Nevertheless, a bundled approach will not suit every project. Specific questions associated with a bundled approach include: Are there efficiency gains to be obtained from bundling? What particular services can sensibly be combined into one contract? Or would a number of separate (unbundled) contracts be preferable? These are matters that can benefit from a systematic analysis. The next section outlines a framework for guiding the decision-making process associated with the adoption, or otherwise, of a bundled PPP.

The framework developed builds on and extends earlier work by the authors on PPPs and traditional procurement (Grimsey and Lewis, 2004c, 2005a), risk management (Grimsey and Lewis, 2002), VFM (Grimsey and Lewis, 2004a, 2005b) and contractual governance (Grimsey and Lewis, 2004b). It also draws on ongoing work at Ernst & Young Transaction Advisory Services (Ernst & Young, 2006a,b).

21.3   Developing a Methodology

Public procurement of any service or facility must begin with an analysis of the need, or rationale, for the project as defined by the preliminary business

Part Three

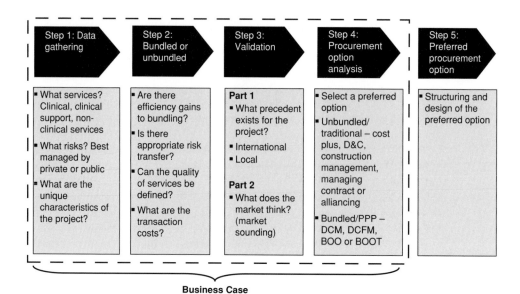

Figure 21.1   A framework for procurement options analysis.

case. Questions to be asked include: Is there a pressing need for change in the service? Is now the time for a review of the service and its delivery? Is the *status quo*, with incremental rather than substantial change, an option? If change is required, is it likely to involve significant new investment? An assessment of such questions will be supported by the use of appraisal techniques such as cost–benefit analysis, environmental impact statements and perhaps even a triple bottom line assessment.

Once this analysis is completed, there is a need to consider what procurement option is appropriate for the project. Obviously, this decision must be made on a case-by-case basis. In doing so, however, it is important that there be a proper appreciation of the risks, and who might bear them, and that the comparisons between publicly and privately financed options be fair, realistic and comprehensive. It is the authors' belief that the process can be aided by establishing a deliberate, step-by-step approach to selecting the preferred procurement model for the project concerned.

Developing a systematic approach to procurement options analysis involves constructing a decision-making framework to select the most appropriate procurement model and financing mode for a particular project. In this section, a decision-making process is outlined and elaborated on, which is based around five stages: data gathering; bundling analysis; procurement validation; procurement options analysis; and selecting the preferred procurement approach. The key elements involved in this process are depicted in Figure 21.1.

21.3.1   Data gathering

Data gathering includes performing a diagnostic review of the existing services and understanding the way in which the current service arrangements are being supplied. For example, are the services being delivered in-house or are they outsourced? The key tasks involve identifying the relevant services for bundling, determining the key procurement risks associated with following a bundled or unbundled approach, and understanding the unique service provision characteristics of the project.

Selecting the most appropriate procurement option for the project consequently requires a sound understanding of the following issues:

- Services. What are the key services that must be delivered by the infrastructure? What (if any) part or parts of the proposed service mix is a service that the government itself should deliver to its citizens (termed a 'core' service)? In the case of a prison, for example, the core service might be custodial services. Are there other services that might fall into this category (e.g. education, vocational training, medical services to prisoners)? What are the non-core services that could be delivered by the private sector (e.g. cleaning, laundry, security systems, facilities maintenance, transport of prisoners)?
- Risks. What are the project risks? A risk-management process can usefully be run in parallel and identify many of the risks. This will inform the procurement options analysis, and help highlight specific risks that might be better managed by the public or private sector.
- Characteristics. What are the characteristics of the project or business? What is unique about the infrastructure and what features make this facility different from others of its type? Is it a 'greenfield' project, redevelopment or a combined redevelopment and new facility?

The data collected from this exercise provides the base from which objective decisions can be taken in the subsequent steps.

21.3.2   Bundling analysis

Considering a bundled or unbundled approach involves assessing the data gathered in step 1 and involves, *inter alia*, articulation of the qualitative benefits (e.g. efficiency gains) and risks of bundling services, assessment of which risks are better managed by the private or public sectors and a quantification (where possible) of the incremental costs or cost savings of bundling services. The result is a decision as to whether any, or all, of the services should be bundled and procured as one package. This decision requires an objective analysis on the following:

- Efficiency. Are there efficiency gains from bundling services together? What are they?
- Quality. Can the services be adequately defined (in terms of quality) and specified in a contract?
- Cost. What are the transaction costs?

Bundling means that only one party is in charge of building, maintaining and operating (providing core and/or non core services) an asset. This would mean that the government writes a contract that defines the quality

Part Three

Table 21.1 Bundled and unbundled procurement models.

Bundled models	Unbundled models
Bundled approach includes the following procurement models: • Design, construct, maintain (DCM) • Design, construct, maintain, finance (DCMF) • Build, own, operate (BOO) • Build, own, operate, transfer (BOOT)	The unbundled approaches are centred on construction based procurement models, e.g.: • Cost plus • Design and construct • Construction management • Management contracting • Alliance

of these services. Bundled procurement models (see Table 21.1) are those such as serviced infrastructure models (design, construct, maintain (DCM), design, construct, maintain, finance (DCMF)) with performance based payments, concession models including full operations and franchise arrangements (build, operate, transfer (BOT), build, own, operate, transfer (BOOT), build, own, operate (BOO)), and 'privatisation' models involving market-based payments. Privatisation models are usually considered to fall outside of what is generally regarded as PPP-type procurement. However, we have included them here for completeness and also to illustrate the full risk spectrum when considering private sector involvement in infrastructure services.

By contrast, the unbundled approach means that the government would need to write at least two, or more, contracts, whereby one party would build the asset and another would operate the services. This could include various parties, including the public sector itself. Unbundled models include cost-plus reimbursement with fees based on fixed amounts, target cost arrangements involving alliance contracting, management contracts and construction management, and traditional contracting methods based on fixed price, bill of quantities and schedule of rates.

The main rationale for bundling is that, by putting one party in charge of all stages of the production chain, cost savings can be made over the whole life-cycle which could result from innovation, risk pricing and trade-offs between higher initial costs and lower operating costs. The government can extract the benefit of these savings by running a competitive process for the contract. However, efficiency savings can come at a cost that manifests itself as either a reduction in quality, e.g. a social cost, or an increase in the initial contracting costs. What needs to be considered is whether the services can be adequately defined and contracted such that the risk of a reduction of quality is minimised or removed. If this is possible, such efficiency gains may be counterbalanced by the increased costs of contracting for the bundled services.

21.3.3   Procurement validation

Procurement validation occurs at two levels:

■ Benchmark projects: the aim of these projects is to challenge the assumptions underpinning the bundling analysis with reference to other projects procured under a PPP framework or other similar bundled approaches. A 'desk top' study of precedent projects is conducted to identify major issues that may impact upon the project and shape the procurement options analysis.

■ Market sounding: this process seeks an independent confirmation of the assumptions made by the project team. It involves meetings with private sector constructors, facilities managers and financiers to establish interest and ascertain likely issues prior to the options analysis. This market intelligence and initial 'testing' is particularly important for bundled PPP projects for which the number of potential bidders can be lower than for the unbundled approach.

21.3.4   Procurement options analysis

Having decided whether it is worthwhile going down the bundled or unbundled approach, the framework should enable a consideration of the various procurement models available. Both bundled and unbundled approaches are associated with a number of different procurement models. Even where a relatively standard procurement model is chosen, this will inevitably require tailoring to the project.

The models can be compared in terms of criteria such as price certainty, flexibility, risk transfer and incentives structures. For example, amongst the bundled models, a design, construct, maintain (DCM) approach offers a high degree of flexibility and price certainty because the risks are relatively well understood and there is a tightly defined service specification. However, there is often less risk transfer to the private sector and less incentive to innovate than a build, own, operate (BOO) model. These trade-offs are illustrated in Figure 21.2.

Amongst the unbundled approaches, traditional contracting approaches, such as fixed-price contracts, have high price certainty, transfer risks to the contractor so long as the specification remains unchanged and create the incentive to keep within the defined scope and contractual terms. However, there exists limited flexibility for design changes and variations. By contrast, alliancing and cost-reimbursed models build in flexibility, but leave risks with the public procurer. They therefore often have lower price certainty, and cost reduction may be secondary. Figure 21.3 depicts the trade-offs between the models.

In considering the different options, there should be a focus upon VFM, affordability and the public interest. VFM relies on risk allocation, whole-of-life costing, innovation, asset utilisation, economies of scale, bid costs and financial skills. Quantitative and qualitative considerations need to be evaluated. Affordability depends on third party revenues, capital receipts, current and future budgets and additional funding sources. The public interest test considers access, equity and project effectiveness.

Any procurement option needs to be measured against some common objectives. The relevant criteria include:

Part Three

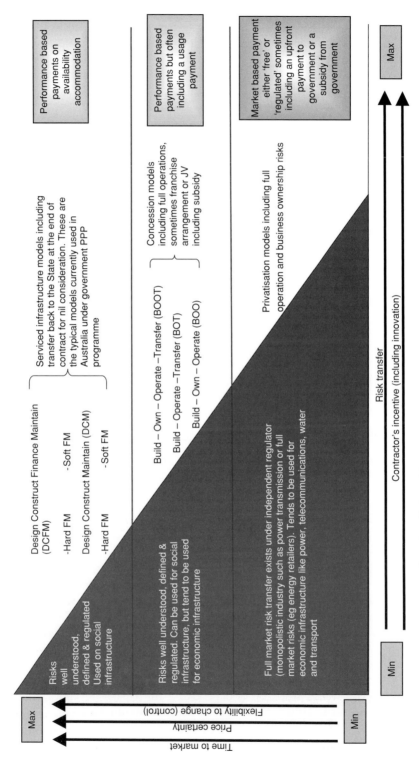

Figure 21.2 Procurement options analysis: bundled models.

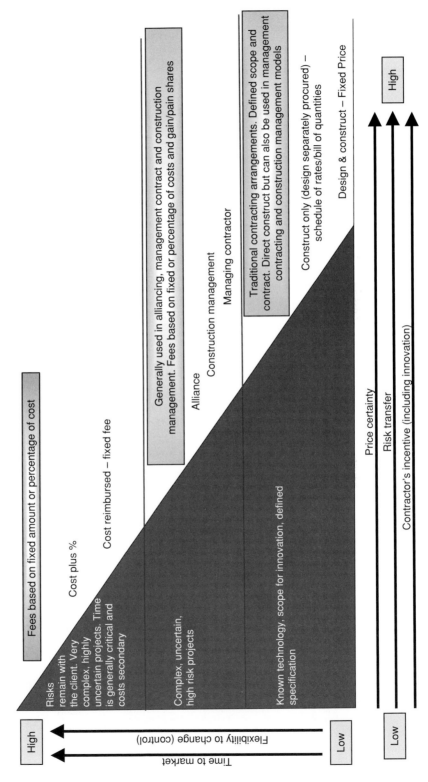

Figure 21.3   Procurement options analysis: unbundled models.

Part Three

- To demonstrate that the procurement option represents best VFM, assessed on a rigorous and objective basis.
- To ensure that the procurement option is consistent with the public interest.
- To retain, where appropriate, the ability to test the competitiveness of elements of the service from time to time.
- To make sure that any procurement option is sufficiently flexible to capture the benefits of changes in industry practices and technology over time.
- To ensure that the procurement option captures the benefits of innovative approaches to the design, construction, operation, maintenance and/or finance of infrastructure facilities.
- To ascertain that any procurement option enables appropriate responses to changing service requirements over the term of the contract.
- To confirm that any procurement option achieves timely provision of infrastructure facilities.
- To be confident that any procurement option provides certainty and continuity of service delivery.
- To verify that any procurement option is consistent with the government's overall social and fiscal objectives.

21.3.5   The preferred procurement option stage

This stage involves the structuring and design of the preferred procurement option in detail. Usually, this step takes place after approval of the full business case. Parameters to consider include cost, time, quality and risk. This step may include the development of proposed commercial structures to ensure appropriate risk allocation. Eventually, the project is developed to a point where it can be taken to market.

As part of this approval process, several questions need to be asked of any chosen project design. These are:

- Is the project as structured affordable?
- Is the project bankable?
- Have the outputs been defined with sufficient clarity?
- Is the risk allocation optimal?
- Are the key terms and conditions defined unambiguously?
- With what degree of certainty is there likely to be VFM?
- Is the timetable achievable?
- Is the project team committed and resourced appropriately?
- Is there the needed stakeholder commitment?

Case Study 21.1: Procurement Analysis for a Hospital Project

The procurement options framework outlined above is illustrated by considering how it would be used in the case of the redevelopment of a large tertiary hospital project. Whilst the case study is primarily meant to illustrate the options procurement process discussed above, it is based on real procurement studies carried out by the authors on several hospital projects in Australia. Complexity is introduced into this case study by allowing for a number of indicative project delivery scenarios

Table 21.2 An assessment of procurement models.

Delivery model	Scope	Risk/uncertainty	Size	Asset life
Fixed price	Can be defined Input based	Risks can be articulated, valued and managed Public sector bears the cost risk of scope expansion and design risk Damages for late completion usually apply, but often contractors are able to argue shared or government blame Government retains quality/fitness for purpose risk Adversarial arrangement that often leads to disputes	The project would be too large for a fixed-price model	N/A
Managing contractor	Can be defined Functional brief supported by concept designs	Risks can be articulated, valued and managed Contractor manages the risks for the public sector Government keeps the cost risk of scope, design and delay risks Government bears quality/fitness for purpose risk Incentives can be applied to align the managing contractor with the procurer on cost, time and quality objectives	No issues	N/A
Alliancing	Can be defined Functional brief supported by concept designs The public sector has the flexibility to make changes at any time, but will bear the costs	Risks can be articulated, valued and managed, but alliancing is based on a risk sharing Alliancing is not appropriate where risks can be identified and understood. It is more suited to high risk projects that have significant uncertainties	No issues	N/A
PPP, i.e. a bundled approach	Can be defined Output based for the asset and ancillary infrastructure services, e.g. cleaning Can require more time and effort than input specifications, but applies greater discipline to ensure the public sector receives what it needs over the life of the operating concession period granted	Risks can be articulated, valued and managed Risks are allocated between the parties Generally, the government is able to transfer more of the risks associated with cost, time and quality to the private sector Risk transfer includes whole-life risks associated with maintaining the assets to a specified level of performance	No issues	The asset life is sufficient

Part Three

under a range of development options, which include a staging of government funding for the redevelopment.

There is added complexity in the case study because, while the hospital visualised for redevelopment is part of a regional health service in a metropolitan environment, it is the main facility and the key referral point for other hospitals in the region on a 'hub-and-spoke' model of health service delivery in which various management and support services are supplied to the other district hospitals. Allowing the hospital to continue to function is a prerequisite of the project. Accordingly, the project involves the relocation of some facilities, the building of a new hospital, in which construction is sequenced to enable continued operation of the hospital, the transfer of services to the new facilities and the demolition of the old buildings not being retained.

Because of the need for a sequencing of construction work, the project considers two redevelopment options, both of which envisage a complete rebuilding of the hospital. The first option contemplates delivering the project in a single stage, i.e. within a single sequenced construction programme, whereas the second option contemplates delivery in two, or more, stages. The time between stages is unknown and depends on future spending priorities. In both options, construction is staggered in order to allow for the continued operation of the hospital. This entails an ordered moving programme to keep the hospital operational during construction. However, from the viewpoint of the procurement options analysis, the impact of staging, as it relates to the second option, needs to be considered in the procurement study when determining a preferred procurement strategy.

Data gathering

As the major facility within the regional health area, the hospital provides key services to two other district hospitals. They, in turn, provide services to other healthcare facilities under the regional *aegis*. Continuity of operations is a critical aspect of the project.

In both, the single-stage, and two-stage options for the construction and delivery of redevelopment, a progressive relocation programme is envisaged to keep the hospital operational during construction. However, the risks of a single- and two-stage redevelopment are very different. In almost all cases, the risks are increased when the project proceeds in two stages rather than a single stage. Planning is more protracted and the outcomes more uncertain. Site availability and site access are less certain. Disruption is increased and potentially prolonged. Design risks and changes in scope are more likely. Due to the age of the existing facilities, operational risks are increased by staging. There is also the risk that the government will consider the later stages of the project to be less urgent and will therefore re-order procurement priorities to other sectors.

As part of the data gathering exercise, a complete list is made of all services provided. Altogether 24 service categories can be identified and classified into, first, the type of service and, second, how they are currently disposed. They include:

- Clinical support or non-clinical services provided by the existing hospital
- Those services provided on a region-wide basis
- Those services currently outsourced

This information is preliminary to the bundling analysis.

Bundling analysis

For this project, the decision 'to bundle or not to bundle' is conditioned by a number of factors:

- Government policy, resulting from protocols with public sector unions, mandates that 'core' healthcare services be delivered by the public sector. Clinical services and clinical support services are consequently excluded from the bundle. This includes doctors and nurses, diagnostic imaging, health information services, pharmacy, pathology and collection services, library, medical gases, theatre transport.

- As a corollary, the healthcare services considered for bundling relate only to the provision of non-clinical services, i.e. 'hard' and 'soft' facility maintenance services.
- Linen and laundry services are currently outsourced on a region-wide basis. Since the hospital in question is presumed unlikely to generate volumes that would encourage innovation, such as on-site facilities, and since there are insufficient synergies with other services, these are also excluded from the bundling analysis.
- Car parking, with its associated construction, operation and associated risk, is also excluded. This service has the potential to generate a separate revenue flow to the health authority sufficient to finance repayment of a government loan for construction, with additional returns to fund other activities, such as purchases of medical equipment.
- The services considered suitable for bundling are engineering maintenance and facility management, utilities, medical gases services, installation and maintenance, food services, non-theatre patient and general transport, waste management, patient services assistants, cleaning, security, grounds and garden maintenance, pest control and retail facilities. The analysis of the various services suggests that there are considerable synergies, efficiencies and quality improvements likely to be gained by grouping these services and adopting a common platform for their delivery. All of these services are considered suitable for bundling and the preferred option is for all these services to be bundled together.

Project validation

Step 3 of the methodological framework incorporates testing the assumptions underlying the analysis, particularly with respect to bundling, with reference to other comparable projects, and seeking an independent perspective through a market-sounding exercise.

As concerns benchmarking, desk top research reveals few projects relevant for benchmarking purposes in two respects. First, there are examples of services being provided on a bundled basis to retained as well as refurbished or new facilities on the same site, but either the retained buildings are relatively uncomplicated, as exemplified by Hereford Hospital, or the refurbishment is of a much more limited scope or different performance indicators were applied to the retained buildings, as exemplified by Manchester Super Hospital or Barts and Royal London Hospital. Second, no examples are found of a multi-staged redevelopment of a major teaching hospital using a PPP approach, as is envisaged in the second procurement scenario. There are projects that adopt a staged approach, such as the NHS's Local Improvement Finance Trust (LIFT) programme for primary healthcare facilities in the UK (see Grimsey and Lewis, 2007), but these new hybrid PPPs involve different locations, less complex designs and competitively test VFM at each stage rather than use upfront pricing.

As concerns market sounding it is felt that it is valuable to ascertain whether the market, particularly the facilities maintenance contractors, agree that efficiency gains identified as potentially being achievable through a bundled PPP approach, are realisable. The key cost components of a PPP scheme are:

- Construction
- 'Soft' facilities management (e.g. catering, portering etc.)
- 'Hard' facilities management (e.g. boiler maintenance and major refits)
- The value placed on risk

VFM can only be achieved by reducing one or more of these costs. This can be achieved in two ways. One is by reducing construction and facilities maintenance costs through value engineering the design, innovation and by efficiency savings. The other is if the private sector assumes risks which would otherwise have been borne by the public sector.

Another key consideration for the project relates to the impact of staging on the viability of the project and therefore its attractiveness to the market. In considering a staged construction (and budget allocation) option it is useful to determine whether there are commercial constraints that will

Part Three

affect the project being delivered as a PPP. This potentially has three elements: market constraints, i.e. whether the structure is sufficiently attractive to potential private sector participants who are capable of delivering a commercial response to the envisaged structure; technical constraints, i.e. whether the scale of the investment required is deliverable in the timeframe envisaged; financial constraints, i.e. whether the market has the capacity to fund the investment under the proposed risk sharing arrangements.

The market-sounding process is an opportunity for industry to engage with the project team while the project is in the detailed business case stage. There is no obligation for any industry participant to engage in the process, but from the viewpoint of the project, it is a cost-effective way to test the bundling scenarios developed and the staging options by discussing their attractiveness with potential private sector participants and financial investors. The aims of this market-sounding process are: to identify issues about which the private sector has concerns; to ensure their appropriate consideration during the development of the project; to obtain feedback from the private sector on specific commercial, technical and procedural matters which can be used to further develop the project; and to raise industry awareness of the project

In this case study a limited market sounding is performed with well-respected representatives of:

- The financial industry, with particular reference to the issue of a staged approach to funding in the context of a PPP
- The facilities maintenance and soft service industry with respect to the issues of bundling services and potential efficiency gains that might be achieved through such an approach
- The construction industry, on the issue of a staged approach to redevelopment, within a PPP approach framework

Organisations that are identified as trustworthy and as possessing the relevant expertise are issued with a letter of invitation and an initial meeting explains the scope of the project and the purpose of the market-sounding exercise, the need for confidentiality, as well as setting out caveats regarding this not being a procurement exercise. The market sounding then takes the form of semi-structured interviews based around a 14-point questionnaire. During the meetings and interviews it is emphasised that the exercise is not part of the procurement process, and that they should not regard themselves as chosen bidders. A formal process begins only when the government has approved a business case and funding to proceed under a PPP or an alternative, such as management contracting.

Those who participate generally agree that more opportunity exists for innovation and efficiency gains where greater numbers of non-clinical services are bundled together. In particular, some of the soft services, e.g. cleaning, are seen as providing significant opportunity to realise whole-life trade-offs. No specific exclusions are identified and it is felt that the greater the number of non-core services included in the bundle the better will the project perform. Efficiency gains relate to innovation in design, a whole-life approach to building, operating a facility for a fixed concession period, and being able to manage the risks.

There is a general consensus that a staged approach would introduce significant inefficiencies, as compared to a single-stage redevelopment. This could result in increased construction costs arising from preliminaries and site set-up etc., the retention of inefficient buildings and increased risks of engineering services failure and integration issues. Risks are seen to arise from the scope of the second and subsequent stages, the timing of the second stage and the interface issues thereby created. In particular, in terms of the financing requirements, equity would need to be sized to take into account risks and uncertainties of such a large change to the contract at the time. If the private sector entities were required to build into the bid the risks involved in future stages, then this requirement could result in a potentially significant risk premium being added. This risk premium would increase if pricing for subsequent stages is to be locked in upfront and would be needed to cover timing and scoping risks.

Alternatively, there could be contract change management processes for demonstrating VFM in stage 2, similar to the UK's LIFT/LEP model. These processes could include competitive tendering of some of the construction sub-contracts and open-book accounting and disclosure of the costs with the public sector body. Nevertheless, it was thought that these models may not be appropriate for a large tertiary hospital campus where the single site and interface risks are significantly greater than those associated with the UK's LIFT programme. Also, two VFM 'drivers' could not be realised through contract change management. Specifically innovation would be difficult to demonstrate if the design and integration of stage 2 would not be market tested, and residual risk of service performance and whole-life considerations would have less force.

In general, there is scepticism as to whether a staged approach would represent VFM for the public sector. Funding a staged redevelopment, where the second stage is undefined, could be expensive, owing to uncertainty about future market conditions, the opportunity cost and returns to sponsors of other future investments, and the change in requirements of hospital buildings. In these circumstances, the constructors see some merit in alliance-contracting techniques where VFM is validated by open-book accounting for costs. About 70% of the construction price is sub-let to the various trades and these can still be competitively tendered. In addition, hard facilities maintenance, i.e. those services not benchmarked or market tested, e.g. engineering services, could similarly be validated by open-book accounting and benchmarked against similar facilities.

Procurement analysis

A decision to apply a traditional, or a PPP-based, delivery model requires a detailed consideration of the project characteristics to test whether the project is better suited to a particular delivery model. The analysis below considers the delivery models according to the project's scope, i.e. the extent to which the government can define the scope; risk/uncertainty, i.e. to what extent the government can articulate, value and manage the risks; size, i.e. whether the project is the right size for the delivery model; and asset life, i.e. the degree to which the asset lends itself to a PPP-style whole-life approach to its procurement.

Table 21.2 summarises the results of this assessment. On this basis, it is concluded that the fixed-price and alliance models are unlikely to be suitable. In the former case, the size of the project and the adversarial nature of fixed-price contracting suggest that the public sector would retain significant risks and associated management responsibilities. In the latter case, the government has a good understanding of the project risks and can define and quantify them to support allocation under a contract. As a general rule, alliancing would not be appropriate where risks can be identified and understood. It is more suited to high-risk projects which have significant uncertainties and risks that cannot be appropriately defined and quantified.

On balance, both managing contractor and PPP delivery models would seem to be more suitable for this project. PPP delivery is a whole-life approach that tends to transfer facilities-based risks to the private sector. Evidence to date suggests that it is a good option for major hospital redevelopment projects and is used extensively in the UK, Australia and Canada, as compared to the more traditional managing contractor approach. By contrast, the managing contractor approach could give the government more control during construction, although this control is typically achieved at a cost.

Preferred procurement option

The choice between these procurement options is governed by whether a single- or multiple-stage development strategy is pursued. For a single-stage redevelopment the provision of support services to facilities on the same site under a bundled approach is likely to realise efficiency gains. This would point to a PPP approach. Opportunity exists to deliver VFM from synergistic relationships between design, construction, whole-life facilities maintenance and risk transfer.

However, there are a number of difficulties in using the PPP route for a staged redevelopment, when the public sector authorities keep control over the time and scope of future changes. Changes to interfaces and service levels are likely to be significant, throwing open the appropriateness of the risk allocation and pricing. Ultimately the performance risk, design and services innovation cannot be competitively market tested for future stages as these ultimately fall back on the incumbent sponsor. In addition, the staged option presents the service provider with significant opportunity to open up risk pricing issues and renegotiate its position. For example, the service provider may try to renegotiate its key performance indicators to reduce its performance risk by arguing that the impact of staging has increased its risk exposure. The public sector bodies involved may be able to manage and control these issues better under a managing contractor approach. This approach would enable the government to manage its construction and service contracts in a way that maximises competition and controls the future stages.

21.4   Conclusion

There now exists a variety of delivery models, embracing traditional construction-based procurement methods and PPPs of various forms, along with hybrids of these, that can accommodate different infrastructure service needs. The decision as to which one to use is conditioned by the specific project, but the choice can be aided by adopting a systematic framework to procurement analysis that is capable of being applied to a wide range of different projects. A five-stage approach is outlined here embracing data gathering, bundling analysis, procurement validation, procurement option analysis and the preferred procurement option.

This framework is elucidated by the case study of the procurement options analysis for an illustrative large hospital redevelopment based on and consequently representative of actual redevelopment projects. The example given demonstrates how the project risks and characteristics can be used to analyse objectively what is the best procurement model for the project. Once a procurement model is identified it can be designed in detail such that it is tailored to the project. This task is usually carried out at the next stage once the business case is approved and the project can be developed to an appropriate level to be taken to market.

References

Ernst & Young (2006a) *Impact of Property on Public Infrastructure*, a study by Ernst & Young on the impact of property in various infrastructure projects including PPPs. Ernst & Young Project Finance and Real Estate Advisory Services, Australia.

Ernst & Young (2006b) *PPPs in Education*. Ernst & Young Project Finance Advisory Services, Australia.

Grimsey, D. and Lewis, M.K. (2002) Evaluating the risks of public private partnerships for infrastructure projects. *International Journal of Project Management*, 20(2), 107–118.

Grimsey, D. and Lewis, M.K. (2004a) Discount debates: rates, risk, uncertainty and value for money in PPPs. *Public Infrastructure Bulletin*, 3, 4–7.

Grimsey, D. and Lewis, M.K. (2004b) The governance of contractual relationships in public private partnerships. *Journal of Corporate Citizenship*, 15, 91–109.

Part Three

Grimsey, D. and Lewis, M.K. (2004c) *Public Private Partnerships: The World-wide Revolution in Infrastructure Provision and Project Finance.* Edward Elgar, Cheltenham.

Grimsey, D. and Lewis, M.K. (eds.) (2005a) *The Economics of Public Private Partnerships. The International Library of Critical Writings In Economics.* Edward Elgar, Cheltenham.

Grimsey, D. and Lewis, M.K. (2005b) Are public private partnerships value for money? Evaluating alternative approaches and comparing academic and practitioner views. *Accounting Forum*, 29(4), 345–378.

Grimsey, D. and Lewis, M.K. (2007) Public–private partnerships and public procurement. *Agenda, A Journal of Policy Analysis & Reform*, 14(2), 171–188.

Part Three

22

The Payment Mechanism in Operational PFI Projects

Jon Scott and Herbert Robinson

22.1   Introduction

Public sector bodies put forward a VFM case for procuring a project through the PFI route which rests upon risk transfer and efficiency in service delivery. The payment mechanism puts into financial effect the allocation of risk and service performance when PFI projects become operational. However, there are several factors affecting the role of the payment mechanism as an incentive for the service provider to improve performance, or as a tool for financial deductions when services are not delivered in accordance with the PFI contract.

This chapter discusses the function of the payment mechanism in the delivery of public services procured through PFI. It starts with an outline of the key principles underpinning PFI projects and the VFM arguments. Key components of the payment mechanisms, such as the output specification which defines the services required by the public sector client, and the performance measurement system to monitor the level of services delivered by the private sector service provider, are then examined. Using a case study approach, findings from public sector clients and private sector operators on specific issues affecting the effectiveness of the payment mechanism in improving service performance and providing VFM in PFI projects are analysed and discussed.

22.2   The Key Principles

The policy objective of PFI is to improve public services and is underpinned by a theory focusing on the delivery of services rather than the ownership of assets. The contestability of public services, i.e. whether the private sector can deliver the equivalent services cheaper or at better quality, is at the heart of PFI theory. There is a number of key principles associated with the delivery of PFI projects. First, the PFI option must demonstrate VFM and risk transfer. PFI is the UK government's preferred procurement route where it is shown to provide VFM when compared to the traditional public sector funded route adjusted to include a realistic pricing of all services and the value of risks. Second, payments to the private sector are based on the successful supply of services linked to quality of assets or physical infrastructure produced (Grout, 1997). Certain elements of contract payment are therefore at risk as the link between quality of services and payments provides a powerful incentive for PFI contractors to deliver the standard of services required by the public sector client. Payments received by the PFI contractor cover the project capital costs, the operating costs involved in providing facilities management service and associated financing costs (Ball *et al.*, 2000), usually referred to as capital expenditure (CAPEX) and operating expenditure (OPEX). However, payments are not received until the asset is ready for use and is fully operational. Therefore, the VFM case for PFI cannot be truly tested until projects become operational because it is at this stage that the effectiveness of the payment mechanism can be assessed in terms of risk allocation and as an incentive to improve service delivery.

22.3   Value for Money Arguments

Public assets have not been properly maintained in the past, as public sector bodies under tight financial constraints often cut back on maintenance spending (Ball and King, 2006). PFI projects, due to the long-term nature of the contracts, encourage both the private contractor and the public sector department to consider costs over the whole life of an asset rather than considering the design, construction and operational periods separately. It is argued that this integrated and whole-life approach can lead to efficiencies through synergies between design and construction and its later operation and maintenance. The outcome should lead to a reduction in costs, both for the contractor and the public sector client due to innovation and the improved allocation of risk resulting in better VFM (ACCA, 2002).VFM, defined as 'the optimum combination of whole-life cost (capital and operating costs) and quality of services to meet the requirement of the public sector', is therefore central to the PFI debate.

Davies (2006) further argued that by internalising 'project maintenance costs post-construction, PFI contractors may have an incentive to install more efficient types of technology and deliver the project at a lower cost'. Also, as PFI contracts specify the condition in which a building is to be handed back to the public sector at the end of the contract, the contractor is incentivised to ensure the building is well maintained (NAO, 2003). The lower costing from the PFI consortium is due to the strong incentives to 'reduce costs but not to jeopardise quality' or services through innovation and better risk-management practices from the private sector.

A key benefit of PFI is the opportunity for innovation in terms of funding packages, design, construction, technology and the asset delivery of services. The perceived wisdom dictates that innovation in terms of design and construction leads to operational cost savings (Ball *et al.*, 2000). However, this is often the subject of intense debate. Sussex (2003) argued that whilst PFI probably leads to more projects being completed on time and better maintained hospitals, it may or may not offer design improvements and lower construction costs and probably does not lead to more cost-effective support services. Another key benefit relates to risk management. Problems have occurred with conventionally procured projects often because of a failure to identify all the potential risks and to manage them. Traditionally procured public projects tend to be prone to what is often referred to as 'optimism bias', usually associated with the tendency to underestimate risks, particularly cost and time overruns due to a culture of predicting lowest cost and earliest completion. Typically, projects seemed to value risk transfer at around 30–35% of construction costs (ACCA, 2004). PFI route is selected if it is lower than the hypothetical risk-adjusted costing known as the public sector comparator (PSC) when expressed in net present value terms. Pollock and Vickers (2002) highlighted a case argued where the cost of a PFI hospital became lower than the publicly funded hospital only after including risk transfer. In other words, the VFM case rested upon risk transfer at the design, construction and operational stages.

Operational risks are directly related to the payment mechanism in a PFI project. For example, volume risk, availability, performance, maintenance, lifecycle, legislation and technology risks will all affect the revenue or the unitary payment received by the private sector operator. Private sector firms tend to reduce their exposure to volume risk such as the demand for their facilities. In prisons PFI, the private sector consortia are often unwilling to take on the risk of a facility being unoccupied because of a change in sentencing policy and in the education sector there is a risk of falling school roll as a result of a change in population parameters (Ball et al., 2000).

Grout (1997) reported evidence that volume risk is often borne by the public sector but argued that usage is dependent upon quality of assets and associated risks ought to be borne by the builder or owner. In other cases, risk transfer in PFI projects is less problematic. For example, if the maintenance cost of a hospital turns out to be higher than expected the PFI contractor has no other option but to bear the burden. For risks relating specifically to service performance and non-availability of a facility, penalties are applied to the private sector. However for this to be effective, penalties should be set at an appropriate level and information about service performance and availability should be collected (Ball and King, 2006). It is therefore important to understand the financial consequences of risks at contract negotiation to ensure that the payment mechanism is seen as an effective risk-allocation tool to improve service performance during contract monitoring.

22.4   Key Components of the Payment Mechanism

The payment mechanism is at the heart of the operational PFI contract, as it puts into financial effect the allocation of risks, particularly the operational risks and responsibility of the private sector operator relating to service performance and availability of facilities.

There are various payment models for PFI projects, and in all cases deductions are made if facilities are unavailable or services delivered are not to acceptable standards. In model A below, the unitary payment is based on the number of available places (e.g. prison, school or hospital places) which includes associated core services such as heating, cleaning, mail delivery, food. The payment structure is non-separable so there is a single payment for availability of facility and services as they are included in the definition of available place. This is illustrated in the model below.

Example of model A payment structure

$P = (F + I) - Z$

P = unitary payment per place
F = fixed amount per available place per day
I = indexed amount per available place per day (e.g. increased by retail price index – RPI)
Z = performance deductions

In model B, the unitary payment is based on the full provision of overall accommodation divided into units and includes associated core services such as heating, mail delivery and food. This is another example of non-separable or single payment structure but there are separate deductions for unavailability and performance. However, the level of deductions will reflect the importance of each unit or type of accommodation if the service provider fails to provide an available place.

Example of model B payment structure

$P = (F \times I) - (D + E)$

P = unitary payment per place/day
F = price per day for overall accommodation requirement
I = indexation factor
D = deductions for unavailability
E = performance deductions

However, model C is an example of a payment structure that is separable where the unitary payment is divided into separate availability payment stream and facilities management services payment stream. The availability payment is for the provision of assets such as buildings and equipment, and the service payment is for the provision of facilities management.

Example of model C payment structure

$P = (A + Q) - (D + E)$

P = unitary payment per unit
A = availability payment
Q = indexed facilities management payments
D = deductions for unavailability
E = performance deductions

Part Three

In some contracts there may be a variable element or charge that depends on usage, volume or demand factors such as occupancy rate of a hospital ward, or use of sport facilities, which is what Handley-Schachler and Gao (2003) refer to as a VFM risk arising from the danger that a service which is very expensive will be used very little. The availability payment usually forms a significant part of the unitary charge which is fixed for the concession period but the PFI contract allows for an annual adjustment for inflation and periodic adjustments for the service component of the charge through benchmarking and market testing. Market testing is used to adjust the payment of services to ensure that VFM principles are followed throughout the operational stage (Boussabaine, 2007).

The split between the availability and services payments in model C is crucial in terms of performance risks. Whilst one of the key principles of PFI is that payment of the unitary charge or payment is conditional upon supplying the required services, and should in theory be reflected in non-separable or single payment structure. In practice, lenders seek to minimise their credit risk by ensuring that there is a separate availability payment stream for their capital investment.

The extent to which deductions are made from availability payments is also minimised to protect revenues and ensure that debt service cover ratio (DSCR) reflecting the level of credit risk is acceptable to lenders. For this reason, the availability payment is sometimes seen as a fixed cost that changes slightly, and PFI transactions are often seen as a three-way relationship between the public sector client, private sector PFI contractor and lenders, who want to safeguard their investment by requiring the PFI contractor to maintain a certain level of DSCR. Failure to maintain the minimum level of DSCR due to the unavailability of the facility will result in a breach of the agreement between the PFI contractor and the lenders who provided the capital.

The payment mechanism is therefore based on the interaction of several elements: core assets or the type of facility (e.g. operating theatre/laboratory space or school place); associated facilities management services (e.g. heating, air conditioning, lighting, other environmental factors); or as non-core FM services (e.g. catering, cleaning and mail delivery). For example, a school classroom with inadequate lighting or that is not properly cleaned will have failed to meet the performance standard but could be used. In a hospital ward if the temperature falls below the level stated in a performance specification, the ward becomes unavailable and a penalty is imposed, which will increase steeply if the situation continues (ACCA, 2004). Payments are deducted for unavailability and failure in service performance with the level or amount of deductions reflecting the severity of the failure. But the payment mechanism is sometimes viewed as complex, containing separate lagged variables for availability and service performance; there have been issues relating to the appropriateness of weightings that are applied to different aspects of service elements. The payment mechanism therefore establishes the incentives for the contractor to deliver exactly the service required in the manner that provides VFM (HM Treasury, 2004).

The operation and effectiveness of the payment mechanism depends, however, on the output specification setting out the services performance level

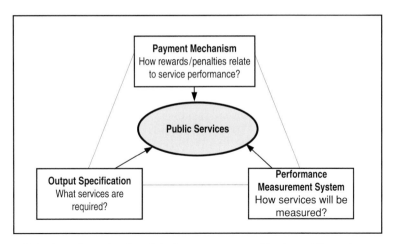

Figure 22.1 Interdependencies of the key components affecting public services.

required by the public sector authority and the performance measurement system measuring the performance of the PFI contractor, both in terms of availability and the standard of service (Figure 22.1).

22.5   The Output Specification

The output specification has two elements, accommodation and service performance standards, which are linked to the payment mechanism or payment received by the private sector operator. The accommodation standard relates to the physical condition and the design and performance of the building and services within the affordability limits set out in the outline business case. The service performance standards reflect the scope and level of requirement for each service category, priority for service delivery, the pass or fail criteria for assessing performance and rectification periods if the service fails.

According to McDowall (1999), the introduction of output specification has helped to change attitudes to specifying buildings and services by concentrating on those aspects of performance which are important to clients. Heavisides and Price (2001) noted that there is a significant debate generated by output-based systems. Unlike a technical specification which focuses on 'how' a facility should be delivered by specifying the dimensions, materials and workmanship, an output specification focuses on 'what' services are required. It sets out the operational requirements of the project in terms of accommodation standards and a wide range of services from hard facilities management (FM) services (e.g. building maintenance, groundwork, landscaping etc.) to soft FM services (e.g. cleaning, catering, security etc.). A well-drafted output specification is therefore fundamental to the operation of PFI projects and the successful delivery of long-term services (4Ps, 2005). The output specification provides an opportunity for bidders to be flexible, to think about the long-term implications of the service and to offer innovative solutions in PFI projects. But developing an output specification is an

Part Three

extremely difficult process and the public sector authorities have the challenging task of specifying a wide range of services in a manner that allows innovation but is not open to misinterpretation. As one senior partner from a top legal firm involved in PFI recently put it 'you have to be extremely clever to develop an output specification'. An example of the operational requirements for PFI prisons from the National Audit Office (NAO, 2003) is shown below under seven broad headings:

- Keeping prisoners in custody, e.g. the number and type of searches to be carried out.
- Maintaining order, control, discipline and a safe environment, e.g. the provision of a system of incentives and earned privileges for prisoners.
- Providing decent conditions and meeting prisoners' needs, e.g. safeguarding prisoners' personal property.
- Providing positive regimes, e.g. provision of education and counselling services.
- Preparing prisoners for their return to the community, e.g. pre-release courses.
- Delivering prison services, e.g. selection and recruitment policies of prison staff and provision of probation and healthcare staff.
- Community relations, e.g. facilitating access to the prison for invited members of the community.

Pitt and Collins (2006) argued for output specifications to provide bidders with the opportunity to prioritise the service by defining the client's required level of criticality (relating to the event impacting on the asset) and functionality (relating to the asset's importance). However, there is a danger in the preparation of output specifications that require services far in excess of what is intended leading to affordability problems (Heavisides and Price, 2001). A key issue in operational PFIs is therefore the need for concise definition in the output specification and clarity of the performance standards. Sometimes the precise definition of a high-quality service may be elusive, which allows different interpretations and can result in post-contract disputes (Akintoye *et al.*, 2003). Subjectivity in output specifications creates different interpretations and disagreements between parties with the public sector client having one view on the performance requirement and the service provider having another (4Ps, 2005).

Output specifications are not always comprehensive to cover all the services required. For example, in the Darent Valley Hospital, the NAO reported that the trust had been in disagreement with the service provider regarding circumstances that were not foreseen or explicitly stated in the output specification. The disagreement was over whether the contractor was responsible for de-icing the car park when there was an exceptionally heavy snowfall (NAO, 2005). Changes in the provision of core services provided by the public sector can also affect the requirements set out in the output specification. A feature of many PFI projects including hospitals and schools is that the core services, the delivery of clinical services or education, is not part of the PFI contract so any change in the core services can affect the provision of facilities management services specified in the output specification. Getting changes agreed involving a number of parties can be difficult, however, unless everyone is committed to the process. Partnership UK (2006) argued that things only move at the pace of the slowest party involved and a disproportionate amount of personal involvement is needed to make fairly basic changes to the output specification.

22.6   Performance Measurement System

There are several aspects involved in measuring the performance of, or monitoring, a PFI project, setting the standards, establishing measurement metrics and monitoring methods (McDowall, 2000). The accommodation and performance standards set out in the output specification discussed in the previous section determine availability and the level of services which are both critical in the development of a performance measurement system. The service provider is not paid if a hospital ward, a classroom or a prisoner place is unavailable for use. In addition, many PFI projects also require a system that measures the level of service against a percentage scale with a minimum standard, and a scale for applying penalties if performance falls below the threshold. The NAO (2003) report on PFI prisons provides details on the standard requirements to be met for a prisoner place which includes access to healthcare, the opportunity for exercise, and the availability of clean bedding, clothes and three meals a day. Provided these standard requirements are met, the service provider will be paid for the prisoner place whether or not the Prison Service allocates a prisoner to it. Figure 22.2 is an example of a percentage scale used at Darent Valley Hospital.

The percentage scale is applied to the individual FM service areas of:

- Estates and maintenance, grounds and garden
- Domestic, window cleaning and pest control
- Portering, transport and internal security
- Linen and laundry
- Catering
- Switchboard and telecommunications
- External security
- Car parking

The minimum service standard required for the service provider to be paid in full in each FM service area is 95%. Furthermore, if the standard for a particular service falls below 75% for 4 consecutive months then the trust is able to insist that the service provider changes its sub-contractor responsible for that particular service (NAO, 2005).

McDowall (1999) argued that the output specification specifies 'levels of services which can be more robustly measured and ultimately offer better value for money'. However, Grimshaw et al. (2002) predicted that the lack of experience and absence of prior measures of productivity and performance will make specification within PFI contracts difficult. Developing a robust performance measurement system with appropriate metrics to facilitate the monitoring of service performance is therefore very challenging. The NAO (2003)

Part Three

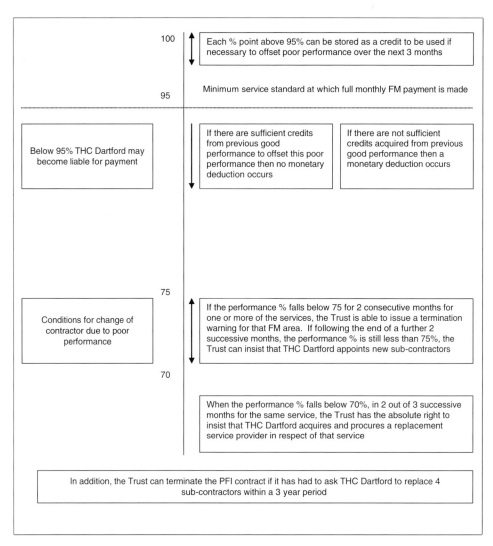

Figure 22.2   Darent Valley performance measurement scale (NAO, 2005).

report on PFI prisons describes the difficulty in developing effective performance measurement systems for the provision of custodial services in terms of availability and the performance of FM services. It is relatively straightforward to have a contractual requirement that penalises contractors for failing to prevent prisoners escaping but it is less clear how a contract can measure the extent to which a prison has contributed to reducing the likelihood of re-offending (NAO, 2003). There is also a balance to strike between ensuring the PMS providing relevant information to monitor service performance without it becoming unmanageable. Partnership UK (2006) cited the case of an operational PFI providing accommodation and training facilities, where the performance measurement was based on self-monitoring by the service provider against 361 KPIs. As a result of these problems, the trust combined

Table 22.1   Unitary charge payments and performance deductions 2000–2004 (NAO, 2005).

	2000–01 £000	2001–02 £000	2002–03 £000	2003–04 £000	April–July04 £000
Total basic charge	9990	17 941	18 306	16 636	5662
Other charges	1037	2341	2168	2423	937
Payments *before* deductions	**11 027**	**20 282**	**20 474**	**19 059**	**6599**
Deductions for availability	0	−4	0	0	0
Deductions for FM services	−10	−1	−0	−7	0
Waste (discount)	0	−19	−4	−5	−1
Total deductions	**−10**	**−24**	**−4**	**−12**	**−1**
Payments *after* deductions	**11 017**	**20 258**	**20 470**	**19 047**	**6598**

the generic and specific monitoring forms to produce around ten key indicators for each FM service area. These key indicators then had weightings attached and were used as the basis for scoring the FM areas (NAO, 2005). However, the NAO also noted that the weightings applied to each aspect of service delivery do not always appear appropriate and are sometimes not consistent across services. For example, in catering aspects for a hospital, hygiene has a weighting of 40 out of 485 points, which seems low compared to the weighting of 90 out of 485 points for presentation.

There is also a need for more objectivity in determining performance metrics or scores. Whilst some degree of subjectivity is inevitable, too many subjective elements can make it harder to agree performance scores to determine the level of performance deductions central to the operation of the payment mechanism. The recent surveys by 4Ps (2005) and Partnership UK (2006) found that in many projects there had been minimal or zero deductions. In the Partnership UK (2006) survey, it was noted that in about one third of projects performance deductions had not been applied and another 50% have had deductions imposed on less than 25 occasions which totalled less than £50 000 per project. Table 22.1 provides details on the deductions made to the unitary charge in a PFI hospital.

Deductions are based on the level of criticality and the variables used should reflect materiality and proportionality in the operation of the payment mechanism. For example, in a hospital any shortfall in the standard of basic facilities could have greater consequences compared to other buildings because of adverse effects on patients receiving healthcare. The level of services such as cleanliness and availability of facilities influence infection rates, whether operating theatres are available or whether patients can be moved promptly to receive care. In a housing PFI project examined it was found that there was a lack of proportionality in the current payment mechanism. Availability measures fail to bite with a daily availability deduction of £3 per property per day compared to a £3000 deduction for failure to provide reports on time. An NAO (2003) report noted that whilst the service provider can lose up to 100% of its availability fee in any month, not all hospital areas are liable to a financial deduction for unavailability. Greater emphasis should therefore be placed on areas considered most critical to patient care or support. Some of the areas including pathology, the fracture clinic, physiotherapy and

Part Three

medical records where the trust cannot make deductions if the facilities are unavailable could cause considerable difficulties for the treatment of patients if facilities cannot be used. The fact that the service provider cannot suffer a financial deduction for these areas could, in theory, reduce the incentive to keep these areas well maintained and to remedy any problems promptly.

Establishing monitoring methods is also crucial for the public sector client and private sector operator to ensure that the right deductions are made. Partnership UK (2006) found that a range of methods are used for assessing user satisfaction with 40% of the projects surveyed using formal customer service satisfaction surveys. Other methods used for monitoring performance include using regular meetings with stakeholders, feedback from the helpdesk, real-time information systems such as building management systems (BMS), spot checks, site visits and third party audits (McDowall, 2000). For some PFI projects user satisfaction is also relevant to the determination of the overall level of service performance. However, there are problems associated with user satisfaction surveys as multiple layers of users often create problems. For example one hospital manager commented that more complaints were received from hospital staff than from patients, but saw this as positive because they felt that faults should be prevented or rectified before they affected the end user (Partnership UK, 2006).

22.7   Case Studies

This section presents the findings following a series of semi-structured interviews with the key stakeholders of operational PFI projects. The experience of two public sector or local authority PFI contract managers and two managers from private sector facilities management companies involved in PFI projects are presented and discussed below.

Case Study 22.1: Building Asset Manager for a Local Authority

The local authority has a number of operational PFI projects which includes road infrastructure project providing road improvement and maintenance, schools and care homes. The asset manager has been involved in the procurement of PFI projects and offers advice on the preparation of output specifications and tender documents.

Output specification

The local authority is encouraged to describe services in output terms to allow the private sector consortia room to innovate and use their experience to find the most cost-effective solution. In some PFI projects, particularly where services were of a sensitive nature, there were difficulties in specifying soft service delivery. For example in a care home it had been difficult to put into words the amount of cleaning that would be required. The operational PFIs have also encountered problems in areas where the private sector did not have previous experience, such as in a special needs school. Potential bidders for PFI projects were encouraged to visit existing facilities providing similar services as bids that show a good understanding of the service required and potential problems are viewed favourably.

Services in the PFI projects were generally delivered in accordance with the output specification but there were disputes over interpretation and subjectivity with reference to levels of service and standards. For example, the furniture required in a care home for young people with behavioural problems was specified to be 'of a high standard capable of withstanding a high level of daily wear and tear'. The furniture provided by the private sector proved not to be robust enough as a result of a misinterpretation of the phrase 'a high level of daily wear and tear'. A further example was provided with regards to vehicles for a care home for the elderly. The output specification specified that vehicles must have an electronic tail gate to enable access for passengers in wheelchairs. An electronic tail gate was provided which required a higher level of manual operation than anticipated by the driver. It was felt that the contractor had provided the cheapest option to meet the specification of an electronic tail gate but had not provided consideration for the extra labour involved. To provide an output specification across so many services that is comprehensive, but not too prescriptive, has proven extremely difficult.

Performance measurement system

Generally, performance measurement systems within the authority are viewed to be complex and there is a lack of understanding on how they operate amongst public sector staff. Each PFI project contains approximately 100–120 performance targets and there is even a case to make systems more comprehensive, covering eventualities not anticipated during contract negotiations. Performance targets are weighted against the disruption that a service lapse would cause. User satisfaction surveys are conducted but are not linked to the payment mechanism. As with most PFI projects, contractors are self-monitoring and their performance subject to audit by the authority. The local authority is invoiced each month, with deductions for performance calculated by the contractor. Quarterly meetings are held between the authority and the contractor to discuss performance but there were issues with the resources available on the public sector side for contract monitoring.

Contract management

Very few performance deductions have been made in operational PFIs to date. The level of deductions stipulated in contracts is thought to be an inadequate compensation for disruption caused by service lapses. In some projects the administrative cost of imposing a deduction outweighed the actual deduction the contractor could be liable for. It was felt that service delivery in PFI projects tends to have less flexibility than where the delivery of services is under the direct employment of the local authority. For example, there was lack of flexibility in cleaning regimes when a room is required for alternative use. There have also been problems with the division of responsibility between different sub-contractors and complaints from passengers over the cleanliness of public transport vehicles. Drivers refused to clean up litter at the end of journeys, as this was left to the cleaning sub-contractor to be carried out at end of the day. Outside PFI, the local authority reviews contracts every 3 years and can re-tender if not satisfied with service. However, under PFI the authority felt that it may not be possible or too restrictive if issues over service delivery continued.

The authority placed a high level of importance on building working relationships and was keen to avoid a 'them and us' culture. The payment mechanism was not seen as a way of punishing service providers and did not provide a hindrance to building partnerships. The general view was that partnerships were not easy when commercial interests influence working roles and practices.

Case Study 22.2: Assets and Capital Manager for a Local Authority

The local authority's operational PFI projects consist of three secondary schools, all completed in 2003. The asset and capital manager acted as a project manager during the construction phase of project and fulfils the role of an intermediary between the local authority and the SPV during the operational phase.

Output specification

It is acknowledged that the output specification for the schools' PFI contains subjective elements. The cleaning and waste management specification states that there must be 'minimal instances of staining or marks' to the interior and exterior surfaces of the school buildings. However the interpretation of 'minimal instances' is problematic and as a consequence there have been disputes between the local authority and service provider. Examples were provided of performance standards relating to cleaning hours and the content of vending machines. The service provider is adamant that they are meeting the minimum requirements set out in the output specification and any changes to the service will have a cost implication, however minor. For example, there have been arguments about the provision of tools to open windows in classrooms, but it is recognised that there would be costs implication to provide an opening tool per classroom.

The main issue in the schools relates to the design of the buildings and problems with acoustics and ventilation. The design was compliant with the government department's regulations at the time the contract was agreed. However these regulations have recently been updated and changes were required that could not have been foreseen at the time of developing the output specification. It was felt that discussions over resolving this issue were detracting attention away from problems with service delivery. There have been specific changes to the contract since financial close which have included an additional fitness facility to one school, an additional pupil entrance to another and changes to the use of rooms, for example, from an IT room to a store room. The changes resulted in additional costs and the level of fees associated with each change, even if minor, could have a significant cost implication.

Performance measurement system

It is felt that the self-monitoring by the contractor is not rigorous enough. Due to the complexity of the performance measurement system, the authority cannot challenge the reported performance from the contractor. The local authority is also unable to monitor performance adequately due to a lack of resources. There have been issues with staff not reporting faults and it is believed that a 'culture change' is required to encourage staff to report any problems with services.

A helpdesk is located within one of the schools and the other two schools contact the helpdesk by telephone. It is noticeable that more faults are reported at the school where the helpdesk is located due to convenience or visibility and it has been suggested that the location of the helpdesk is rotated around all three schools.

Contract management

There have been deductions for availability in the PFI schools, however only a few deductions relate to soft service performance. It is felt that the availability of rooms/facilities is easier to monitor but the standard of cleaning or catering is subjective. Any deductions are taken from FM budget of the service provider but it is argued that this could lead to a 'vicious circle' resulting in even poorer performance as a result of budget constraints. The interviewee also commented on the lack of flexibility in PFI contracts. PFI also represents a 'culture change' for head teachers as FM staff are no longer under the direct control of the school or the local authority. A high level of importance is placed on partnership but there were initial problems encountered with frequent changes in the FM manager. The current FM manager is developing a good relationship with teaching staff.

Case Study 22.3: General Manager of an FM Service Provider

This case study is an interview with the general manager responsible for the delivery of both hard and soft FM services at a PFI hospital project. The hospital specialises in the treatment of patients with mental illness and has been operational since March 2003. The service provider is not part of the SPV but was involved during the procurement phase of the project and contract negotiations. The service provider is directly responsible for many of the services such as catering and cleaning, however certain specialist services such as window cleaning and security are sub-contracted.

Output specification

After financial close the service provider discussed with the trust the practicalities of service delivery. As a result changes to the output specification were agreed that provided cost savings. For example, it was agreed that the requirement for cooked breakfasts for staff and patients was not necessary and was subsequently removed from the specification. It was also felt that the output specification include subjective elements on the soft services side. The service provider has tried to be flexible where there are differences in interpretation or changes required. They have agreed to alterations to service delivery and not pursued increases to the unitary charge where the changes are minor and do not have a significant cost or time implications. Any change to service delivery is always formally recorded to provide clarification and to avoid disputes in the future.

There has also been a number of changes to the contract post-financial close which have had cost implications and affected the unitary charge. Additionally, there have been variations that the trust has paid for outside the PFI contract. The service provider also felt that the public sector staff responsible for the operation of the hospital should be involved in procurement and contract negotiations as this will increase their understanding of contract and permit input from staff with knowledge of how a facility works in practice.

Performance measurement system

As with other PFI contracts the service provider self-monitors performance and carries out a daily internal audit but their performance is subject to monthly audits and random checks by both the trust and the SPV. The service provider has developed its own performance measurement system which supports the system in the contract between SPV and trust. Monthly performance reports are produced for both the SPV and the trust. The hospital is split into functional areas (wards) and the provider can incur penalty points if there is a reported non-compliant incident that is not rectified within the time stipulated in the contract. The level of penalty points can escalate if there is more than one failure in the same service or functional area in the same day or week. Performance deductions are imposed if a threshold of penalty points is exceeded. Faults are reported via a helpdesk which is in the form of an intranet and telephone number. Each ward in the hospital has access to this intranet via an icon on the hospital's computers and provides a more cost-effective method of obtaining feedback. The service provider encourages the trust staff to use the intranet to avoid congestion on the switchboard.

Customer satisfaction surveys are carried out every 4 months but this is not part of the contracted PMS and the service provider cannot therefore incur penalty deductions for a low score in a survey. However the results of the surveys are reported to both the trust and SPV who expect the service provider to investigate and act upon any areas that receive low scores.

The service provider noted that the performance measurement system is complex and argued that simplification could only be brought about by changes to the payment mechanism. Performance is scored against functional areas and the payment mechanism calculates deductions based on the unavailability of functional areas. Furthermore, each functional area is made up of functional units. For example a ward is made up of a number of bedrooms, staff room, store room, cleaner's room etc. Therefore a method of measuring the performance of each functional unit has to be found. The Manager provided an example of a larger PFI hospital with 49 functional areas and 1200 functional units which his company is involved in.

Contract management

There have been minimal deductions on this project to date relating mainly to construction defects. The deductions incurred are claimed back from the contractor responsible for the construction of the hospital. The interviewee considered that the payment mechanism certainly acts as an incentive to deliver the standard of service stipulated in the contract, as it incentivises the service provider or contractor to avoid deductions. However, performance above the contract standard would require

Part Three

an increase in the unitary charge but the trust would not be able to afford a scheme that provided bonuses for service that was over the agreed level of service in the contract. The service provider was sceptical about the government's proposal to leave soft services outside PFI contracts and to let these contracts separately on a 5-year basis.

The service provider placed a great deal of importance on developing good working relationships with both trust staff and the SPV. The interviewee commented that he was aware of other PFI contracts that have full-time public sector contract managers which are funded from imposing performance deductions. Such a situation can create an adversarial relationship. There have been issues with staff changes in the trust's monitoring team and new staff not possessing an understanding of the level of services that are specified in the contract. The provider now makes a conscious effort to manage the expectations of the trust staff and carry out inductions for all new members to ensure that they are familiar with both the nature and standards of service that are set out in the contract.

Case Study 22.4: PFI Contract Manager for an FM Company

This case study is an interview with a facilities management company that provides hard services for a PFI project comprising three local authority care homes for the elderly. The FM company is a subsidiary of the contractor that was responsible for the construction of the homes and is part of the SPV. Soft service delivery is provided by a separate FM company that specialises in such facilities. The interviewee has a wide range of experience in other PFI projects including a library, a school and a hospital where he oversees the provision of both soft and hard services.

Output specification

There have been some problems with output specifications, as there were 'grey areas' that were open to interpretation and instances that were not foreseen when the specifications were drafted. It was felt that it would be a near impossible task to draft an output specification that would cover all eventualities.

The interviewee described the output specification as 'the bible' and it was referred to when any disputes arose. The service provider does try to show some flexibility and goodwill. In the library PFI project, cleaning is sub-contracted out, but there have been some issues with the adequacy of the contracted number of cleaning visits. This was resolved by the service provider agreeing to take on some cleaning duties at no extra cost. A further example was provided in the care homes PFI – portering in care homes was not part of duties of the service provider, however certain tasks such as moving furniture are carried out by the provider's on-site staff when time permits, again at no extra charge. The process for agreeing changes to the output specification can be cumbersome. For any variation to the contract a change notice has to be issued, and lifecycle implications are calculated which will then have to be agreed with the funders of the project. The local authority or public sector client must then agree to the extra cost which results in an increase in the unitary charge.

Performance measurement system

Performance measurement systems for each of the PFI projects are complex and consist of a large number of KPIs. If any of these KPIs fall below a certain percentage threshold then the service provider can be liable for penalties. Each of the projects carry out customer satisfaction surveys, however these surveys are not part of the payment mechanism and a low score does not trigger any penalties. Any faults are reported directly to on-site staff and users and public sector staff are encouraged to use a standard form for clarity purposes. Faults can also be reported by telephone which is mainly used 'out of hours' when there are no on-site staff available.

Contract management

Very few performance deductions have been imposed in the projects to date but there have been some penalties for unavailability in a care home due to a leak. There were also penalties for the failure of a chiller unit in a care home but this was recouped from the installer of the chiller unit. The interviewee felt that the standard of building is higher in PFI projects than buildings that are procured via traditional methods due to the fact that PFI contracts contain requirements that stipulate the building condition when handed back to the local authority at the end of the project. The service provider places a high level of importance on partnership and building a good working relationship which is emphasised during the recruitment of staff for the service providers. Each project has on-site staff acting as caretakers, seen as essential in fostering a good working relationship. There had been some issues relating to staff changes on the local authority side with new staff not aware of the responsibilities of the FM provider.

22.8   Analysis and Discussion of Findings

The key issues that emerged during the case study interviews with local authorities or public sector clients and service providers are summarised in Tables 22.2 and 22.3. In some areas the representatives from local authorities

Table 22.2   Summary of key issues on the output specification and performance measurement system.

Local authorities	Service providers
Output specification	
Difficulties in specifying soft service delivery. Subjectivity with reference to levels of service and standards, disputes over interpretation	Near impossible task to draft an output specification that would cover all eventualities. Subjectivity in soft services specification
Difficulties in assessing the cost implication of any changes to the output specification	Flexibility where there are differences in interpretation or changes required
Level of fees associated with each change can result in even a minor change having a significant cost implication	Variations that have been paid for outside the PFI contract
Problems where the private sector did not have previous experience and services need to be delivered in difficult environments, e.g. special needs schools, care homes for young people with behavioural problems	Practicalities of the output specification after financial close
	Involvement of operational staff in the procurement and drafting of the output specification, to allow user input and develop an understanding of the contract
Performance measurement system	
Complex and a lack of understanding amongst public sector staff	Performance measurement system is complex and is a function of the payment mechanism
Self-monitoring by the contract is not rigorous enough	Performance subject to own internal daily audit
Inadequate resources available for contract monitoring	Developed most cost effective methods of receiving feedback including an intranet and standard forms
Staff not reporting faults, 'culture change' required to encourage staff to do this. Location of helpdesk an issue of convenience and visibility	

Part Three

Table 22.3   Summary of key issues relating to contract management from local authorities and service providers.

Local authorities	Service providers
Very few performance deductions with most deductions relating to availability which is less subjective and easier to monitor	Minimal deductions have been passed on to subcontractors
Administrative cost of imposing a deduction can sometimes exceed the actual amount deducted	Deductions 'eat into' the contractors' profit margins; there is therefore a big incentive to avoid these deductions
Less flexibility in service delivery than where staff are under the direct control of the local authority	Provides an incentive to deliver the contractual level of service only
Problems with the division of responsibility for various tasks between different subcontractors	Payment mechanism is not seen as a way of punishing the service provider and does not hinder partnership working
Payment mechanism is not seen as a way of punishing the service provider and does not hinder partnership working	Efforts to manage the expectations of staff and carry out inductions for all new members of staff
Partnerships are crucial but not easy when commercial interests influence working roles	
Frequent changes in staff can make building effective working relationships difficult	

or public sector clients share similar objectives to private sector service providers, notably in the desire to form effective partnerships. However, in other areas there were different views and perceptions. For example, local authorities or public sector clients felt that PFI contracts lacked flexibility and any required changes had a cost implication. Service providers, on the other hand, maintained that they try to adopt a flexible approach to PFI contracts in the interest of building good working relationships with public sector bodies. The findings from the local authority clients and service providers are discussed in the following sections.

22.8.1   Interpretation and changes to output specification

The case studies provided examples of differences between the public and private sector in the interpretation of the output specifications. However in all of the case studies there were changes to the contract/unitary charge since financial close as a result of the interpretation or changes to the output specification and the projects became more expensive than originally anticipated. The public sector found the change process in PFI both cumbersome and time consuming. Case study 22.2 commented that getting change agreed involves a number of parties including all members of the SPV, and funders of the project need to assess the impact of change on the risk profile of the project. The level of fees associated with changes can also mean that even a minor variation has a significant cost implication. The government recognised the difficulty of incorporating variations into highly detailed PFI contracts and are setting up a PFI operational taskforce to advise on how to negotiate contract variations (HM Treasury, 2006a). Partnership UK (2006) also argued for the need to involve both operational public sector staff and end users during the drafting of the output specification. All case study participants agreed that the involvement of these parties was a good idea as it allows input from staff that possess knowledge of the practicalities of service delivery and the needs of users. Case Study 22.3 in particular demonstrated the value of involving the service provider during the drafting of the output specification.

22.8.2   Scope of FM services

The government has announced that future PFI projects in health will not automatically include soft service delivery so that public authorities must provide the case for including soft services. This is likely to mean that, unless a VFM case can be proved, soft services in a PFI project will be let under separate short-term contracts. Case Study 22.3 was sceptical about the government's proposal to leave soft services outside PFI contracts. Under such a scenario it was felt that during the early phase of the contract there would be high level of service, followed by cost cutting during the middle phase and in the final year of the contract there would be either a high level of service to encourage renewal or a poor level of service if provider knows the contract will not be renewed. The service provider argued that benchmarking would be a better option of providing VFM, as it would provide a more consistent approach to improve level of service and be less disruptive.

22.8.3   Performance measurement systems

The study found that there were difficulties in developing effective performance measurement systems in PFI projects. Whilst it is important for PMS to be comprehensive, it is recognised that there is a need to strike a balance between ensuring the performance measurement system provides relevant information without it becoming unmanageable. As the contractor is essentially self-monitoring, measurement systems need to be transparent to allow the public sector to audit performance. When drafting payment mechanisms it is also important to consider the implications in terms of its relevance to service standards defined in the output specification. Case Study 22.3 felt that the complexity of the performance measurement system is a function of the payment mechanism. There is evidence of public sector staff sympathising with the service provider which has influenced the amount of performance failures reported. In Case Study 22.2 it is believed that a 'culture change' is required amongst school staff in order to encourage the reporting of problems with services. A helpdesk is located within one of the schools and it was noted that more faults are reported at the school where the helpdesk is located due to convenience or visibility. It has been suggested that the location of the helpdesk is rotated around all three schools. The service providers in Case Studies 22.3 and 22.4 deliver the most cost-effective solutions to fault reporting via an intranet and standard forms. If public sector staff are reluctant to report faults and the methods available are not the most convenient

Part Three

then the reported performance of service provider may not fully reflect the actual standard of service delivery.

22.8.4   Service performance and deductions

In general service providers are meeting their contractual obligations in terms of service delivery standards, as reflected by the low level of performance deductions. In Case Studies 22.3 and 22.4, the service providers indicated that on the few occasions that deductions had been imposed they have been passed on to sub-contractors. A key issue is whether the payment mechanism provides an incentive for service provider to deliver a standard of service above that set out in the contract. Case Studies 22.3 and 22.4 agreed that payment mechanisms do not incentivise the service provider to deliver a level of service higher than stipulated in the contract, as the focus is on avoiding deductions. The service provider in Case Study 22.3 felt that the public sector would not be able to afford a scheme that provided bonuses for service that was above contract. The interviewee could foresee arguments about the level of performance and the amount of bonus payable if such a scheme was adopted. Both the public sector clients and private sector service providers recognised that it is important that performance deductions are not viewed as a punishment and a hindrance to the building of effective working partnerships.

22.8.5   Contract monitoring and resource implications

There were problems relating to monitoring and performance measurement. The monitoring undertaken by the public sector was seen as unnecessary because of the repetition of the processes already implemented by the service provider. The service provider monitors the performance of its own sub-contractors which is then subject to monthly audits and random checks by both the SPV and the public sector body. The case studies have shown that different monitoring methods are used including customer surveys. The government has recently announced that it will seek to create an acceptable mechanism for linking user satisfaction with payment under future PFI contracts and to align the incentives of service providers more closely with user expectations (HM Treasury, 2006a).

The neutrality of the public sector in performance measurement can also be questioned. There are dangers of the public sector sympathising with the service provider and not reporting all performance failures, or stringently applying the contract, which could adversely affect working relationships in a project. The government intends to trial a project delivery organisation that would be responsible for the auditing of performance in operational PFIs but this may also be questionable if their fees are paid solely by either the public sector body or the SPV. The need for an independent organisation or third party to audit and certify performance and the importance of this role is increasingly recognised and paid for jointly by the SPV, service provider and public sector client.

There is evidence to suggest that the public sector has not fully assessed the resource implications of performance monitoring and has not set aside sufficient resources for it. It seems very little focus was given to the practical issues of contract management resourcing during the procurement process. The local authority in Case Study 22.1 needed to devote more resources to contract monitoring than anticipated and Case Study 22.2 did not have enough resources to carry out the level of monitoring required. Adequate resources from the public sector are crucial for effective contract monitoring of services delivered by the private sector service providers otherwise VFM may not be achieved throughout.

22.8.6   Building relationships and knowledge sharing

The service providers recognised the importance of good working relationships with the public sector and on occasions have adjusted their service provision without a formal contract change. The case studies demonstrated the importance that all parties placed on partnership. However, one public sector client did feel that the commercial interests of service providers could adversely affect working relationships. Staff changes were raised as an issue that can interrupt efforts to build effective working relationships as it can cause problems in contract monitoring, particularly if there is subjectivity in output specification and the performance measurement system.

Both the public sector and private sectors are undergoing a learning process in operational PFI which should lead to greater knowledge sharing and improvements in the delivery of future PFI contracts (Carrillo *et al.*, 2006). Long-term relationships between service providers and public sector clients can provide a powerful stimulus if partnerships are built to facilitate learning, knowledge sharing and innovation. There are also organisations such as the 4Ps and Partnerships UK to offer support and share best practice.

22.9   Concluding Remarks

The payment mechanism ensures that the public sector client's objectives for PFI projects are delivered as set out in the output specification and monitored through the performance measurement system. However subjectivity in the output specification and complexity in the performance measurement system affect the effectiveness of the payment mechanism as a risk allocation tool and also raise questions as to whether the low level of deductions truly reflect the actual level of service that is being delivered. Subjectivity and interpretation of the output specifications increase the unitary charges or payments and result in low level of performance deductions in operational PFIs. The findings from the case studies also suggest that there have been added costs during the operational phase of PFI projects due to additional public sector resources for contract monitoring. There is also some evidence of the public sector foregoing entitled deductions in the 'spirit of partnership' and in exchange for minor contract variations.

Part Three

Current payment mechanisms provide an incentive for contractors to deliver a contractual level of service but do not incentivise them to deliver a higher standard of service as the focus is on avoiding deductions. There is a need for improving output specifications to reduce subjectivity, simplifying performance measurement systems so that they are more transparent, and more significantly to strengthen the logic and link between the output specification, performance measurement system and the payment mechanism. Both the public and private sector are undergoing a learning process which should lead to improvements in the drafting of future PFI contracts and monitoring operational PFI projects to ensure that VFM is achieved throughout.

References

4Ps (2005) *4ps Review of Operational PFI and PPP Projects*. November, www.4ps.gov.uk.

Association of Chartered Certified Accountants (ACCA) (2002) *PFI: Practical Perspectives*. Certified Accountants Educational Trust, London.

ACCA (2004) *Evaluating the Operation of PFI in Roads and Hospitals*, Research Report No. 84. Certified Accountants Educational Trust, London.

Akintoye A., Hardcastle C., Beck M. *et al.* (2003) Achieving best value in Private Finance Initiative project procurement. *Construction Management and Economics*, 21, 461–470.

Ball, R., Heafey, M. and King, D. (2000) Private Finance Initiative – a good deal for the public purse or a drain on future generations? *Policy and Politics*, 29, 95–108.

Ball, R. and King, D. (2006) The Private Finance Initiative in local government. *Economic Affairs*, 26(1), 36–40.

Boussabaine, A. (2007) *Cost Planning of PFI and PPP Building Projects*. Taylor and Francis, Oxford.

Carrillo, P.M., Robinson, H.S., Anumba, C.J. and Bouchlaghem, N.M. (2006) Knowledge transfer framework: the PFI context. *Construction Management and Economics*, 24(10), 1045–1056.

Davies, J. (2006) *Risk Transfer in Private Finance Initiatives (PFIs) – An Economic Analysis*. DTI, Industry Economics And Statistics Directorate (IES) Working Paper.

Froud, J. and Shaoul, J. (2001) Appraising and evaluating PFI for NHS hospitals. *Financial Accountability And Management*, 17(3), 247–270.

Gaffney, D. and Pollock, A. (1999) Pump priming the PFI: why are privately financed hospital schemes being subsidised. *Public Money and Management*, 17(3), 11–16.

Grimshaw, D., Vincent, S. and Willmott, H. (2002) Going privately: partnership and outsourcing in UK public services. *Public Administration*, 80(3), 475–502.

Grout, A.P. (1997) Economics of the Private Finance Initiative. *Oxford Review of Economic Policy*, 13(4), 53–66.

Handley-Schachler, M. and Gao, S.S. (2003) Can the Private Finance Initiative be used in emerging economies? Lessons from the UK's successes and failures. *Managerial Finance*, 29, 36–51.

Heavisides, B. and Price, I. (2001) Input versus output-based performance measurement in the NHS – the current situation. *Facilities*, 10, 344–356.

HM Treasury (2004) *Value for Money Assessment Guidance*. HMSO, London.

HM Treasury (2006a) *PFI: Strengthening Long-Term Partnerships*. HMSO, London.

Part Three

McDowall, E. (1999) Specifying performance for PFI. *Facilities Management*, June, 10–11.

McDowall, E. (2000) Monitoring PFI contracts. *Facilities Management*, December, 8–9.

National Audit Office (NAO) (2003) *The Operational Performance Of PFI Prisons*, Report Of Comptroller And Auditor General, HC 700, Session 2002–3. The Stationery Office, London.

NAO (2005) *Darent Valley Hospital: The PFI Contract in Action*, Report Of Comptroller And Auditor General, HC 209, Session 2004–5. The Stationery Office, London.

Partnership UK (2006) *Report On Operational PFI Project*. www.partnershipsuk.org.uk.

Pitt, M. and Collins, N. (2006) The Private Finance Initiative and value for money. *Journal of Property Investment and Finance*, 24(4), 363–373.

Pollock, A. and Vickers, V. (2002) Private finance and value for money in NHS hospitals: a policy in search of a rationale? *British Medical Journal*, 324, 1205–1208.

Sussex, J. (2003) Public-private partnerships in hospital development: lessons from the UK's Private Finance Initiative. *Research in Health Care Financial Management*, 8(1), 59–76.

Part Three

23

Concession Period Determination for PPP Infrastructure Projects in Hong Kong

Xueqing Zhang

23.1   Introduction

PPPs have been practised for many years in Hong Kong. For example, the build, operate, transfer (BOT) approach has been used in the development of major road tunnels and the design, build, operate (DBO) approach in the development of sophisticated solid waste management facilities. In the recent public sector reform, the Hong Kong government has been seeking innovative and flexible financing strategies to stimulate economic activities and increase competitiveness in pubic works and services in order to provide better public services. In June 2001, the Hong Kong government set the policy principles in the Private Sector Involvement Program, *Serving the Community by Using the Private Sector*, and in August 2003, the government released a guideline for implementing PPPs in Hong Kong, *Serving the Community by Using the Private Sector – An Introductory Guide to Public Private Partnership* Consequently, a wide range of public works and services are proposed to be delivered through PPPs, ranging from an international exhibition centre, to prisons, sewage treatment services and massive cultural district projects. For example, ten recreational and cultural facilities projects were proposed with an estimated total value of about HK$2.5bn. The government also earmarked approximately HK$29bn per year over a 5-year period following 2003 for direct expenditure on infrastructure works.

The large scale of private investment and the long-term contract periods associated with PPP projects often lead to considerable public debate and conflicting interests. There is substantial controversy in public opinion about PPPs, particularly on how to ensure accountability, transparency, efficiency and cost effectiveness. On the one hand, there is a public concern that the private sector may gain unreasonable windfall profits due to the lack of adequate competition, which sacrifices the interests of the public sector and could lead to social and political risks to the government. On the other hand, there are various risks associated with PPP projects: social, political, environmental, technical as well as economic risks. They may emerge at different stages of the project lifecycle and have a combined impact on the project company's profitability and sustainability. Therefore, PPPs are not merely a device for the governments to develop infrastructure projects by transferring all the risks to the private sector and thus shedding all of its own responsibilities. Rather, PPPs require appropriate allocation of risks, assigning risks to those best placed to control them.

The Hong Kong government needs to address two critical aspects. One is to successfully attract private funds to infrastructure projects that are particularly needed. The other is to ensure that the projects be developed efficiently and provide an acceptable service to the public. This necessitates a comprehensive set of win–win regulatory rules, procedures and methodologies for successful adoption and management of PPP projects. In this regard, one important issue is the determination of the appropriate length of the concession period for a particular PPP project. The length of the concession period, to some extent, demarcates the rights and obligations between public and private sectors in a project's lifecycle and it is also critical to the project's sustainable development.

This chapter introduces an innovative methodology and consequently develops a simulation-based framework for concession period determination based on a win–win principle for public and private parties involved in a PPP project.

23.2   PPP Projects in Hong Kong

A wide range of infrastructure projects has been developed through PPPs in Hong Kong in the past 35 years.

23.2.1   Road tunnels

Five major road tunnels have been developed through BOT contracts. They are the Cross Harbor Tunnel, Eastern Harbor Crossing, Tate's Cairn Tunnel, Western Harbor Crossing and Route 3 Country Park Section – Tai Lam Tunnel and Yuen Long Approach Road. Some comparative information on these tunnels is provided in Table 23.1.

23.2.2   Port works

In 1999, a river trade terminal was built in Tuen Mun to satisfy the increasing demand of river trade cargo shipment in the Pearl River Delta. Being the first purpose-built container terminal in Hong Kong for river trade cargo, this terminal was intended to be a logistics hub in the Pearl River delta. A private company was granted the land to build and operate the terminal.

23.2.3   Railways

The Mass Transit Railway Corporation, a government-owned corporation established in 1975, had been responsible for the construction, operation and

Table 23.1 BOT tunnel projects in Hong Kong.

Project name	Tunnel length (m)	Immersed tube length (m)	Number of lanes	Traffic design capacity (v/d)	Planned construction period (months)	Actual construction period (months)	Concession period (years)	Construction start date	Opening date	Approximate cost HK$ (million)	US$ (million)
CHT	1852	1064	Dual 2	90 000	47	36	30	09/69	08/72	320	56
EHC	2255	1860	Dual 2, + 2 tracks	90 000	42	37.5	30	07/08/86	21/09/89	4400	564
TCT	4000		Dual 2	90 000	37	34	30	11/07/88	01/06/91	2150	277
WHC	2000	1360	Dual 3	135 000	48	44	30	02/08/93	01/04/97	7500	969
R3(CPS)	3800		Dual 3	135 000	38	38	30	31/05/95	30/07/98	7250	936

CHT – Cross Harbor Tunnel.
EHC – Eastern Harbor Crossing.
TCT – Tate's Cairn Tunnel.
WHC – Western Harbor Crossing.
R3(CPS) – Route 3 Country Park Section.

management of the Hong Kong mass transport system. The Mass Transit Railway Corporation was succeeded by the Mass Transit Railway Corporation Limited on 30 June 2000, which was listed on the Hong Kong Stock Exchange on 5 October 2000. The Mass Transit Railway Corporation Limited has been actively developing properties close to railway stations in addition to rail lines. For example, some recently developed stations are incorporated into large housing estates or shopping complexes.

23.2.4   Waste management

Eight refuse transfer stations (RTSs) and three strategic landfills have been built for waste management. The eight RTSs are Kowloon Bay Transfer Station, Island East Transfer Station, Island West Transfer Station, Shatin Transfer Station, North Lantau Transfer Station, Outlying Islands Transfer Facilities, West Kowloon Transfer Station and North West New Territories Transfer Station. The three strategic landfills are West New Territories Landfill, South East New Territories Landfill and North East New Territories Landfill. Waste collected in major urban centres of population is delivered to the RTS where the waste is compacted and containerised for onward transportation to the strategic landfills. The RTS is managed by the private collector under the DBO contract for 15 years.

23.2.5   Highway and bridge maintenance

Under the current PPP programme for highway/bridge maintenance in Hong Kong, the private sector is involved in the whole process of maintenance and is paid based on the performance standards of the highway system under its maintenance. For example, the Tsing Ma Control Area is managed and maintained by a private consortium. The Tsing Ma Control Area covers the Lantau Link and related road networks in Hong Kong, including Tsing Ma Bridge, Ting Kau Bridge, Rambler Channel Bridge, Kap Shui Mun Bridge, Cheung Tsing Tunnel, Tsing Kwai Highway and North Lantau Highway.

23.2.6   Tourism projects

The Disneyland Theme Park on Lantau Island, Hong Kong, the Walt Disney Company's third international theme park, opened in 1999. The project company is a joint venture between the Walt Disney Company (43% equity) and the Hong Kong government (57% equity). With a concession period of 40 years, the US$1.8bn project includes a world-class international theme park, a Disney-themed resort hotel complex and a retail, dining and entertainment centre.

Another PPP tourism project is the HK$1bn Ngong Ping 360 project (formerly known as the Tung Chung Cable Car project). Under a 30-year concession award by the Hong Kong government, the Mass Transit Railway Corporation Limited was granted the right to design, construct, operate and

Part Three

maintain the cable car system and pay a royalty to the government. The Ngong Ping Skyrail provides a spectacular 5.7 km cable car journey between Tung Chung town centre and Ngong Ping on Lantau Island, within the natural setting of the Lantau North Country Park.

23.2.7   AsiaWorld-Expo

AsiaWorld-Expo is a world-class exhibition venue, located at the centre of an extensive and efficient air, land and marine transport network connecting Hong Kong with China's Pearl River delta and the world's business capitals. It offers over 70 000 m² of rentable space for exhibitions, conventions, concerts, sports and entertainment events. Developed at a cost of HK$2.35bn, AsiaWorld-Expo is a PPP involving funding from the Hong Kong government and a private sector consortium including Dragages Hong Kong Limited and Yu Ming Investments Limited, with the Hong Kong Airport Authority contributing the land.

23.2.8   Information technology and property development

In 2000, the Hong Kong government and the Hong Kong Cyberport Management Company Limited (HKCMCL) signed an agreement to establish a Cyberport to nurture the development of information technology and multimedia, with the aim of helping the rebound of Hong Kong's economy after the East Asian financial crisis in 1997. The project was developed on a 24-hectare site in the southern district of Hong Kong Island, including four office buildings, a five-star hotel, a retail entertainment complex and a deluxe residential development. The government's capital contribution to the project was the land of the residential portion of the project and the associated infrastructure facilities. The HKCMCL was responsible for the construction costs of both the Cyberport portion and the residential portion. The Cyberport was intended to be home to a strategic cluster of about 100 IT companies and 10 000 IT professionals.

23.3   Build, Operate, Transfer Scheme

Many public-private partnership (PPP) models have been explored in international infrastructure development, including Hong Kong. Among these PPP models, the BOT scheme is a typical approach or a popular procurement methodology underlying different PPP scenarios. The BOT concept has generated a number of related acronyms that reflect variations of governmental interest/preference and industrial characteristics in procurement approaches (Palaneeswaran *et al.* 2001): BBO (buy, build, operate), BLT (build, lease, transfer), BOO (build, own, operate), BOOM (build, own, operate, maintain), BOOT (build, own, operate, transfer), BT (build, transfer), BTO (build, transfer, operate), DBFO (design, build, finance, operate), DBOM (design, build, operate, maintain), DOT (develop, operate, transfer), LDO (lease,

Part Three

develop, operate), MOT (modernise, operate, transfer), ROO (rehabilitate, own, operate), ROT (rehabilitate, operate, transfer) and TOT (transfer, own, transfer).

Under the BOT scheme, an infrastructure project is developed through a concession agreement between a public authority and a private consortium (the concessionaire). In this agreement, the public authority grants the concessionaire the rights to build and operate the project for a certain period (the concession period). The concessionaire pays back the loan and recovers its investment with a certain level of profit through revenues from the project during the concession period, and at the end of the concession transfers the project to the public authority.

BOT projects usually require a substantial upfront construction investment, the recovery of which is through revenues from the project over the concession period. One important issue for the government considering using the BOT scheme to develop infrastructure facilities is the determination of the appropriate length of the concession period. Different projects will incur different cash flow profiles during their lifecycles. There are many uncertainties and risks in the construction and future operation of the project, which have significant impacts on the length of the concession period.

23.4   Concession Period

23.4.1   Fixed vs. flexible concession period

The length of concession is critical to the project's sustainable development and it, to some extent, demarcates the rights and responsibilities between public and private sectors in the project's lifecycle. The length of concession is usually determined based on 'normal' or 'expected' conditions, which are subject to various changes that may cause extension to the original concession or its early termination. In practice, a long-term fixed concession period is commonly used, although there may be a mechanism to extend the concession for unexpected risks, such as a *force majeure* event or a market demand that is far below the expected level. However, a flexible concession period may be preferable where (1) the scope of the project has not been clearly defined, (2) the concessionaire is financially high-leveraged, (3) construction activities of the project are very complex with substantial risks (e.g. cost and duration overruns), and (4) the cash flows in future operation are very difficult to predict.

23.4.2   One-period concession vs. two-period concession

In the concession arrangement, some projects include the construction phase as part of the concession period while others do not. The former is called the 'one-period' concession, in which the concession starts when construction begins. The latter is called the 'two-period' concession, in which the concession begins at the completion of the construction. The one-period concession

Part Three

combines the construction period and the operation period. This transfers the construction time overrun risk to the concessionaire: the operation period is shorter if the construction period is longer, and vice versa. The two-period concession has a fixed operation period regardless of the actual completion time of construction.

23.4.3   Factors affecting length of concession

The length of concession depends on a number of factors, such as project type, scope, asset specificity, construction complexity, project lifespan, project development costs, combination of financing instruments, opening asset value, depreciation, operation and maintenance costs, market demand (price and quantity) of the services provided by the project, interest rate, inflation rate, foreign exchange rate (if foreign currency is involved) and governmental regulation practices.

The concession period should be short to permit frequent competition without jeopardising the incumbent concessionaire's return on socially desirable investment if no substantial sunk investments are involved (Kwoka, 1996). However, a long concession period is desirable if a project involves large initial sunk costs in construction, construction/operation equipment and other project-specific assets. In general, the concession period should not be longer than the designed life of the project. Furthermore, whether a fixed-term or flexible-term concession, or whether a one-period concession or two-period concession, it should satisfy conditions specified in the concession agreement and required by relevant laws, for example, the allocation of risks specified in the concession agreement, and the maximum allowable length of concession limited by the law or regulations if any.

23.4.4   Concession extension and termination

Various project variables may happen to be quite different from those assumed before or at the time of the award of the concession, unexpected situations may appear, and the public client's objectives in the concession may change. The changed conditions necessitate the modifications and changes of the original concession agreement to reflect. For example, the Argentinian government suspended the intercity road concessions and renegotiated with concessionaires only 5 months after the concessions had been in operation, leading to a major overhaul in the design of the concession:

1. The number of toll booths was reduced and their locations adjusted.
2. Tolls were reduced by more than 50%.
3. The government withdrew the 'canon' requirement and granted a total annual subsidy of US$57m to compensate the concessionaires (Estache and Carbajo, 1996).

The concession may be extended in order to compensate the concessionaire for the impact of risks that are beyond the control of the concessionaire or not assigned for the concessionaire to bear. For example, in the more than

US$13bn programme of concession toll roads under the Puebla Panama Plan in Mexico, a clause for concession extension was provided for traffic levels falling below government forecasts, cost overruns resulting from government-imposed delays or design modifications, and cost overruns in excess of 15% of the original project budget (Vazquez and Allen, 2004).

23.5   Concession Period Determination Methods

The following methods may be used to determine the length of concession in light of specific project conditions.

23.5.1   Concession period integrating construction and operation

It is a common practice to include the construction phase as part of the concession period to encourage early project completion and early opening of services to the public. In this method, the concessionaire is often required to design and build the project facilities by a specified date. The concessionaire would have to pay liquidated damages if the works were not completed on time. There may also be a 'backstop' date (e.g. 1 year from the target completion date) on which the client would be entitled to terminate the concession agreement if completion has still not been achieved (Guislain and Kerf, 1995).

23.5.2   Short concession with high service price

This approach allows the concessionaire to recover development costs in a short time while still maintaining the efficiency from frequent competition. However, this mechanism may not always be feasible unless the government pays the concessionaire. Otherwise, unaffordable prices may reduce market demand to a degree that the initial project development costs might not be recovered at any price level over the short concession period (Guislain and Kerf, 1995). High prices may also cause strong public opposition and consequent social and political problems. For example, in Mexico, the loans to BOT projects were characterised by high floating interest rates due to a lack of mature domestic financial market. The government adopted the shortest concession period length as a key award criterion to address the difficulty in obtaining long-term fixed-rate financing. This encouraged the concessionaires to charge the maximum allowable toll with the aim of reducing the payback time. The combination of high floating interest rate and short maturity period resulted in prohibitively high tariffs (Vazquez and Allen, 2004).

For a service that is traditionally free to the public, or where there is an alternative option that is free, the users may not use the tolled facilities. This would result in the project's being financially non-viable and 'congestion' to free facilities. The use of shadow tolls would be suitable for projects where there is a perception of end-users being resistant to paying tolls. Shadow tolls are 'per vehicle' amounts paid to the concessionaire by a sponsoring governmental entity rather than the end-users.

Part Three

23.5.3   Staged lifespan concession and pricing system

A staged concession system with variable prices may be explored in the designed lifespan of a project. For example, the construction costs of a project that has a life of 3X years before major repairs are needed may be recovered in an X-year concession. The competition in the second X-year concession would cause prices to fall to the level needed to operate and maintain the project. In the third X-year concession, the prices are set to a level enough to cover the operation and rehabilitation costs. This approach has some weaknesses. In addition to the 'feasibility' and 'opposition' problems in the first and third concessions resulting from high prices due to huge construction or rehabilitation costs as discussed in the 'short concession with high service price' mechanism, the periodic significant changes of prices result in an unstable toll regime that may not be socially desirable.

23.5.4   Bidding-driven concession period

It is a common international practice that the concession period is fixed by the public client before advertising the request for proposals. However, there is another option, i.e. listing the concession period as one of the factors to be bid for by the private sector. This approach was taken in the Talca-Chillan stretch of route 5 in Chile (Engel *et al.* 1996).

23.5.5   Condition-dependent (flexible) concession period

The length of the concession may be determined by the actual occurrence of endogenous factors according to a pre-defined formula. For example, it is determined over time by reference to the date of recovery by the lenders of their principal and interest, the date by which equity holders have achieved a certain level of return, or the date by which the project has achieved a certain level of production/usage (Clement-Davies, 2001). One case in point is the concession of the Queen Elizabeth II Bridge in Dartford, the United Kingdom, which will end when the concessionaire's cumulative revenue has reached the level of outstanding debt or after 20 years, whichever comes first (HM Treasury, 1995). The flexible and condition-dependent concession leaves more space for dealing with risks and uncertainties.

23.6   Simulation-Based Concession Period Determination Methodology

23.6.1   Reasonable concession period

The concession period divides the project's revenues over its lifecycle between the public and private sectors. Normally, a longer concession period will allow the concessionaire to collect more revenues with reduced interests to the public sector, and vice versa. A PPP project should allow the concessionaire to obtain a 'reasonable but not excessive' level of return. This necessitates that the length of the concession be long enough to allow the concessionaire to achieve a 'reasonable' return. But it should not be too long such that the concessionaire's return would be 'excessive'.

23.6.2   Mathematical definition of concession period

The length of the concession period is determined by two time variables: construction period and operation period. According to the 'reasonable but not excessive' principle, the concession period T is defined as (Zhang and AbouRizk, 2006):

$$T = T_c + T_o \tag{23.1}$$

where T_c = project completion time; T_o = operation period; and T_c and T_o satisfy conditions 2 to 4:

$$T_c \leq T_{c\,\mathrm{max}} \tag{23.2}$$

$$T_o \leq T_{oe} \tag{23.3}$$

$$NPV_I(1 + R_{\mathrm{min}}) \leq NPV\,|_{T_o=t} \leq NPV_I \times (1 + R_{\mathrm{max}}) \tag{23.4}$$

where $T_{c\,\mathrm{max}}$ = maximum allowable project completion time; T_{oe} = designed economic operation life of the project; NPV_I = net present value of the total project development cost; R_{min} = minimum rate of return required by the private sector in the development of a certain type of projects; R_{max} = maximum rate of return to the total project development cost that is acceptable to the public sector; and $NPV\,|_{T_o=t}$ = net present value of net revenues generated from a operation period $T_o = t$.

23.6.3   Simulation-based risk analysis

A PPP infrastructure project is subject to a variety of risks and uncertainties. The determination of an appropriate concession period T requires a good estimation of the construction period T_c and the operation period T_o. T_c is dependent on the durations of various construction activities and their relationships. Various construction risks may occur in the project site, relationships of contractual parties, contractual arrangements, technical specifications and other areas. These risks have significant impacts on the project completion time. T_o depends on the project development cost (NPV_I) and the net present value of the net revenues in the operation period ($NPV\,|_{T_o=t}$). NPV_I depends on the costs of various construction activities. The various construction risks may also greatly increase the project development cost. $NPV\,|_{T_o=t}$ depends on the construction period T_c and many risks that may be encountered in the future operation of the project.

Computer simulation is a useful tool for decision making under uncertainties and risks. In this chapter, Monte Carlo simulation is used to quantify and reason with the risks affecting the length of the concession period of a BOT-type project. Project development parameters are assumed to be random variables following certain statistical distributions. Major risk variables

Part Three

considered here are construction period T_c, project development cost NPV_I, market demand, sale price, project operation and maintenance (OM) costs and discount rate.

A simulation-based framework for concession period determination has been developed based on Zhang and AbouRizk (2006), as shown in Figure 23.1. Details of each step are discussed below.

23.7.1   Developing work breakdown structure

The work breakdown structure (WBS) is a progressive hierarchical break-down of the project into smaller pieces to the lowest practical level at which work activities are carried out or costs controlled. The WBS can be used to manage the project from a time, cost and quality perspective. There are some basic guidelines in establishing the WBS (Halpin, 2006):

1. Work packages must be clearly distinguishable from one another.
2. Each work package must have unique starting and ending dates.

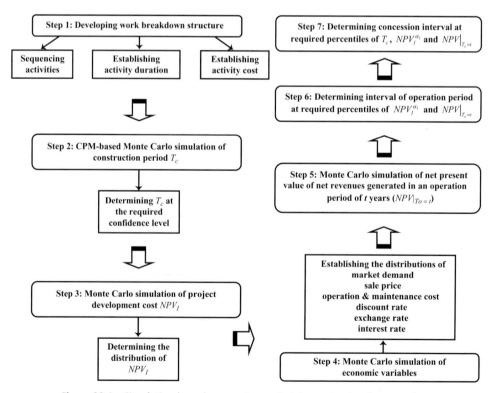

Figure 23.1 Simulation-based concession period determination framework.

3. Each work package should have its own unique budget.
4. Work packages should be small enough to allow measurement of work progress.

In sequencing the work activities for time control, the location of each work package, its construction methods and the material and resources required should be considered in terms of how these aspects will impact the order of these work activities. For example, the location of a work package may determine its sequence. However, in some cases, the sequence is driven by management logic rather than physical requirements.

23.7.2   CPM-based Monte Carlo simulation of project completion time, T_c

The critical path method (CPM) and Monte Carlo simulation can be combined to simulate the construction duration under risks and uncertainties. Firstly, the construction project is broken down into distinct work activities that are logically sequenced by a precedence diagram, arrow diagram or conditional diagram. Secondly, the time distribution of each activity in the diagram is established based on historical data and/or expert knowledge. Thirdly, Monte Carlo simulation is used to establish the statistical distribution of the project completion time using the CPM method based on a randomly generated set of durations of all work activities. Instead of determining the path criticality of a construction project as in the conventional CPM method, Monte Carlo simulation examines activity criticality based on the statistical distribution of the duration of each activity (Ahuja *et al.*, 1994). Finally, the project completion time at a particular percentile can be calculated using this established distribution. The distribution of the project completion time also provides a basis on which the maximum allowable project completion time ($T_{c\,max}$) is determined.

23.7.3   Monte Carlo simulation of project development cost NPV_I

The probability of achieving an estimate of the total project development cost NPV_I that is within a certain range is determined by Monte Carlo simulation based on the statistical cost distributions of major project development activities. This is done through the following procedures (Zhang, 2005):

1. Define the project scope and establish its work breakdown structure.
2. Classify the work items of each work package into two groups: group one – work items with high degree of cost certainty; and group two – work items with uncertain costs.
3. Establish or assume the statistical cost distributions of uncertain work items.
4. Establish the statistical cost distribution of each work package.
5. Establish the statistical distribution of the total construction cost of the project.
6. Calculate the total project development cost at a required percentile.

23.7.4   Monte Carlo simulation of economic variables

Major economic risk variables in a PPP project include market demand (price and quantity), OM costs, interest rate, currency exchange rate, inflation rate and discount rate. Statistical distributions of these economic risk variables can be established using Monte Carlo simulation based on their sample data. In this regard, sample data of OM costs can be generated from historical data of similar projects with appropriate adjustments, while sample data of other economic variables can be derived by analysing the historical economic data of the country where the project is located.

23.7.5   Monte Carlo simulation of $NPV \mid_{T_o=t}$

$NPV \mid_{T_o=t}$, the net present value of the net revenues generated in a specific operation period $T_o = t$, is calculated using the following formula:

$$NPV \mid_{T_o=t} = \frac{1}{(1+r)^{T_c}} \sum_{i=1}^{t} \frac{NCF_i^o}{(1+r)^i} = \frac{1}{(1+r)^{T_c}} \sum_{i=1}^{t} \frac{(I_i^o - C_i^o)}{(1+r)^i} \quad (23.5)$$

$$I_i^o = Q_i^o \times P_i^o \quad (23.6)$$

where NCF_i = net cash flow; I_i^o = income; C_i^o = operation and maintenance cost; Q_i^o = quantity of demand; P_i^o = sale/service price in the ith year of operation; and r = annual discount rate.

$NPV \mid_{T_o=t}$ is dependent on T_c, I_t^o, C_t^o and r. The distribution of T_c is established in step 2 and the distributions of I_t^o, C_t^o and r are established in step 4. Therefore, the distribution of $NPV \mid_{T_o=t}$ can be established using Monte Carlo simulation based on the distributions of T_c, I_t^o, C_t^o and r. $NPV \mid_{T_o=t}$ can be reasonably assumed as a normal distribution with mean μ_o and standard deviation $\sigma_o \cdot \mu_o$ and σ_o can be determined by a large number of simulation runs. Finally, $NPV \mid_{T_o=t}$ corresponding to a specific percentile α_o can be calculated based on this established normal distribution.

23.7.6   Determining the interval of operation period T_o

T_o must satisfy the condition $NPV_I(1 + R_{\min}) \leq NPV \mid_{T_o=t} \leq NPV_I \times (1 + R_{\max})$. $NPV \mid_{T_o=t}$ corresponding to different percentiles can be calculated based on the established distributions of $NPV \mid_{T_o=t}$. Let $(T_o^l \ T_o^u) \mid_{\alpha_o}^{\alpha_I}$ denote the interval of the operation period T_o at α_I percentile of NPV_I and α_o percentile of $NPV \mid_{T_o=t}$. Then, T_o^l is the minimum t that satisfies $NPV_I^{\alpha_I}(1 + R_{\min}) \leq NPV^{\alpha_o} \mid_{T_o=t}$ and T_o^u is the maximum t that satisfies $NPV^{\alpha_o} \mid_{T_o=t} \leq NPV_I^{\alpha_I}(1 + R_{\max})$, where $NPV_I^{\alpha_I}$ is the net present value of the total project development cost at α_I percentile and $NPV^{\alpha_o} \mid_{T_o=t}$ is the net present value of the total annual net cash flows from operation year 1 to t at α_o percentile.

23.7.7   Determining the interval of concession period T

Let $T_c^{a_c}$ be the project completion time T_c at the required percentile of α_c, then the concession interval at α_c percentile of T_c, α_I percentile of NPV_I and α_o percentile of $NPV \mid_{T_o=t}$ can be calculated as $(T_c^{a_c} + T_o^l, \; T_c^{a_c} + T_o^u)$.

Case Study 23.1: Calculation of a Concession Interval

A hypothetical BOT infrastructure project is used to demonstrate the application of the proposed methodology, mathematical model and simulation-based approach discussed in the above. Please note that this project is intentionally simplified for the purpose of demonstration. In this case study, the package *CRYSTAL BALL* was used for conducting Monte Carlo simulations. A total of 20 000 simulation analyses was conducted in each required simulation variable, such as construction time, project development cost, and the accumulative net present value of the net revenues up to a particular operation year in the designed economic operation life of the project.

Statistical distributions of key project variables

The estimates on key project variables are given probability distributions. These variables are project development cost, activity duration, market demand, sale price, operation and maintenance (O&M) cost and discount rate.

Activity costs and durations

The project is divided into four major work activities (1, 2, 3 and 4). It is assumed that the distributions of the costs (in million dollars at the beginning of the first year of construction) and durations of the four activities are already established based on historical data, using the methods mentioned in Sections 23.7.3 and 23.7.2 respectively. These distributions are shown in Table 23.2.

Market demand and price

The designed annual production capacity of the project is 10×10^8 units. In the operation period, the annual market demand of the product follows a normal distribution, with mean value of 8×10^8 units and standard deviation of 2×10^8 units. The sale price of the product follows a normal distribution with a mean of $0.4/unit and a standard deviation of $0.04/unit.

Table 23.2   Construction cost and duration distributions of different activities.

Activity	Cost distribution	Duration distribution
1	Normal distribution, with mean $150m and standard deviation $15m	Triangular distribution, with most likely duration of 1.5 years, minimum duration of 1 year, and maximum duration of 2 years
2	Normal distribution, with mean $200m and standard deviation $30m	Uniform distribution, with minimum duration of 1 year, and maximum duration of 2 years
3	Triangular distribution, with most likely value of $200m minimum value of $100m and maximum value of $300m	Normal distribution, with mean of 1.5 years and standard deviation of 0.2 years
4	Uniform distribution, with minimum value of $100m and maximum value of $300m	Triangular distribution, with most likely duration of 1 year, minimum duration of 0.5 year, and maximum duration of 1.5 years

Operation and maintenance cost

The designed economic operation life of the project is 30 years. It is assumed that the O&M cost increases over this operation life. For simplicity, it is assumed that the annual O&M cost is 20% of the total annual sales revenue in the first 10 years of operation, 30% in the second 10 years and 40% in the third 10 years. As the annual quantity of demand and sale price are random variables, the annual O&M cost is also random.

Annual discount rate

Discount rate can be seen as the interest rate charged by financial institutions for the use of their money. It is used to discount cash flows to reflect risks and the time value of money. The discount rate r can be calculated in the following formula (Brealey et al. 2003):

$$r = (1 + r_r)(1 + r_l) - 1 \qquad (23.7)$$

$$r \approx r_r + r_l \qquad (23.8)$$

where r_r = real interest rate; and r_l = inflation rate. Here it is assumed that the annual discount rate r follows a normal distribution with mean of 10% and standard deviation of 1%.

Simulation of project completion time T_c

Assume that the four activities follow finish–start relationships from activity 1 to activity 4, then, T_c is a stochastic variable whose value is the summation of the randomly generated values of the durations of activities 1 to 4. The statistics of T_c are shown in Table 23.3. Figure 23.2 and Figure 23.3 are the frequency and cumulative charts of T_c. Based on the statistics and shapes of the frequency and cumulative charts, it is reasonable to assume that T_c follows normal distribution, with a mean of 5.83 years and standard deviation of 0.48 years.

Let $T_c|_a$ denote the ath percentile of the random variable T_c, then

$$T_c|_a = \overline{T_c} + z_a \sigma \qquad (23.9)$$

where $\overline{T_c}$ = mean of T_c; z_a = critical value of standard normal distribution at the specified percentile value a; and σ = standard deviation of T_c.

The project completion time can be derived according to equation 9 based on the risk tolerance of the decision maker. For example, if a decision maker of low risk tolerance sets the project completion time at the 95% percentile, denoted by $T_c|_{a=95\%}$, then

$$T_c|_{a=95\%} = \overline{T_c} + z_a \sigma = 5.83 + 1.645 \times 0.48 = 6.62 \text{ years}$$

Table 23.3 Statistics of total construction time (year).

Statictics	Value
Mean	5.83
Median	5.83
Standard deviation	0.48
Variance	0.23
Skewness	0.08
Kurtosis	2.78
Coeff. of variability	0.08
Range minimum	4.25
Range maximum	7.56
Range width	3.31
Mean std. error	0.00

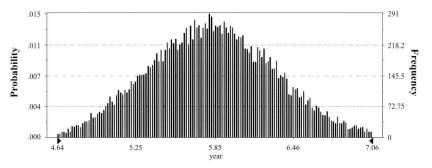

Figure 23.2   Frequency chart of total construction time.

Simulation of NPV_I

The total project development cost NPV_I is a stochastic variable, whose value is the summation of the randomly generated values of the costs of the four activities. The statistics of NPV_I are shown in Table 23.4. Figure 23.4 and Figure 23.5 are the frequency and cumulative charts of NPV_I. Based on the statistics and shapes of the frequency and cumulative charts, it is reasonable to assume that NPV_I follows normal distribution, with mean of $751.04m and standard deviation of $78.97m.

If the total project development cost is set at the 95% percentile, denoted by $NPV_I|_{a=95\%}$, then

$$NPV_I|_{a=95\%} = 751.04 + 78.97 \times 1.645 = \$880.95\ m.$$

Simulation of $NPV|_{T_o=t}$

As shown in equation (5), $NPV|_{T_o=t}$ is a stochastic variable that depends on stochastic variables T_c, I_t^o, C_t^o and r. Here, T_c is set at the 95% percentile, that is, 6.62 years as calculated in a previous section. According to the assumption made in the section 'Operation and Maintenance Cost', for year 1 to year 10 of the operation period, $NCF_i^o = I_i^o - C_i^o = I_i^o - 0.2I_i^o = 0.8I_i^o$; for year 11 to year 20 of the operation period, $NCF_i^o = I_i^o - C_i^o = I_i^o - 0.3I_i^o = 0.7I_i^o$; and for year 21 to year 30 of the operation period, $NCF_i^o = I_i^o - C_i^o = I_i^o - 0.4I_i^o = 0.6I_i^o$.

In the simulation process, the following condition is satisfied:

$$Q_i^o = q_i^o \text{ if } q_i^o \le 10 \times 10^8$$
$$Q_i^o = 10 \times 10^8 \text{ if } q_i^o > 10 \times 10^8$$

where q_i^o = the randomly generated quantity of demand for the ith year of operation.

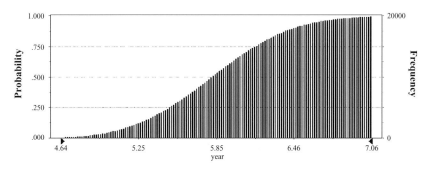

Figure 23.3   Cumulative chart of total construction time.

Table 23.4   Statistics of total project development cost ($ million).

Statistics	Value
Mean	751.04
Median	750.27
Standard deviation	78.97
Variance	6236.21
Skewness	0.01
Kurtosis	2.59
Coeff. of variability	0.11
Range minimum	494.91
Range maximum	994.96
Range width	500.05
Mean std. error	0.56

For simplicity, it is assumed that there is no penalty to the concessionaire for not being able to satisfy a total demand that is beyond the designed capacity of the project. The mean, standard deviation, minimum, maximum, range width and 75% percentile of $NPV \mid_{T_o=t}$ for $t = 1$ to 20 are shown in Table 23.5. Figure 23.6 shows the mean, minimum and maximum of $NPV \mid_{T_o=t}$.

Determination of concession interval

Assume the government decides to use the 95% percentile value of T_c and NPV_I, and the 75% percentile value of $NPV \mid_{T_o=t}$. As the project completion time $T_c \mid_{a=95\%}$ is already derived, the concession interval is known if the lower and upper limits (T_o^l and T_o^u) of the operation period are known.

Lower limit of operation period T_o^l

Assume $R_{min} = 12\%$, then the minimum total net revenue required by the concessionaire as discounted at the beginning of the first year of construction is calculated as follows:

$$NPV_I \mid_{a=95\%} (1 + R_{min}) = 880.95 \times (1 + 0.12) = \$986.67\$m.$$

From Table 23.4, it is known that $NPV \mid_{T_o=11} = \$956.60m$ and $NPV \mid_{T_o=12} = \$1,004.59m$. Therefore, T_o^l is between 11 and 12 years. Assume there is a linear relationship between $NPV \mid_{T_o=t}$ and

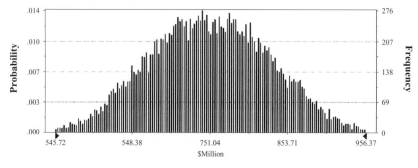

Figure 23.4   Frequency chart of total construction cost.

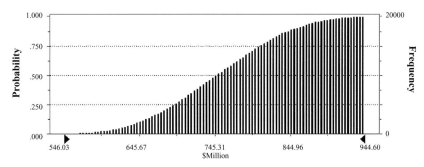

Figure 23.5 Cumulative chart of total construction cost.

t in this short duration, then T_o' is calculated as follows

$$NPV_I|_{a=95\%}(1+R_{min}) = NPV_I|_{T_o=11} + \frac{(T_o'-11)}{(12-11)}(NPV_I|_{T_o=12} - NPV_I|_{T_o=11})$$

$$T_o' = 11 + \frac{NPV_I|_{a=95\%}(1+R_{min}) - NPV_I|_{T_o=11}}{NPV_I|_{T_o=12} - NPV_I|_{T_o=11}} = 11 + \frac{986.67 - 956.60}{1004.59 - 956.60} = 11.63 \text{ (years)}$$

Upper limit of operation period $T_o{}^u$

Assume $R_{max} = 20\%$, then the maximum total net revenue allowed by the government as discounted at the beginning of the first year of construction is calculated as follows:

$$NPV_I|_{a=95\%}(1+R_{max}) = 880.95 \times (1+0.2) = \$1057.14\ m.$$

Table 23.5 Statistics of $NPV_I|_{T_o=t}$ ($ million).

Year	Mean	Standard deviation	Minimum	Maximum	Range width	75 percentile
1	122.02	31.20	11.82	247.79	235.97	143.92
2	232.61	44.92	71.60	437.51	365.90	262.48
3	333.56	56.39	151.30	604.66	453.36	370.56
4	425.35	66.59	209.30	769.43	560.13	469.39
5	508.70	76.35	258.80	928.91	670.11	558.33
6	584.84	85.88	296.79	1096.07	799.28	640.09
7	654.16	95.25	350.30	1224.02	873.72	715.70
8	717.30	104.43	401.43	1318.71	917.28	784.36
9	774.58	113.03	428.91	1468.52	1039.61	846.30
10	826.85	121.50	446.48	1590.76	1144.28	902.97
11	874.64	129.93	464.06	1711.03	1246.96	956.60
12	918.15	138.01	486.70	1825.46	1338.76	1004.59
13	957.73	145.89	499.07	1931.43	1432.37	1048.79
14	993.77	153.37	512.72	2067.69	1554.97	1089.07
15	1026.82	160.60	526.52	2134.81	1608.29	1125.71
16	1056.82	167.52	537.69	2227.87	1690.17	1159.10
17	1084.12	174.20	545.47	2268.51	1723.03	1190.33
18	1109.11	180.59	551.05	2330.29	1779.24	1218.37
19	1131.84	186.70	556.37	2382.94	1826.57	1245.13
20	1152.63	192.55	562.83	2441.72	1878.88	1268.99

Part Three

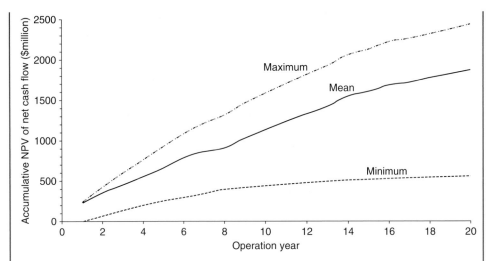

Figure 23.6   Mean, minimum and maximum of $NPV(T_o)$.

From Table 23.5, it is known that $NPV \mid_{T_o=13} = \$1{,}048.79m$, and $NPV \mid_{T_o=14} = \$1{,}089.07m$.
    Therefore, T_o^u is between 13 and 14 years. Again, assume there is a linear relationship between $NPV \mid_{T_o=t}$ and t in this short duration, then T_o^u is calculated as follows:

$$NPV_I \mid_{a=95\%} (1 + R_{max}) = NPV \mid_{T_o=13} + \frac{(T_o^u - 13)}{(14 - 13)} (NPV \mid_{T_o=14} - NPV \mid_{T_o=13})$$

$$T_o^u = 13 + \frac{NPV_I \mid_{a=95\%} (1 + R_{max}) - NPV \mid_{T_o=13}}{NPV \mid_{T_o=14} - NPV \mid_{T_o=13}} = 13 + \frac{1057.14 - 1048.79}{1089.07 - 1048.79} = 13.21 \text{ years}$$

Therefore, the concession interval is $(T_c + T_o^l, \ T_c + T_o^u) = (6.62 + 11.63, \ 6.62 + 13.21) = (18.25, \ 19.83)$.

23.8   Conclusions

The Hong Kong government has been seeking innovative and flexible financing strategies to stimulate economic activities and increase efficiency and cost-effectiveness in the provision of pubic works and services. A wide range of public works and services has been delivered or proposed to be delivered through PPPs. There is substantial controversy in public opinion about PPPs, particularly on how to ensure accountability, transparency, efficiency and cost effectiveness. PPPs require appropriate allocation of risks.

The length of concession is an important issue in infrastructure development through PPPs. In practice, both fixed and flexible concession periods have been used. Concessions can also be differentiated in terms of one-period concessions or two-period concessions. Several methods have been identified, from which a suitable one may be chosen to determine the appropriate length of the concession of a particular PPP project, taking into consideration the characteristics of the project and the environment in which the project operates.

The essence of the concession period methodology proposed in this chapter is that the concession should integrate construction and operation to encourage innovations, efficiency, cost savings and early project completion. The project completion time should allow a competent contractor to complete the project on schedule and the operation period should be long enough to enable the concessionaire to achieve a 'reasonable' return, but not too long such that the concessionaire's return is 'excessive' and the public sector's interests are sacrificed.

Informed assessments and analysis of risks and uncertainties are a prerequisite to the determination of an appropriate length of concession. Monte Carlo simulation is a useful tool to measure uncertainties and reason with construction and economic risks, including project development cost, project completion time, market demand and price of project services/products, operation and maintenance cost, interest rate and inflation rate.

The proposed methodology, mathematical model and simulation-based approach would facilitate the public sector in the determination of a suitable concession period for a particular infrastructure project, and the private sector in determining whether to bid for a concession solicited by a public client. It would also facilitate the private sector in developing unsolicited concession proposals for potential infrastructure projects and the public sector in evaluating such unsolicited proposals.

References

Ahuja, H.N., Dozzi, S.P. and Abourizk, S.M. (1994) *Project Management: Techniques in Planning and Controlling Construction Projects.* John Wiley & Sons, NJ.

Brealey, R.A., Myers, S.C., Marcus, A.J. *et al.* (2003) *Fundamentals of Corporate Finance.* McGraw-Hill, Toronto.

Clement-Davies, C. (2001) Public-private partnerships on central and eastern Europe: structuring the concession agreement. *Business Law International*, 1, 18–37.

Engel, E., Fischer, R., and Galetovic, A. (1996) *Highway Franchising in Chile.* Center for Applied Economics, Universidad de Chile, Santiago.

Estache, A. and Carbajo, J. (1996) Designing toll road concessions – lessons from Argentina. *Public Policy for the Private Sector*, Note No. 99. The World Bank Group, Washington, DC.

Guislain, P., and Kerf, M. (1995) *Concessions – The Way to Privatize Infrastructure Sector Monopolies.* Public Policy for the Private Sector, Note No. 59. The World Bank Group, Washington, DC.

Halpin, D.W. (2006) *Construction Management.* John Wiley & Sons, NJ.

HM Treasury (1995) *Private Opportunity, Public Benefit: Progressing The Private Finance Initiative.* London.

Kwoka, J. E. (1996) *Privatization, Deregulation, and Competition: A Survey of Effects on Economic Performance.* Private Sector Development Department Occasional Paper 27. World Bank, Washington, D C.

Palaneeswaran, E., Kumaraswamy, M.M. and Zhang, X.Q. (2001) Reforging construction supply chains: a source selection perspective. *European Journal of Purchasing and Supply Management*, 7(3), 165–178.

Vazquez, F., and Allen, S. (2004) Private sector participation in the delivery of highway infrastructure in Central America and Mexico. *Construction Management and Economics*, 22, 745–754.

Part Three

Zhang, X.Q. (2005) Financial viability analysis and capital structure optimization in privatized public infrastructure projects. *Journal of Construction Engineering and Management*, 131, 656–668.

Zhang, X.Q. and AbouRizk, S. M. (2006) Determining a reasonable concession period for private sector provision of public works and services. *Canadian Journal of Civil Engineering*, 33, 622–631.

Part Three

Index